Universal's
MATCHBOX®
TOYS

Schiffer Publishing Ltd

The Universal Years, 1982-1992
with Price Guide   ***Revised 2nd Edition***

4880 Lower Valley Road, Atglen, PA 19310 USA

"Models of Australia Art" Yesteryears Y-21E "Pro Hart" & Jenny Kee"

Revised Price Guide: 1999
Copyright © 1999 by Charlie Mack
Library of Congress Catalog Card Number: 98-83293

ISBN: 0-7643-0771-1
Printed in China

Published by Schiffer Publishing Ltd.
4880 Lower Valley Road
Atglen, PA 19310
Phone: (610) 593-1777; Fax: (610) 593-2002
E-mail: Schifferbk@aol.com
Please visit our web site catalog at
www.schifferbooks.com

This book may be purchased from the publisher.
Include $3.95 for shipping. Please try your bookstore first.
We are interested in hearing from authors
with book ideas on related subjects.
You may write for a free printed catalog.

In Europe, Schiffer books are distributed by
Bushwood Books
6 Marksbury Avenue
Kew Gardens
Surrey TW9 4JF England
Phone: 44 (0)181 392-8585; Fax: 44 (0)181 392-9876
E-mail: Bushwd@aol.com

Contents

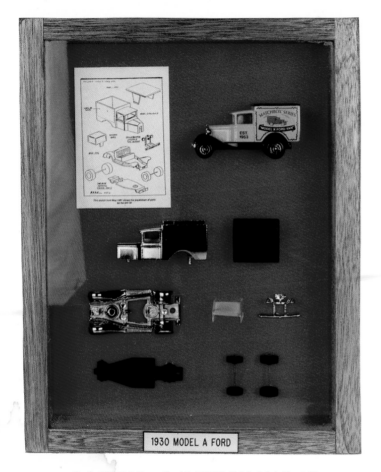

Code 2"PC-1 Fame" with MB38-E Model A Ford Van

Acknowledgements

Thanks to the those individuals who helped proofread this book, loaned models from their collection or provided their guidance in the production of this book. Jim Gallegos, Shaun Gallegos, Craig Hill, Everett Marshall III, Neil Waldmann, Wayne Wilson, Jeff Snyder and Leslie Bockol.

Freddy Kreuger Talking Doll

Dinky Toys miniature MB9-D/74-E Fiat Abarth

History 1982-1992

It was the end of an era. Lesney Products, in business since 1947, had gone bankrupt on June 11, 1982. On that day, R.D. Agutter and G.T.E. Ped joint receivers. The company was reformed as "Matchbox Toys Ltd." and was looking for a buyer. Both Fisher Price and Mattel were interested but Universal Toys, owned by David Yeh, came up as the buyer on September 24, 1982. It took several years for "Lesney Products" to be removed from all the molds. Both the miniatures and Yesteryears especially carried on the Lesney name up until 1985! The models produced in 1982 and 1983 are not completely defined in books 2 and 3 and there is an overlap between the Lesney and Universal issues.

Jack Odell, one of the founders of Lesney Products, bought much of the Lesney tooling and machinery and began his own new diecast company called "Lledo" (Odell spelled backwards).

In 1977, Kidco toys was formed to market Universal Toys in the U.S. and was merged with Matchbox in the early 1980's. In 1978 Universal bought 80% of LJN Toys of New York. When Matchbox was bought by Universal, the entire Kidco and LJN lines were marketed as Matchbox product in Europe. In 1980, the Universal Group bought space in Macau for a company called "Macau Toys Ltd." and in 1981 "Macau Diecasting Toys Ltd." was also formed.

Within three months of the Lesney takeover, much of the tooling for Matchbox Toys began moving to Macau with the first Macau-made Matchbox being seen in May, 1983. It was decided though to retain production in Rochford, England. In 1985 however, even the Yesteryear production moved to Macau.

In 1986, Matchbox Toys began negotiations with Kenner Parker to buy the Dinky trademark. By 1987, the deal was consummated. To protect the Dinky trademark, six miniatures were produced in special versions and packaged in Dinky blisterpacks. In 1989, the first Dinky prototypes were shown at toy fairs in the U.K.

Matchbox Toys was not just a diecast toy company. Many "unusual" products started appearing with the Matchbox brand through the 1980's. Dolls continued to be manufactured with stuffed animals being produced also. In 1988, Matchbox hit a licensing winner with their Pee Wee Herman line in the United States. In 1989, Matchbox hit upon their most controversial offering, a Freddy Kreuger talking doll. Freddy was the star character of the *Nightmare On Elm Street* movies. Due to parental and religious groups requests, Freddy soon disappeared from the shelves. As the 1990s hit, Matchbox was taking a down-turn, probably due to the fact that the line had expanded into too many different areas not all of them moneymakers. By early 1992, David Yeh, chairman of Universal Toys, began shopping around the company to larger toy companies for a buy-out. In another earmark in Matchbox history, in May of 1992 Tyco Toys gave their intentions to buy the Universal Group. After several months of negotiations it was official, on October 2, 1992 Tyco Toys now owned Matchbox Toys.

As 1992 was almost over and very little product was still left to be released by Matchbox Toys, it was decided to include all items issued through December 31, 1992 in this book. The year 1993 will begin the "Tyco Years."

LICENSED ITEMS
Row 1: Porsche martini glass; paper cup; Leslie Smith figure & box; watch
Row 2: "happy" sack; "Direct Line Insurance" telephone; puzzle by Nordevco
Row 3: greeting cards by Gibson

ON-PACK OFFERS
Row 1: PG tips; Heinz pickles; Swarfega; Dairylea
Row 2: Rowntree's Jelly; Jammie Dodgers; Three Torches matches
Row 3: Mars; Chef Boyardee; Tyne Brand; Band Aid Plasters

WHEEL PATTERNS

(1) Five Spoke

(2) Four Spoke

(12) Racing Special

(13) Starburst

(14) Laser

(3) Maltese Cross

(4) Spiro

(5) Five Arch

(15-A)

(15-B)

(15-C)

GRAY DISC WITH RUBBER TIRES

(6) Five Crown

(7) Dot Dash

(8) Five Spoke With Center Cut

(16) Six Spoke

(17) Goodyear Slicks

(18) Nine Spoke Goodyear Slicks

(9) Five Spoke Star

(10) Eight Spoke

(11) Eight Dot

(19) Six Spoke Ringed

MINIATURES
Row 1: MB1D7 **Dodge Challenger**; MB1F1 & 2 **Jaguar XJ6 Police Car**; MB2E1 & 6 **Pontiac Fiero**; MB2G/15G1 **1965 Corvette Grand Sport**
Row 2: MB2G/15G8 **1965 Corvette Grand Sport**; MB3C41 & 33 **Porsche Turbo**; MB4D/43H6 **'57 Chevy**; MB4E1 & 6 **Taxi FX4**
Row 3: MB5D/56F4 **4X4 Jeep**; MB6C/7E1 **I.M.S.A. Mazda**; MB6E/72H 2 & 5 **Ford Supervan II**; MB7C23 **V.W. Golf**; MB7D7 **Rompin Rabbit**
Row 4: MB7F1, 6& 10 **Porsche 959**; MB8G1O **Vauxhall Astra Police Car**; MB8H/39F1 **Mack CH600 Aerodyne**; MB8I/24I1 **Airport Tender**

Row 1: MB8G2 & 1 **Vauxhall Astra Police Car**; MB8F/71E1 & 3 **Scania T142**; MB8D15 **Rover 3500**; MB8E1 **Greased Lightning**
Row 2: MB9D/74E12 **Fiat Abarth**; MB9F/74G1 **Toyota MR2**; MB9H/53E1 & 5 **Faun Dump Truck**; MB10C2 & 8 **Gran Fury Police Car**
Row 3: MB10C11 & 15 **Gran Fury Police Car**; MB10D1, 7 & 10 **Buick LeSabre**; MB11E/67E1 **I.M.S.A. Mustang**
Row 4: MB11E/67E3 **I.M.S.A. Mustang**; MB12E/51E6, 8 & 23 **Pontiac Firebird SE**; MB12H/33H1 & 4 **Mercedes 500SL Convertible**

Row 1: **MB13D/63D9 4X4 Open Back Truck;** MB14E/69E2 & 4 **1983/84 Corvette;** MB14G/28G1O **1987 Corvette;** MB15D/55E8 & 15 **Ford Sierra**
Row 2: MB15D/55E16, 33 & 38 **Ford Sierra;** MB15E7 & 9 **Peugeot 205 Turbo;** MB15F/22F1 **Saab 9000 Turbo**
Row 3: MB15F/22F5 **Saab 9000 Turbo;** MB15H/6G1 & 4 **Alfa Romeo;** MB15I/41I1 **Sunburner;** MB16D/6D2 **F.1 Racer;** MB16E/51H1 **Ford LTD Police Car**
Row 4: MB17C/28F/51F5, 20, 29 & 46 **London Bus;** MB17D/9E2 & 3 **AMX Prostocker**

Row 1: MB17E/37G1, 5 & 10 **Ford Escort Cabriolet**; MB17F/50F3 **Dodge Dakota**; MB18C1 & 4 **Fire Engine**
Row 2: MB18C19 **Fire Engine**; MB19D9, 11 & 19 **Peterbilt Cement Truck**; MB20C/14F5 & 13 **4X4 Jeep**
Row 3: MB20C/14F2 **4X4 Jeep**; MB20D/23G/62I1, 7 & 26 **Volvo Container Truck**; MB20E1 & 7 **Volkswagen Transporter**
Row 4: MB21D1 **Corvette Pace Car**; MB21E/53F1 & 4 **Breakdown Van** ; MB21F/71F1 & 7 **GMC Wrecker**; MB22D/35E8 **4X4 Mini Pickup**

Row 1: MB22E3 & 5 **Jaguar XK120**; MB22G/41G1 **Vectra Cavalier GSi 2000**; MB22H/49F1 **Lamborghini Diablo**; MB23D/25F6 & 10 **Audi Quattro**
Row 2: MB23D/25F15 & 13 **Audi Quattro**; MB23F1 **Honda ATC**; MB24E1 & 2 **Datsun 280Z**; MB24F1 **Nissan 300ZX**
Row 3: MB24F3 **Nissan 300ZX**; MB24H/70F1 **Ferrari F40**; MB25El & 3 **Ambulance**; MB26D6 **Cosmic Blues**; MB26F/49G1 **Volvo Tilt Truck**
Row 4: MB26F/49G2, 3, 5, 6 & 13 **Volvo Tilt Truck**; MB26G/31H1 **BMW 5 Series**

Row 1: MB26G/31H3 **BMW 5 Series**; MB27Dl, 3, 4, 14 & 17 **Jeep Cherokee**
Row 2: MB28E2, 7 & 13 **Dodge Daytona**; MB28H/61F1 & 3 **Fork Lift Truck**; MB29C24 **Tractor Shovel**
Row 3: MB29C31 **Tractor Shovel**; MB30E/23E1O & 13 **Peterbilt Quarry Truck**; MB30F1, 7 & 4 **Mercedes G Wagon**
Row 4: MB30F15 **Mercedes G Wagon**; MB31G/2F1, 6 & 8 **Sterling**; MB31I/21G1 & 2 **Nissan Prairie**

Row 1: MB31E1 & 2 **Mazda RX7**; MB32D/6F12 **Atlas Excavator**; MB32E/12G4 **Modified Racer**; MB33D/56E/63G1 & 3 **Volkswagen Golf GTi**
Row 2: MB33D/56E/63G9 & 10 **Volkswagen Golf GTi**; MB33E/43F4 **Renault 11**; MB33G/74I1 & 8 **Utility Truck**; MB34C13 **Chevy Prostocker**
Row 3: MB34C16 **Chevy Prostocker**; MB34D1, 2 & 7 **Ford RS200**; MB34E/72J1 **Sprint Racer**; MB35C/16C6 **Pontiac T-Roof**
Row 4: MB35F/16F1, 2, 4 & 12 **Land Rover 90**; MB36D15 & 18 **Refuse Truck**

Row 1: MB36D19 **Refuse Truck**; MB37F4 & 3 **4X4 Jeep**; MB38E8, 7 & 19 **Model A Ford Van**
Row 2: MB38E285 **Model A Ford Van**; MB38F2 **Ford Courier**; MB38H/39G1 **Mercedes 600**; MB39C/60E2 & 7 **Toyota Supra**; MB39D/28I1 **BMW Cabriolet**
Row 3: MB39D/28I15, 14, 7 & 16 **BMW Cabriolet**; MB39E/35G1 & 4 **Ford Bronco II**
Row 4: MB39E/35G9 **Ford Bronco II**; MB40C/62E/58F9, 11, 12 & 18 **Corvette T-Roof**; MB40D/60H1 **NASA Rocket Transporter**

Row 1: MB40F/68H1 & 2 **Road Roller**; MB41D5, 6 & 7 **Kenworth Aerodyne**; MB41F/1E1 **Jaguar XJ6**
Row 2: MB42D4 & 7 **1957 Ford Thunderbird**; MB42E5 & 7 **Faun Crane**; MB43C/63H22 **0-4-0 Loco**; MB43D1O **Peterbilt Conventional**
Row 3: MB43E1, 7, 2, 4 & 8 **AMG Mercedes 500SEC**; MB43G/24H1 **Lincoln Town Car**
Row 4: MB43G/24H7 **Lincoln Town Car**; MB44G1 **Skoda LR**; MB44E1 & 7 **Citroen 15CV**; MB44D/68D/26H6 & 18 **4x4 Chevy Van**

Row 1: MB44H1 **1921 Model T Van**; MB45C3 & 7 **Kenworth Aerodyne**; MB45E/69G1 & 3 **Highway Maintenance Vehicle**; MB46D3 **Beetle Streaker**

Row 2: MB46C14 & 13 **Ford Tractor & harrow**; MB46F/57F1, 2, 3 & 13 **Mission Helicopter**

Row 3: MB47D6 **Jaguar SSIOO**; MB47E1 **School Bus**; MB48D4 **Red Rider**; MB48F1, 4 & 8 **Vauxhall Astra/Opel Kadette**

Row 4: MB49D1 & 3 **Sand Digger**, MB49E4 & 7 **Peugeot Quasar**; MB50E2 **Chevy Blazer**; MB51G/68F1 **Chevy Camaro**

Row 1: **MB51G/68F7, 9 & 15 Chevy Camaro; MB52D1, 2 & 3 BMW M1**
Row 2: **MB52D7 BMW M1; MB52E1 & 3 Isuzu Amigo; MB53D6 & 7 Flareside Pickup; MB53G/56G1 Ford LTD Taxi**
Row 3: **MB54F1, 14 & 18 Command Vehicle; MB54-H8 & 35 Chevy Lumina; MB55F/41E1 Racing Porsche**
Row 4: **MB55F/41E6 & 10 Racing Porsche; MB55G/33F1 Mercury Sable Wagon; MB56D/5E6, 10 & 12 Peterbilt Tanker**

Row 1: MB56D/5E9, 17, 14, 13, 16 & 25 **Peterbilt Tanker**
Row 2: MB57D3 **4X4 Mini Pickup**; MB57E3 **Carmichael Commando**; MB57H/50G1 & 2 **Mack Auxiliary Power Truck**; MB58D1 & 2 **Ruff Trek**
Row 3: MB58E1 & 3 **Mercedes 300SE**; MB59D30 & 29 **Porsche 928**; MB59D/61D/28J1 & 11 **T-Bird Turbo Coupe**
Row 4: MB60F/12F1, 2 & 10 **Pontiac Firebird Racer**; MB60G/57G3, 1 & 5 **Ford Transit**

Row 1: MB60G/57G29 & 13 **Ford Transit**; MB61E/37H1 **Nissan 300ZX**; MB61C5, 8 & 14 **Peterbilt Wrecker**

Row 2: MB62D11 **Chevrolet Corvette**; MB62F/31F2 **Rolls Royce Silver Cloud**; MB62G1 & 3 **Volvo 760**; MB62H/64F1 & 3 **Oldsmobile Aerotech**

Row 3: MB13E/63E3, 6 & 9 **Snorkel**; MB64D/9G25 & 27 **Caterpillar Bulldozer**; MB65D2 **F.1 Racer**

Row 4: MB65D3 **F.1 Racer**, MB66E/46E2, 4, 12 & 13 **Sauber Group C Racer**, MB66F/55H1 **Rolls Royce Silver Spirit**

Row 1: MB66F/55H2 & 3 **Rolls Royce Silver Spirit**; MB67D2 **Flame Out**; MB67G1, 4 & 5 **Ikarus Coach**

Row 2: MB67G6 & 13 **Ikarus Coach**; MB67F/11F1, 5, 10 & 15 **Lamborghini Countach**

Row 3: MB68E/64E1, 2, 5, 7 & 9 **Dodge Caravan**; MB68G/73D1 **TV News Truck**

Row 4: MB68G/73D6 **TV News Truck**; MB69D5 **'33 Willys Street Rod**; MB69E3 **Volvo 480ES**; MB70D1O & 16 **Ferrari 308 GTB**; MB70E/45D1 **Ford Cargo Skip Truck**

Row 1: MB70E/45D6 **Ford Cargo Skip Truck**; MB71D8 & 16 **1962 Corvette**; MB71G/59E1, 4 & 8 **Porsche 944**

Row 2: MB72E9 **Dodge Delivery Truck**; MB72F1 **Sand Racer**; MB72G/65E1 & 2 **Plane Transporter**; MB72I/65F1 & 3 **Cadillac Allante**

Row 3: MB72I/65F9 **Cadillac Allante**; MB73C/55I8, 9 & 10 **Model A Ford**; MB73E/27E1 **Mercedes Tractor**; MB74F2 **Mustang GT**

Row 4: MB74F5 & 6 **Mustang GT**; MB74H/14H1 & 2 **Grand Prix Racer**; MB75D19 **Helicopter**; MB75E1 **Ferrari Testarossa**

GIFT SET MINIATURES

Row 1: MB32D/6F1O **Atlas Excavator**; MB60G/57G11 **Ford Transit**; MB75D24 **Helicopter** (from G-3H1); MB30F2 **Mercedes G Wagon**; MB59D35 **Porsche 928**; MB65B28 **Airport Coach** (from G-11B1)

Row 2: MB15D/55E19 **Ford Sierra**; MB65B31 **Airport Coach**; MB68E/64E1O **Dodge Caravan**; MB75D23 **Helicopter** (from G-6E1); MB48F5 **Vauxhall Astra/Opel Kadette**; MB60G/57G7 **Ford Transit** (from CY-206A1)

Row 3: MB33D/56E/63G5 **Volkswagen Golf GTi**; MB60G/57G6 **Ford Transit** (from G-5E1); MB75D27 **Helicopter**; MB46F/57F1O **Mission Helicopter** (from MC-12A1); MB58D1O **Ruff Trek** (from James Bond License to Kill)

Row 4: MB21E/53F3 **Breakdown Van** (from MP1O1A1); MB1OC13 **Gran Fury Police Car** (from 11O1-1A); MB7F18 **Porsche 959** (from MP804-A); MB9A13 **Boat & Trailer**; TP21A **Motorcycle Trailer** (from MP805A)

Row 1: MB16E/51H4 **Ford LTD Police Car;** MB18C9 **Fire Engine;** MB30F11 **Mercedes G Wagon** (from MC-15A1); MB35F/16F11 **Land Rover 90;**
MB68E/64E14 **Dodge Caravan;** MB72E24 **Dodge Delivery Truck** (from MC-17A1)
Row 2: MB3C45 **Porsche Turbo;** MB7F22 **Porsche 959;** MB55F/41E12 **Racing Porsche** (from MC-23A1); MB35F/16F18 **Land Rover 90;** MB68E/64E15 **Dodge
Caravan;** MB75D36 **Helicopter** (from MC-24A1)
Row 3: MB35F/16F21 **Land Rover 90;**MB31C21 **Caravan;** MB40C42 **Horse Box;** MB72K1 **Dodge Zoo Truck;** MB73C/55H17 **Model A Ford** (from MC-803A1)
Row 4: MB42E4 **Faun Crane** (from MC-8A1); MB7F15 **Porsche 959;** MB6E/72H9 **Ford Supervan II** (from MC-9A1); MB51C8 **Combine Harvester** (from MC-7A1)

25

Row 1: MB1D13 **Dodge Challenger**; MB11E/67E7 **I.M.S.A. Mustang**; MB15E8 **Peugeot 205 Turbo**; MB15F/22F6 **Saab 9000 Turbo**; MB15H/6G3 **Alfa Romeo**; MB22G/41G4 **Vectra Cavalier GSi 2000** (from MB835A1)

Row 2: MB31G/2F7 **Sterling**; MB34D6 **Ford RS200**; MB48F9 **Vauxhall Astra/Opel Kadette** (from MB835A1); MB37E19 **Matra Rancho**; TP11OA **Inflatable**; MB54F19 **Command Vehicle** (from EM90)

Row 3: MB39E/35G1O **Ford Bronco II**; MB35F/16F20 **Land Rover 90**; MB21E/53F9 **Breakdown Van** (from EM90); **Volvo Container Truck** (from SB810); MB41F/1E4 **Jaguar XJ6** (from K-803)

Row 4: MB25C19 & 18 **Flat Car & Container** (from Freight Yard playset); MB5D/56F8 **4X4 Jeep** (from Jeep Jamboree); MB66E/46E8 **Sauber Group C Racer** (from Carguantua); MB7F11 **Porsche 959** (from K-159A1)

Row 1: MB1D14 **Dodge Challenger**; MB15F/22F7 **Saab 9000 Turbo**; MB15H/6G6 **Alfa Romeo**; MB15I/41I2 **Sunburner**; MB17F/50F6 **Dodge Dakota**; MB20D/23G/62I28 **Volvo Container Truck (from MB963A1)**

Row 2: MB25E8 **Ambulance**; MB31I/21G4 **Nissan Prairie**; MB32E/12G27 **Modified Racer**; MB34E/72J16 **Sprint Racer**; B36D20 **Refuse Truck**; MB45E/69G4 **Highway Maintenance Vehicle** (from MB963A1)

Row 3: MB47E14 **School Bus**; MB57H/50G4 **Mack Auxiliary Power Truck**; MB63E/13E11 **Snorkel** (from MB963A1); MB6B25 **Mercedes Tourer**; MB23B12 **Atlas** (from 6pc. Value Pack)

Row 4: MB14G/28G11 **1987 Corvette**; MB43G/24G4 **Lincoln Town Car**; MB52E2 **Isuzu Amigo** (from 32900A1); MB5D/56F9 **4X4 Jeep**; MB22E1O **Jaguar XK120**; MB70D19 **Ferrari 308GTB** (from 32900A2)

ENGLISH PROMOTIONALS
Row 1: MB17C/28F/51F9, 10, 11, 13, 14 & 15 **London Bus**
Row 2: MB17C/28F/51F17, 18, 21, 22, 25 & 30 **London Bus**
Row 3: MB17C/28F/51F31, 32, 33, 36, 37 & 45 **London Bus**
Row 4: MB44H2, 4, 5, 6, 7 & 8 **1921 Model T Ford Van**

Row 1: MB44H1O, 13, 16 & 17 **1921 Model T Ford Van;** MB38G1 **Ford Courrier;** MB38F1 **Ford Courier**
Row 2: MB41F/1E2 **Jaguar XJ6;** MB4D/43H14 **'57 Chevy;** MB22D/35E9 **4X4 Mini Pickup;** MB35C/16C5 **Pontiac T-Roof;** MB71G/59E6 & 3 **Porsche 944**
Row 3: MB7F26 **Porsche 959;** MB75E20 **Ferrari Testarossa;** MB3C32 **Porsche Turbo;** MB55F/41E3 **Racing Porsche;** MB33E/43F1 **Renault 11;** MB62F/31F1
Rolls Royce Silver Cloud
Row 4: MB47D7 **Jaguar SS;** MB73C/55I14 **Model A Ford;** MB54H11 **Chevy Lumina;** MB66E/46E11 **Sauber Group C Racer;** MB33D/56E/63G6 **Volkswagen
Golf GTi;** MB53D12 **Flareside Pickup**

Row 1: MB20D/23G/62I3, 4, 9, 13, 15 & 24 **Volvo Container Truck**
Row 2: MB20D/23G/62I25, 27 & 29 **Volvo Container Truck; MB60G/57G1O, 14 & 18 MB60G/57G Ford Transit**
Row 3: MB43C/63H11, 12, 15, 16, 17 & 18 **0-4-0 Loco**
Row 4: MB43C/63H20 **0-4-0 Loco;** MB72E4, 11, 19 & 18 **MB72E Dodge Delivery Truck; MB65B33 Airport Coach**

Row 1: MB38E12, 13, 14, 15, 17 & 21 **Model A Ford Van**
Row 2: MB38E22, 25, 27, 29, 30 & 31 **Model A Ford Van**
Row 3: MB38E32, 34, 36, 37, 39 & 40 **Model A Ford Van**
Row 4: MB38E41, 43, 44, 46, 47 & 49 **Model A Ford Van**

Row 1: MB38E50, 51, 54, 56, 57 & 58 **Model A Ford Van**
Row 2: MB38E62, 63, 64, 67, 68 & 69 **Model A Ford Van**
Row 3: MB38E70, 71, 72, 74, 76 & 80 **Model A Ford Van**
Row 4: MB38E82, 83, 112, 113, 115 & 116 **Model A Ford Van**

Row 1: MB38E121, 156, 157, 159, 160 & 161 **Model A Ford Van**
Row 2: MB38E165, 166, 193, 197, 201 & 264 **Model A Ford Van**
Row 3: MB38E266, 286, 263, 282, 270 & 283 **Model A Ford Van**
Row 4: TP121A2 **Land Rover & Seafire**; SCXX-A1 **Wagon Wheels**; SP1/2A12 **Kremer Porsche**

AMERICAN PROMOTIONALS

Row 1: MB14E/69E7 **1983/84 Corvette**; MB45C4 **Kenworth Aerodyne**; MB47E4 **School Bus**; MB14G/28G8 **1987 Corvette**; MB39E/35G6 **Ford Bronco II**; MB69D11 **'33 Willys Street Rod**

Row 2: MB2G/15G7 **1965 Corvette Grand Sport**; MB4D/43H13 **'57 Chevy**; MB71D19 **1962 Corvette**; MB47E3, 10 & 12 **School Bus**

Row 3: MB12E/51E1O **Pontiac Firebird SE**; MB34C21 **Chevy Prostocker**; MB59B42 **Mercury Fire Chief**; MB12E/51E16 **Pontiac Firebird SE**; MB21F/71F2 **GMC Wrecker**; MB72I/65F7 **Cadillac Allante**

Row 4: MB38E9 **Model A Ford Van**; MB72E21 **Dodge Delivery Van**; MB38E120 & 194 **Model A Ford Van**; MB66E/46E20 **Sauber Group C Racer**

Row 1: MB44H12 **1921 Model T Ford Van**; MB38E38 & 65 **Model A Ford Van**; MB44H3 **Model T Ford Van**; MB38E1O **Model A Ford Van**

Row 2: MB38E5 & 84 **Model A Ford Van**; MB212A3 **Ford Thunderbird** ; MB61E/37H16 **Nissan 300ZX**; MB38E155 **Model A Ford Van**

Row 3: MB17C/28F/51F43 **London Bus**; MB60G/57G35 **Ford Transit**; MB68E/64E11 **Dodge Caravan**; MB74H/14H6 **Grand Prix Racer** ; MB32E/12G28
Modified Racer

Row 4: MB2G/15G9 **1965 Corvette Grand Sport**; **MB14G/28G12 1987 Corvette**; MB40C/62E/58F20 **Corvette T-Roof**; MB71D23 **1962 Corvette**

AUSTRALIAN PROMOTIONALS
Row 1: MB4E2 **Taxi FX4**; MB13E/63E8 **4X4 Open Back Truck**; MB19D 14 **Peterbilt Cement Truck**; MB44D/68D/26H15 **4X4 Chevy Van** ; MB47E13 **School Bus**; MB56D/5Ell **Peterbilt Tanker**
Row 2: MB60G/57G8, 15, 20 & 34 **Ford Transit**; MB65B32 & 30 **Airport Coach**
Row 3: MB66E/46E3 **Sauber Group C Racer**; MB38Ell, 16 & 24 **Model A Ford Van**; CYlllA14 **Racing Team**
Row 4: MB72E8, 16, 17, 14 & 13 **Dodge Delivery Truck**

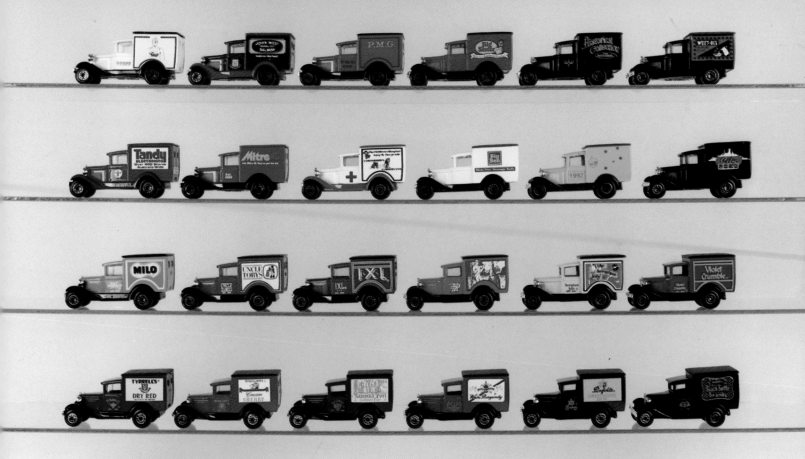

Row 1: MB38E28, 33, 45, 55, 52 & 23 **Model A Ford Van**
Row 2: MB38E75, 158, 259, 262, 265 & 267 **Model A Ford Van**
Row 3: MB38E251, 252, 253, 254, 255 & 256 **Model A Ford Van**
Row 4: MB38E276, 277, 278, 279, 280 & 281 **Model A Ford Van**

JAPANESE PROMOTIONALS

Row 1: MB3C30 **Porsche Turbo**; MB7C22 **V.W. Golf**; MB17C/28F/51F7 & 8 **London Bus**; MB18B19 **Hondarora**; MB20D/23G/62I12 **Volvo Container Truck**

Row 2: MB20B45 **Police Patrol**; MB33C24 **Police Motorcycle**; MB24E3 **Datsun 280Z**; MB36D13 **Refuse Truck**; MB39C/60E4 **Toyota Supra**; MB44F1 **Datsun 280Z Police Car**

Row 3: MB44D/68D/26H7, 8, 9, 10, 11 & 12 **4X4 Chevy Van**

Row 4: MB44D/68D/26H13 & 14 **4X4 Chevy Van**; MB54F9 **Command Vehicle**; MB63E/13E7 **Snorkel**; MB67G7 **Ikarus Coach**; MB71D12 **1962 Corvette**

DUTCH PROMOTIONALS
Row 1: MB3C37 **Porsche Turbo**; MB24F6 **Nissan 300ZX**; MB43E1O **AMG Mercedes 500SE**; MB67F/11F8 **Lamborghini Countach**; MB68E/64E12 **Dodge Caravan**
Row 2: MB6E/72H7 **Ford Supervan II**; MB27D5 **Jeep Cherokee**; MB39D/28I12 **BMW Cabriolet**; MB48F6 **Vauxhall Astra/Opel Kadette**; MB51G/68F11 **Chevy Camaro**
Row 3: MB43C/63H24 **0-4-0 Loco**; MB66E/46E1O **Sauber Group C Racer**; MB60G/57G17 **Ford Transit**; MB38E111 **Model A Ford Van**; MB73C/55I11 **Model A Ford**
Row 4: MB20D/23G/62I1O & 16 **Volvo Container Truck**; CY803-A1 **Scania Lowloader & Dodge Truck**

GERMAN & SWISS PROMOTIONALS
Row 1: MB17C/28F/51F25 & 39 **London Bus**; MB29C26, 28 & 33 **Tractor Shovel**; MB33D/56E/63G11 **Volkswagen Golf GTi**
Row 2: MB38F3 **Ford Courier**; MB60G/57G29 & 19 **Ford Transit**; MB69B8 **Armoured Truck**; MB72E12 **Dodge Delivery Truck**; MB20D/23G/62I19 **Volvo
Container Truck**
Row 3: MB20D/23G/62I20 & 18 **Volvo Container Truck**; MB26F/49G9 **Volvo Tilt Truck**; MB30E/23E11 **Peterbilt Quarry Truck**; MB33D/56E/63G8
Volkswagen Golf GTi; MB38E81 **Model A Ford Van**
Row 4: MB60G/57G27, 24, 26 & 25 **Ford Transit**; MB64D/9G20 **Caterpillar Bulldozer**; MB67G1O **Ikarus Coach**

HONG KONG/GREEK/BELGIAN/SCOTCH PROMOTIONALS

Row 1: MB43E11 **AMG Mercedes 500SEC**; MB2E8 **Ford Fiero**; MB35C/16C7 **Pontiac T-Roof**; MB24F7 **Nissan 300ZX**; MB28E12 **Dodge Daytona**; MB60F/12F12 **Pontiac Firebird Racer**

Row 2: MB9F/74G4 **Toyota MR2**; MB3C38 **Porsche Turbo**; MB75E5 **Ferrari Testarossa**; MB67F/11F9 **Lamborghini Countach**; MB55F/41E9 **Racing Porsche**; MB70D13 **Ferrari 308GTB**

Row 3: MB7F14 **Porsche 959**; MB41F/1E3 **Jaguar XJ6**; MB43E15 **AMG Mercedes 500SEC**; MB75E11 **Ferrari Testarossa**; MB38E35 **Model A Ford Van**

Row 4: MB17C/28F/51F3 **London Bus**; MB20D/23G/62I8, 22 & 2 **Volvo Container Truck**; MB60G/57G31 **Ford Transit**; MB67G9 **Ikarus Coach**

41

CANADA/SPANISH/CHINESE/IRISH/SAUDI/WELSH PROMOTIONALS
Row 1: MB17C/28F/51F34 & 42 **London Bus**; MB20D/23G/62I21 **Volvo Container Truck**; MB70D14 **Ferrari 308GTB**; MB72E23 **Dodge Delivery Truck**;
MB38E154 **Model A Ford Van**
Row 2: MB38E73, 26, 117, 118, 119 & 59 **Model A Ford Van**
Row 3: MB17C/28F/51F35 **London Bus**; MB67G5 **Ikarus Coach**; MB72E15 **Dodge Delivery Truck**; MB23D/25F16 **Audi Quattro**; MB67G12 **Ikarus Coach**;
MB16D/6D3 **F.1 Racer**
Row 4: MB17C/28F/51F38 **London Bus**; MB60G/57G33 **Ford Transit**; MB43E14 **AMG Mercedes 500SEC**; MB38E114 **Model A Ford**; MB35F/16F22 **Land
Rover 90**; MB60G/57G37 **Ford Transit**

SUPERFAST

Row 1: (SF1-6) MB59B39 **Mercury Fire Chief**; MB12E/51E15 **Pontiac Firebird SE**; MB59D32 **Porsche 928**; MB28D9 **Dodge Daytona**; MB43E3 **AMG Mercedes 500SEC**; MB55F/41E4 **Racing Porsche**

Row 2: (SF7-12) MB15D/55E9 **Ford Sierra**; MB71D13 **1962 Corvette** ; MB24E4 **Datsun 280Z**; MB62D14 **Chevrolet Corvette**; MB70D11 **Ferrari 308GTB**; MB34C22 **Chevy Prostocker**

Row 3: (SF13-17) MB14E/69E6 **1983/84 Corvette**; MB14G/28G3 **1987 Corvette**; MB39D/28I1O **BMW Cabriolet**; MB17E/37G9 **Ford Escort Cabriolet**; MB66E/46E6 **Sauber Group C Racer**; MB67F/11F6 **Lamborghini Countach**

Row 4: (SF17-20) MB67F/11F13 **Lamborghini Countach**; MB60E/12E8 & 17 **Pontiac Firebird Racer**; MB2E4 **Pontiac Fiero**; MB24F4 **Nissan 300ZX**

SUPERFAST/LASER WHEELS

Row 1: (SF21-24) MB51G/68F5 **Chevy Camaro**; MB9F/74G2 **Toyota MR2**; MB75E2 & 17 **Ferrari Testarossa**; MB49E2 **Peugeot Quasar**

Row 2: (LW1-6) MB59B41 **Mercury Fire Chief**; MB12E/51E17 **Pontiac Firebird SE**; MB59D33 **Porsche 928**; MB28E1O **Dodge Daytona**; MB43E9 **AMG Mercedes 500SEC**; MB55F/41E5 **Racing Porsche**

Row 3: (LW7-10) MB15D/55E17 **Ford Sierra**; MB71D17 & 18 **1962 Corvette**; MB24E8 & 9 **Datsun 280Z**; MB62D15 **Chevrolet Corvette**

Row 4: (LW10-14) MB62D16 **Chevrolet Corvette**; MB70D12 **Ferrari 308 GTB**; MB34C23 **Chevy Prostocker**; MB14G/28G2 **1987 Corvette**; MB14E/69E9 **1983/84 Corvette**; MB39D/28I11 **BMW Cabriolet**

LASER WHEELS/WORLD CLASS
Row 1: (LW15-20) MB17E/37G8 **Ford Escort Cabriolet**; MB66E/46E7 **Sauber Group C Racer**; MB67F/11F7 **Lamborghini Countach** ; MB60F/12F9 **Pontiac Firebird Racer**; MB2E7 **Pontiac Fiero**; MB24F5 **Nissan 300ZX**
Row 2: (LW21-24) MB51G/68F6 & 10 **Chevy Camaro**; MB9F/74G3 **Toyota MR2**; MB75E3 & 6 **Ferrari Testarossa**; MB49E3 **Peugeot Quasar**
Row 3: (LW25-30) MB1OD2 **Buick LeSabre**; MB72I/65F2 **Cadillac Allante**; MB15F/22F2 **Saab 9000 Turbo**; MB31G/2F2 **Sterling**; MB59F/61D/28J2 **T-Bird Turbo Coupe**; MB69F1 **Volvo 480ES**
Row 4: (World Class) MB2G/15G3 **1965 Corvette Grand Sport**; MB4D/43H18 **'57 Chevy**; MB7F17 **Porsche 959**; MB12H/33H5 & 3 **Mercedes 500SL Convertible**; MB14G/28G7 **1987 Corvette**

WORLD CLASS

Row 1: MB22E6 **Jaguar XK120**; MB22H/49F2 **Lamborghini Diablo**; MB24H/70F2 & 14 **Ferrari F40**; MB40C/62E/5817 & 19 **Corvette T-Roof**

Row 2: MB43E16 **AMG Mercedes 500SEC**; MB43G/24G2 **Lincoln Town Car**; MB52D8 **BMW M1**; MB55F/41E16 & 13 **Racing Porsche**; MB59F/61D/28J8 **T-Bird Turbo Coupe**

Row 3: MB59D36 **Porsche 928**; MB61E/37H5 & 15 **Nissan 300ZX**; MB62F/31F4 **Rolls Royce Silver Cloud**; MB67F/11F16 & 11 **Lamborghini Countach**

Row 4: MB71D20 **1962 Corvette**; MB71G/59E7 **Porsche 944**; MB70D15 **Ferrari 308GTB**; MB72I/65F6 **Cadillac Allante**; MB75E16 & 12 **Ferrari Testarossa**

PRESCHOOL
Row 1: MB4E3 **Taxi FXR**; MB8G7 **Vauxhall Astra Police Car**; MB9H/53E3 **Faun Dump Truck**; MB15D/55E/40E34 **Ford Sierra**; MB16E/51H8 **Ford LTD Police Car**
Row 2: MB17C/28F/51F40 **London Bus**; MB18C8 & 14 **Fire Engine**; MB19D18 **Peterbilt Cement Truck**; MB20C/14F9 **4X4 Jeep**
Row 3: MB21E/53F7 **Breakdown Van**; MB26F/49G1O & 11 **Volvo Tilt Truck**; MB29C27 **Tractor Shovel**; MB31G/2F4 **Sterling**
Row 4: MB33G/74I4, 5 & 2 **Utility Truck**; MB39E/35G7 **Ford Brono II**; MB40C37 **Horse Box**

Row 1: MB40C38 **Horse Box;** MB43C/63H14 **0-4-0 Loco;** TP20A **Side Tipper;** MB43C/63H21 **0-4-0 Loco;** MB44C25 **Passenger Coach**
Row 2: MB43G/24G5 **Lincoln Town Car;** MB47E7 **School Bus;** MB49E8 **Peugeot Quasar;** MB51C1O **Combine Harvester;** MB52B1O **Police Launch**
Row 3: MB53D14 **Flareside Pickup;** MB56D/5E24 **Peterbilt Tanker;** MB64D/9G19 & 24 **Caterpillar Bulldozer;** MB65D12 **F.1 Racer**
Row 4: MB70E/45D2 **Ford Cargo Skip Truck;** MB71C67 **Cattle Truck;** TP19A **Cattle Trailer;** MB75D33 **Helicopter;** MB75E13 **Ferrari Testarossa**

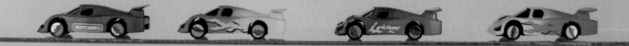

LIGHTNING WHEELS
Row 1: MB24H/70F9, 13, 11 & 12 **Ferrari F40**
Row 2: MB61E/37H8, 10, 12, 11, 19 & 18 **Nissan 300ZX**
Row 3: MB66E/46E15, 16, 19 & 18 **Sauber Group C Racer**
Row 4: MB55F/41E14 & 15 **Racing Porsche**; MB54H14, 13, 39 & 38 **Chevy Lumina**

ACTION PACKS/HOT STOCKS
Row 1: MB1D12 **Dodge Challenger**; MB9H/53E4 **Faun Dump Truck**; MB11E/67E6 **I.M.S.A. Mustang**; MB17F/50F4 **Dodge Dakota**
Row 2: MB21E/53F8 **Breakdown Van**; MB25E6 **Ambulance**; MB32E/12G12 **Modified Racer**; MB33G/74I7 **Utilty Truck**
Row 3: MB34E/72J9 **Sprint Racer**; MB45D/69G2 **Highway Maintenance Vehicle**; MB47E11 **School Bus**; MB63E/13E8 **Snorkel**
Row 4: MB54H22, 23, 21, 20, 18 & 19 **Chevy Lumina**

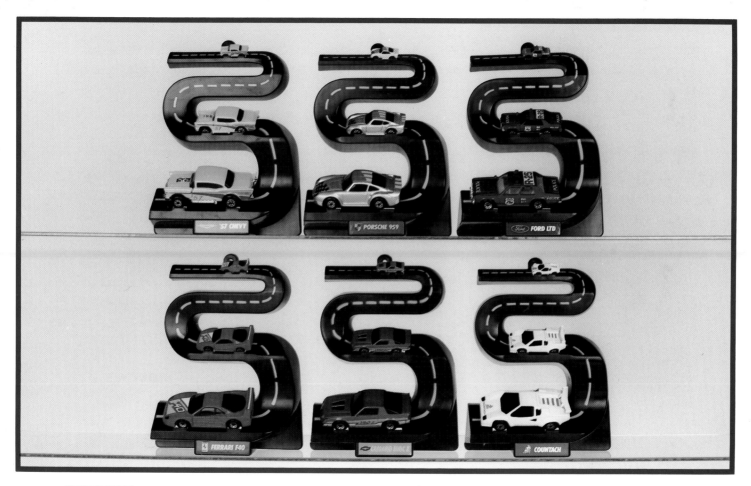

TRIPLE HEAT
Row 1: MB4D/43H19 **'57 Chevy**; MB7F25 **Porsche 959**; MB16E/51H9 **Ford LTD Police Car** with matching SF Minis & World's Smallest
Row 2: MB24H/70F13 **Ferrari F40**; MB51G/68F14 **Chevy Camaro**; MB67F/11F19 **Lamborghini Countach** with matching SF Minis & World's Smallest

LASERTRONICS/INTERCOM CITY
Row 1: MB15D/55E24, 28 & 26 **Ford Sierra**; MB43E17, 21 & 19 **AMG Mercedes 500SEC**
Row 2: MB30FIO, 11 & 12 **Mercedes G Wagon**; MB6E/72H16, 14 & 15 **Ford Supervan II**
Row 3: MB16E/51H11 **Ford LTD Police Car**; MB17F/50F7 **Dodge Dakota**; MB18C19 **Fire Engine**; MB21E/53F13 **Breakdown Van**; MB25E9 **Ambulance**
Row 4: MB30E/23E15 **Peterbilt Quarry Truck**; MB42E8 **Faun Crane Truck**; MB56D/5E31 **Peterbilt Tanker**; MB63E/13E12 **Snorkel**; MB75E38 **Helicopter**

COMMANDO
Row 1: MB2D10 **S.2 Jet**; MB20C/14F6 **4X4 Jeep**; MB26F/49G7 **Volvo Tilt Truck**; MB30C6 **Swamp Rat**; MB30F6 **Mercedes G Wagon**; MB46F/57F6 **Mission Helicopter**

Row 2: MB54F10 **Command Vehicle**; MB56D/5E20 **Peterbilt Tanker**; MB61C13 **Peterbilt Wrecker**; MB70C5 **S.P. Gun**; MB72G/65E **5 Plane Transporter**; MB73B8 **Weasel**

Row 3: MB14D7 **Leyland Tanker**; MB20E2 **Volkswagen Transporter**; MB21E/53F6 **Breakdown Van**; MB26F/49G8 **Volvo Tilt Truck** ; MB27C9 **Swept Wing Jet**; MB35F/16F5 **Land Rover 90**;

Row 4: MB40D/60H3 **Rocket Transporter**; MB46F/57F7 **Mission Helicopter**; MB52B9 **Police Launch**; MB54F12 **Command Vehicle**; MB70C4 **S.P. Gun**; MB73B9 **Weasel**

COMMANDO/INDY
Row 1: **CY15-A6 Peterbilt Tracking Vehicle; CY21-A3 DAF Aircraft Transporter**
Row 2: **32630-A1 Speedway Team Transporter**
Row 3: **32630-A2 Speedway Team Transporter**

INDY
Row 1: MB65D14, 15, 16, 19 & 17 **F.1 Racer**; MB74H/14H4 **Grand Prix Racer**
Row 2: MB74H/14H5, 10, 11, 12, 13 & 14 **Grand Prix Racer**
Row 3: CY111-A6, 10 & 8 **Racing Transporter**
Row 4: CY111-A7, 9 & 11 **Racing Transporter**

Row 1: CY104-A15 & 16 **Kenworth Superstar**
Row 2: CY104-A37 & 38 **Kenworth Superstar**
Row 3: CY104-A39 **Kenworth Superstar**; CY111-A12 **Racing Transporter**
Row 4: MB56D27 **Peterbilt Tanker**; MB72I/65F-**Cadillac Allante**; preproductions of MB21F/71F **GMC Wrecker**; MB68G/73D **TV News Truck**; MB46F/57F
Mission Helicopter

DAYS OF THUNDER
Row 1: CY111-A2 **Racing Transporter;** MB54H3 **Chevy Lumina;** CY104-A8 **Kenworth Superstar**
Row 2: CY111-A4 **Racing Transporter;** MB54H6 **Chevy Lumina;** CY104-A7 **Kenworth Superstar**
Row 3: CY104-A6 **Kenworth Superstar;** MB54H4, 2 & 1 **Chevy Lumina**
Row 4: CY104-A10 & 9 **Kenworth Superstar**

SUPERCOLOR CHANGERS
Row 1: MB4D/43H12 & 11 **'57 Chevy**; MB7F13 **Porsche 959**; MB10D 3 & 6 **Buick LeSabre**; MB12E/51E19 **Pontiac Firebird SE**
Row 2: MB14G/28G4 **1987 Corvette**; MB16E/51H3 **Ford LTD Police Car**; MB27D7 & 9 **Jeep Cherokee**; MB32E/12G2 **Modified Racer**; MB39E/35G2 **Ford Bronco II**
Row 3: MB59F/61D/28J3 & 6 **T-Bird Turbo Coupe**; MB60F/12F14 & 13 **Pontiac Firebird Racer**; MB65D5 **F.1 Racer**; MB75E7 **Ferrari Testarossa**
Row 4: SB1-A10 **Lear Jet**; SB4-A9 **Mirage F1**; SB6-A8 **MIG 21**; SB15-A6 **Phantom F4E**

GRAFFIC TRAFFIC

Row 1: MB1D11 **Dodge Challenger**; MB6E/72H22 **Ford Supervan II**; MB7F24 **Porsche 959**; MB8G9 **Vauxhall Astra Police Car**; MB16E/51H7 **Ford LTD Police Car**; MB17C/28F/51F47 **London Bus**

Row 2: MB17F/50F5 **Dodge Dakota**; MB18C18 **Fire Engine**; MB20E6 **Volkswagen Transporter**; MB21E/53F10 **Breakdown Van**; MB25E7 **Ambulance**; MB31G/2F10 **Sterling**

Row 3: MB31I/21G3 **Nissan Prairie**; MB32E/12G18 **Modified Racer**; MB34D5 **Ford RS200**; MB35F/16F19 **Land Rover 90**; MB37E20 **Matra Rancho**; MB38E271 **Model A Ford Van**

Row 4: MB43C/63H25 **0-4-0 Loco**; MB44C30 **Passenger Coach**; MB44D/68D/26H19 **4X4 Chevy Van**; MB52B12 **Police Launch**; MB56D/5E26 **Peterbilt Tanker**; MB57H/50G3 **Mack Auxiliary Power Truck**

59

Row 1: MB60G/57G32 **Ford Transit**; MB62G5 **Volvo 760**; MB63E/13E10 **Snorkel**; MB66D15 **Super Boss**; MB67F/11F18 **Lamborghini Countach**; MB68G/73D4 **TV News Truck**

Row 2: MB70D18 **Ferrari 308GTB**; MB71D21 **1962 Corvette**; MB75E19 **Ferrari Testarossa**; CY111-A13 **Racing Transporter**; SP13/14A8 **Porsche 959**

Row 3: CY21-A4 **DAF Aircraft Transporter**; SP5/6A6 **Lancia Rally**; SP1/2A11 **Kremer Porsche**; K8-E2 **Ferrari F40**

Row 4: SB3-B4 **NASA Space Shuttle**; SB12-B6 **Pitts Special**; SB24-A11 **F.16**; SB23-A8 **Supersonic Transport**; SP3/4-A8 **Ferrari 512BB**

ROADBLASTERS

Row 1: MB5D/56F5 **4X4 Jeep**; MB6E/72H4 **Ford Supervan II**; MB14E/69E10 **1983/84 Corvette**; MB24E10 **Datsun 280Z**; MB28E11 **Dodge Daytona**; MB29C23 **Tractor Shovel**

Row 2: MB48E7 **Unimog with Plow**; MB49E5 **Peugeot Quasar**; MB50E3 **Chevy Blazer**; MB53D11 **Flareside Pickup**; MB58D9 **Ruff Trek**

Row 3: MB61C12 **Peterbilt Wrecker**; MB66D13 **Super Boss**; Tomy Box; MB66E/46E9 **Sauber Group C Racer**; MB75E4 **Ferrari Testarossa**

Row 1: RB-2522 V.A.R.M.I.T.
Row 2: RB-2532 M.O.R.G.

Row 1: RB-2521 A.L.T.R.A.C.
Row 2: RB-2531 T.R.A.P.P.E.R.

NUTMEG COLLECTIBLES
Row 1: MB34E/72J2, 4, 5, 6, 7 & 8 **Sprint Racer**
Row 2: MB34E/72J10, 11, 12, 13, 14 & 15 **Sprint Racer**
Row 3: MB32E/12G8, 9, 11, 10 & 14 **Modified Racer**
Row 4: MB32E/12G15, 16, 17, 19 & 20 **Modified Racer**
Row 5: MB32E/12G21, 22, 23, 24, 25 & 26 **Modified Racer**

BRAZILIAN MINIATURES

Row 1: MB3C47 **Porsche Turbo**; MB6B28, 26 & 27 **Mercedes Tourer**; MB8D17 **Rover 3500**; MB9D/74E16 **Fiat Abarth**

Row 2: MB15E6 **Peugeot 205 Turbo**; MB19D20 **Peterbilt Cement Truck**; MB21C45 **Renault 5TL**; MB23D/25F17 **Audi Quattro**; MB28D7 **Formula 5000**; MB30E/23E14 **Peterbilt Quarry Truck**

Row 3: MB31E3 **Mazda RX7**; MB40C/62E/58F16 **Corvette T-Roof**; MB40B39 **Horse Box**; MB50E5 **Chevy Blazer**; MB53D13 **Flareside Pickup**; MB55F/41E18 **Racing Porsche**

Row 4: MB56D/5E28 **Peterbilt Tanker**; MB61C17 & 18 **Peterbilt Wrecker**; MB62D17 **Chevrolet Corvette**; MB65B39 **Airport Coach**; MB68E/64E16 **Dodge Caravan**

HUNGARIAN MINIATURES
Row 1: 3X MB20B **Police Patrol**; 3X MB23A **Volkswagen Dormobile**
Row 2: 2X MB17E/9D **AMX Prostocker**; 2X MB25A **Ford Cortina**; 2X MB27A **Mercedes 230SL**
Row 3: 3X MB39B **Rolls Royce Silver Shadow**; 2X MB41A **Ford GT**
Row 4: 2X MB51B **Citroen SM**; 2X MB53C **Jeep CJ6**; 2X MB45B **BMW 3.0 CSL**

BULGARIAN MODELS
Row 1: MB5A **Lotus Europa**; MB6B **Mercedes Tourer**; MB9D/74E **Fiat Abarth**; MB16B **Pontiac**; MB17E/9D **AMX Prostocker**; MB21C **Renault 5TL**
Row 2: MB22C **Blaze Buster**; MB24A **Rolls Royce Silver Shadow**; MB25A **Ford Cortina**; MB27A **Mercedes 230SL**; MB28C **Lincoln Continental MkIV**; MB33E/43F **Renault 11**
Row 3: MB33A **Lamborghini Miura**; MB35C/16C **Pontiac T-Roof**; MB37E **Matra Rancho**; MB39A **Clipper**; MB39C/60E **Toyota Supra**; MB40A **Vauxhall Guildsman**
Row 4: MB41A **Ford GT**; MB45B **BMW 3.0 CSL**; MB46A **Mercedes 300SE**; MB51C **Midnight Magic**; MB51B **Citroen SM**; MB56C **Mercedes 450SL**

Row 1: MB59B **Planet Scout**; MB66B **Mazda RX500**; MB67C **Datsun 260Z**; MB74C **Cougar Villager**; MB75C **Seasprite Helicopter**; MBIX-A **Flamin Manta**
Row 2: K30-A **Mercedes C.111**; K52-A **Datsun 240Z**
Row 3: K59-A **Ford Capri II**; K56-A **Maserati Bora**; K62-A **Doctor's Emergency Car**
Row 4: K66-A **Jaguar XJ12**; K72-A **Brabham**

PREPRODUCTION MINIATURES

Row 1: prototype **Ford Escort**; prototype **Dodge Viper**; plastic MB62F/31F **Rolls Royce Silver Cloud**; MB20D/23G/62I **Volvo Container Truck**; MB38E **Model A Ford Van**; MB58D **Ruff Trek**

Row 2: MB26F/49G **Volvo Tilt Truck**; MB60G/57G **Ford Transit**; MB20E **Volkswagen Transporter**; MB36D **Refuse Truck**; MB40D/60H **NASA Rocket Transporter**; MB49D **Sand Digger**

Row 3: MB38E **Model A Ford Van**; MB24H/70F **Ferrari F40**; MB70D **Ferrari 308GTB**; MB17F/50F **Dodge Dakota**; MB39E/35G **Ford Bronco II**; MB22H/49F **Lamborghini Diablo**

Row 4: MB57E **Carmichael Commando**; MB44H **1921 Model T Ford Van**; MB13E/63E **4X4 Open Back Truck**; MB40C/62E/58F **Corvette T-Roof**; MB57B **Eccles Caravan**; MB67F/11F **Lamborghini Countach**

SUPER GT'S
Row 1: BR1/2 17, 9, 8, 5, 1 & 2 **Iso Grifo**
Row 2: BR1/2 3, 7 & 16 **Iso Grifo**; BR3/4 3, 9 & 1 **Gruesome Twosome**
Row 3: BR3/4 7, 8 & 10 Gruesome Twosome; BR5/6 7, 3 & 9 **Datsun 126X**
Row 4: BR5/6 8 & 1 **Datsun 126X**; BR7/8 12, 9, 8 & 6 **Siva Spyder**

Row 1: BR7/8 7 & 3 **Siva Spyder**; BR9/10 5, 2, 7 & 1 **Lotus Europa**
Row 2: BR11/12 8, 3, 10 & 9 **Saab Sonnet**; BR13/14 11 & 10 **Hairy Hustler**
Row 3: BR13/14 7 & 13 **Hairy Hustler**; BR15/16 16, 15, 14 & 6 **Monteverdi Hai**
Row 4: BR15/16 5, 4, 13, 12 & 7 **Monteverdi Hai**; BR17/18 7 **Fire Chief**

Row 1: BR17/18 8, 9 & 4 **Fire Chief**; BR19/20 4, 7 & 2 **Ford Group 6**
Row 2: BR19/20 3, 9 & 10 **Ford Group 6**; BR21/22 1, 6 & 15 **Alfa Carabo**
Row 3: BR21/22 9 & 16 **Alfa Carabo**; BR23/24 1 & 2 **Vantastic**; BR25/26 4 & 2 **Ford Escort**
Row 4: BR25/26 7 & 8 **Ford Escort**; BR27/28 4, 9, 10 & 17 **Lamborghini Marzal**

Row 1: BR27/28 8 **Lamborghini Marzal**; BR29/30 9, 6, 12, 15 & 1 **Maserati Bora**
Row 2: BR29/30 14 **Maserati Bora**; BR31/32 1, 9 & 3 **Fandango**; BR33/34 3 & 2 **Hi Tailer**
Row 3: BR35/36 9, 1, 13, 12, 14 & 10 **Porsche 910**
Row 4: BR35/36 11 **Porsche 910**; BR37/38 2 & 1 **Ford Capri**; BR39/40 2, 1 & 3 **DeTomasso Pantera**

ORIGINALS

Row 1: MX101-A2 **Road Roller**; MX102-A2 **Tractor**; MX103-A2 **London Bus**; MX104-A2 **Milk Float**; MX105-A2 **Fire Engine**

Row 2: MX106-A1 **Quarry Truck**; MX107-A1 **Wreck Truck**; MX108-A1 **MGA**; MX109-A1 **Concrete Truck**; MX110-A1 **Jaguar XK120**

Row 3: gold plated preproductions of MX101-A, MX102A, MX103-A, MX104-A & MX105-A

Row 4: 40th Anniversary Gift Set includes MX101-A1 **Road Roller**; MX102-A1 **Tractor**; MX103-A1 **London Bus**; MX104-A1 **Milk Float**; MX105-A1 **Fire Engine**

TWIN PACKS

Row 1: TP101-A1 & 4 **Matra Rancho & Pony Trailer;** TP102-A2 **Ford Escort & Glider Transporter**
Row 2: TP103-A3, 4 & 5 **Cattle Truck & Trailer**
Row 3: TP104A2 & 6 **Locomotive & Coach;** TP105-A1 **Datsun & Boat**
Row 4: TP106-A1, 2 & 4 **Renault & Motorcycle Trailer**

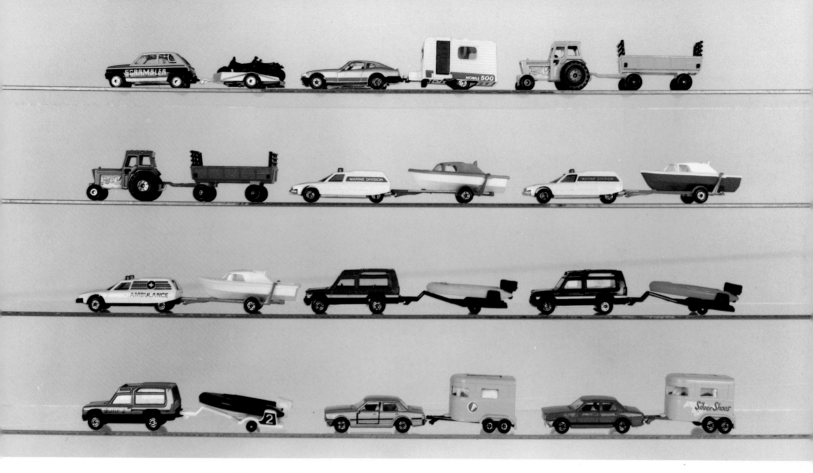

Row 1: TP106-A5 **Renault & Motorcycle Trailer**, TP107-A1 **Datsun & Caravan**; TP108-A3 **Tractor & Trailer**
Row 2: TP108-A1 **Tractor & Trailer**; TP109-A1 & 4 **Citroen & Boat**
Row 3: TP109-A7 **Citroen & Boat**; TP110-A4 & 3 **Matra Rancho & Inflatable**
Row 4: TP110-A5 **Matra Rancho & Inflatable**; TP111-A1 & 2 **Ford Cortina & Pony Trailer**

Row 1: TP112-A1, 2 & 3 **Unimog & Trailer**
Row 2: TP113-A1 **Porsche & Caravan;** TP115-A1 & 2 **Ford Escort & Boat**
Row 3: TP116-A1 **Jeep Cherokee & Caravan;** TP117-A1 & 2 **Mercedes G Wagon & Pony Trailer**
Row 4: TP118-A1 **BMW & Glider Trailer;** TP119-A1 **Flareside Pickup & Seafire;** TP120-A1 **Volkswagen & Inflatable**

Row 1: TP121-A1 **Land Rover & Seafire**; TP122A1 & 2 **Porsche & Glider Trailer**
Row 2: TP123A1 & 2 **BMW & Caravan**; TP125-A1 **Shunter & Side Tipper**
Row 3: TP124-A1 **Locomotive & Coach**; TP128-A1 **Circus Set**; TP126-A1 Tractor & Trailer
Row 4: TP129-A1 **Isuzu Amigo & seafire**; TP130-A1 **Land Rover & Pony Trailer**; TP131-A1 **Mercedes G Wagon & Inflatable**

Row 1: 010301 **Dodge Instant Winner Game**; 010120 **Limited Edition Christmas Offer**
Row 2: 32520 **Days of Thunder Race Car Challenge**; M02-A1 **Ford Matchmates**

WHITE ROSE COLLECTIBLES1992 BASEBALL SET(MB62D Chevrolet Corvette)

WHITE ROSE COLLECTIBLES 1990 FOOTBALL SET (MB38E Model A Ford Van)

WHITE ROSE COLLECTIBLES 1991 FOOTBALL SET (MB38E Model A Ford Van)

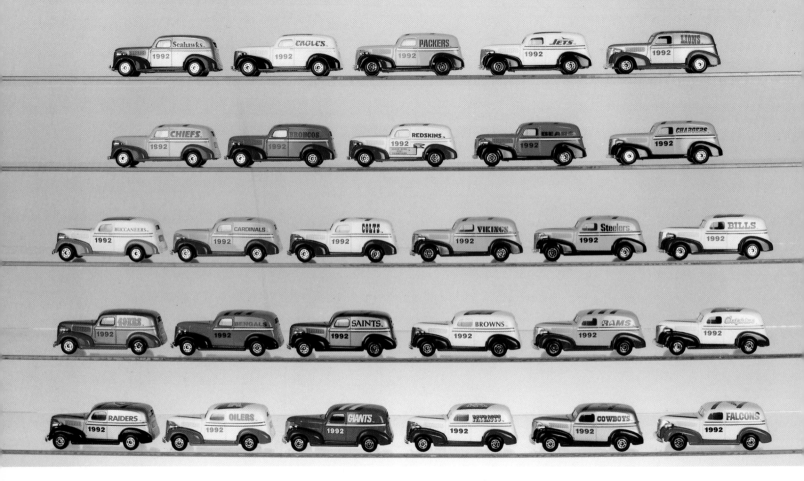

WHITE ROSE COLLECTIBLES1992 FOOTBALL SET (MB215A Chevy Panel Van)

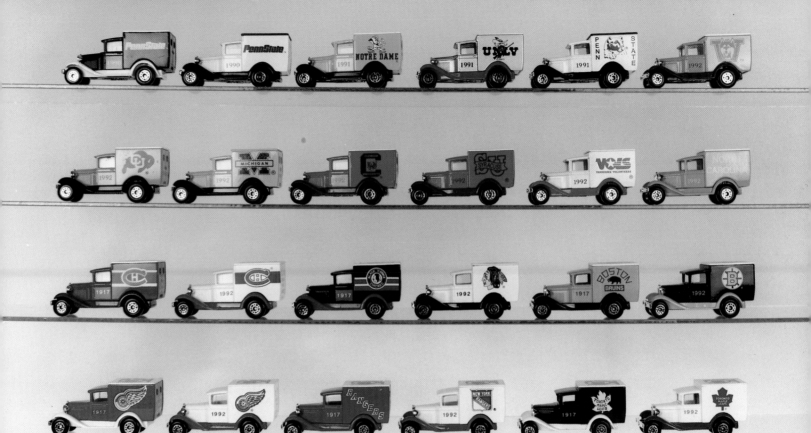

WHITE ROSE COLLECTIBLES
Row 1: MB38E123, 124, 199, 200, 238 & 233 **Model A Ford Van**
Row 2: MB38E234, 235, 236, 237, 258 & 257 **Model A Ford Van**
Row 3: MB38E239, 240, 241, 242, 243 & 244 **Model A Ford Van**
Row 4: MB38E245, 246, 247, 248, 249 & 250 **Model A Ford Van**

Row 1: MB38E78, 125, 204 **Model A Ford Van;** MB212A29 **Chevy Panel Van;** MB73C/55I16 **Model A Ford;** MB47E15 **School Bus**
Row 2: MB38E77, 164, 272, 273, 274 & 275 **Model A Ford Van**
Row 3: MB217A2, 3, 4, 5, 6 & 7 **D.I.R.T. Modified**
Row 4: MB217A1 **D.I.R.T. Modified;** MB62D46 **Chevrolet Corvette;** CY109A12 **Ford Aeromax Superstar;** K-79A-**Gran Fury Police Car**

Row 1: MB54H7, 16, 17, 37, 31 & 36 **Chevy Lumina**
Row 2: MB54H40, 41, 42, 43, 44 & 47 **Chevy Lumina**
Row 3: MB54H48, 50, 52 & 53 **Chevy Lumina**; MB212-A1 & 3 **Pontiac Grand Prix**
Row 4: MB216A10, 8, 6, 12, 13 & 14 **Pontiac Grand Prix**

Row 1: MB212A1, 10, 4, 5, 7 & 8 **Ford Thunderbird**
Row 2: MB212A11, 12, 13, 15, 18 & 19 **Ford Thunderbird**
Row 3: TC63-A1 **J.D. McDuffie Team**; TC65-A1 **Bill Elliot Team**
Row 4: TC60-A1 **Pennzoil Team**; Winross/Matchbox Set including Winross brand truck (background) with MB54-H32 & 33 **Chevy Lumina** (foreground)

Row 1: TC54-A1 & 6 **Goodwrench Team**
Row 2: TC57-A1 **Kodak Team**; TC59-A1 **Schraeder Team**
Row 3: TC56-A1 **Purolator Team**; TC61-A1 **STP (Petty) Team**
Row 4: TC64-A1 **Pontiac Excitement Team**; TC62-A1 **Mello Yello Team**

Row 1: CY104-A1 & 2 Kenworth Superstar
Row 2: CY104-A3 & 4 Kenworth Superstar
Row 3: CY104-A5 & 11 Kenworth Superstar
Row 4: CY104-A13 & 14 Kenworth Superstar

Row 1: CY104-A19 & 20 **Kenworth Superstar**
Row 2: CY104-A22 & 21 **Kenworth Superstar**
Row 3: CY104-A23 & 26 **Kenworth Superstar**
Row 4: CY104-A27 & 28 **Kenworth Superstar**

Row 1: CY104-A29 & 30 **Kenworth Superstar**
Row 2: CY104-A31 & 32 **Kenworth Superstar**
Row 3: CY104-A33 & 34 **Kenworth Superstar**
Row 4: CY104-A18 & 35 **Kenworth Superstar**

Row 1: CY104-A40 & 41 **Kenworth Superstar**
Row 2: CY104-A43 & 42 **Kenworth Superstar**
Row 3: CY104-A44 & 45 **Kenworth Superstar**
Row 4: CY104-A46 & 47 **Kenworth Superstar**

Row 1: CY104-A48 & 49 Kenworth Superstar
Row 2: CY104-A50 **Kenworth Superstar**; CY107-A2 **Mack Superstar**
Row 3: CY107-A3 & 4 **Mack Superstar**
Row 4: CY107-A5 & 6 **Mack Superstar**

Row 1: CY109-A1 & 2 **Ford Aeromax Superstar**
Row 2: CY109-A3 & 4 **Ford Aeromax Superstar**
Row 3: CY109-A5 & 7 **Ford Aeromax Superstar**
Row 4: CY109-A6 & 10 **Ford Aeromax Superstar**

Row 1: CY109-A11 **Ford Aeromax Superstar**; CY107-A7 **Mack Superstar**
Row 2: CY110-A1 **Kenworth Superstar**; CY109-A9 **Ford Aeromax Superstar**
Row 3: CY107-A1 **Mack Superstar**; CY104-A24 **Kenworth Superstar**
Row 4: CY104-A51 **Kenworth Superstar**; CY109-A8 **Ford Aeromax Superstar**

CONVOY
Row 1: CY1-A8 & 10 **Kenworth Car Transporter**
Row 2: CY2-A7 & 8 **Kenworth Rocket Transporter**
Row 3: CY3-A 9 & 11 **Double Container**
Row 4: CY3-A16 **Double Container**; CY3-B **Kenworth Box Truck**

Row 1: CY4-B2 **Scania Box Truck**; CY5-A5 **Peterbilt Covered Truck**
Row 2: CY5-A6 **Peterbilt Covered Truck**; CY7-A9 **Peterbilt Gas Tanker**
Row 3: CY8-A1 & 10 **Kenworth Box Truck**
Row 4: CY8-A13 & 11 **Kenworth Box Truck**

Row 1: CY8-A15 **Kenworth Box Truck**; CY9-A3 **Kenworth Box Truck**
Row 2: CY9-A5 & 11 **Kenworth Box Truck**
Row 3: CY9-A18 & 13 **Kenworth Box Truck**
Row 4: CY9-A14 & 15 **Kenworth Box Truck**

Row 1: CY9-A16 & 19 Kenworth Box Truck
Row 2: CY9-A21 Kenworth Box Truck; CY10-A1 Racing Transporter
Row 3: CY11-A1 & 5 Kenworth Helicopter Transporter
Row 4: CY11-A3 Kenworth Helicopter Transporter; CY12-A1 Kenworth Plane Transporter

Row 1: CY13-A1 & 4 **Peterbilt Fire Engine**
Row 2: CY13-A7 **Peterbilt Fire Engine; CY14-A1 Kenworth Boat Transporter**
Row 3: CY15-A1 & 2 **Peterbilt Tracking Vehicle**
Row 4: CY15-A4 & 7 **Peterbilt Tracking Vehicle**

Row 1: CY15-A11 **Peterbilt Tracking Vehicle**; CY16-A2 **Scania Box Truck**
Row 2: CY16-A4 & 5 **Scania Box Truck**
Row 3: CY16-A6 & 8 **Scania Box Truck**
Row 4: CY16-A9 & 11 **Scania Box Truck**

Row 1: CY16-A12 & 13 **Scania Box Truck**
Row 2: CY16-A16 & 7 **Scania Box Truck**
Row 3: CY16-A15 & 19 **Scania Box Truck**
Row 4: CY17-A1 & 2 **Scania Petrol Tanker**

Row 1: CY17-A3 & 4 **Scania Petrol Tanker**
Row 2: CY17-A7 & 8 **Scania Petrol Tanker**
Row 3: CY18-A1 & 7 **Scania Double Container**
Row 4: CY18-A6 & 5 **Scania Double Container**

Row 1: **CY18-A3 & 4 Scania Double Container**
Row 2: **CY19-A1 Peterbilt Box Truck; CY20-A2 Scania Articulated Truck**
Row 3: **CY20-A3 & 4 Kenworth Articulated Truck**
Row 4: **CY20-A7 & 9 Kenworth Articulated Truck**

Row 1: CY20-A10 **Kenworth Articulated Truck**; CY21-A1 **DAF Aircraft Transporter**
Row 2: CY21-A2 **DAF Aircraft Transporter**; CY22-A1 **DAF Boat Transporter**
Row 3: CY22-A2 & 4 **DAF Aircraft Transporter**
Row 4: CY22-A5 **DAF Aircraft Transporter**; CY23-A2 **Scania Covered Truck**

Row 1: CY24-A2 & 3 **DAF Box Truck**
Row 2: CY24-A5 & 6 **DAF Box Truck**
Row 3: CY25-A3 & 4 **DAF Container Truck**
Row 4: CY25-A1 & 6 **DAF Container Truck**

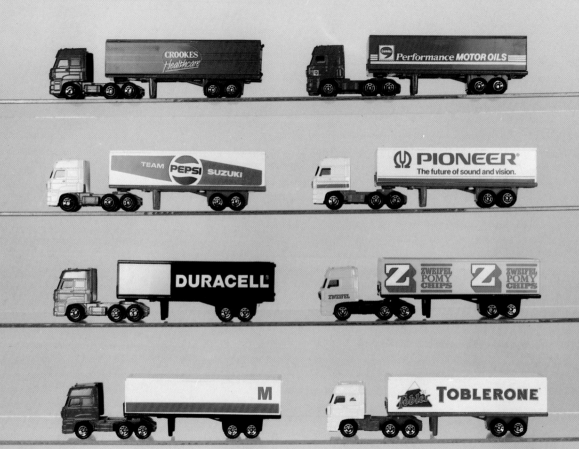

Row 1: CY25-A2 & 5 **DAF Container Truck**
Row 2: CY25-A7 & 9 **DAF Container Truck**
Row 3: CY25-A10 & 11 **DAF Container Truck**
Row 4: CY25-A12 & 13 **DAF Container Truck**

Row 1: CY25-A14 & 17 **DAF Container Truck**
Row 2: CY25-A18 & 19 **DAF Container Truck**
Row 3: CY26-A1 **DAF Double Container; CY27-A1 Mack Container Truck**
Row 4: CY28-A2 & 4 **Mack Double Container**

Row 1: CY29-A1 **Mack Aircraft Transporter**; CY30-A1 **Grove Crane**
Row 2: CY31-A1 **Mack Pipe Truck**; CY32-A1 **Mack Tractor Shovel Transporter**
Row 3: CY33-A2 & 1 **Mack Helicopter Transporter**
Row 4: CY33-A3 **Mack Helicopter Transporter**; CY34-A1 **Peterbilt Emergency Center**

Row 1: CY36-A2 & 1 Kenworth Transporter
Row 2: CY35-A1 **Mack Tanker**; CY105-A1 Kenworth Tanker
Row 3: CY105-A2 **Kenworth Tanker**; CY106-A1 **Peterbilt Tipper Truck**
Row 4: CY108-A1 **DAF Airplane Transporter**; CY203-A1 **Peterbilt Lowloader with Excavator**

Row 1: Preproductions **Peterbilt Container Truck**; CY16-A **Scania Box Truck**
Row 2: Preproductions CY24-A **DAF Box Truck**; **Peterbilt Container Truck**
Row 3: Preproductions CY25-A **DAF Container Truck**; CY8-A **Kenworth Container Truck**
Row 4: Preproductions 2X CY8-A **Kenworth Container Truck**

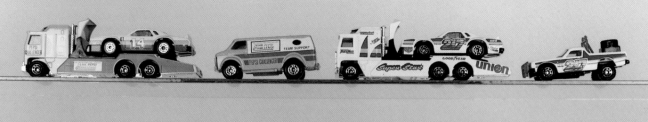

Row 1: TM1-A1(TC7-A1) **Pepsi Team**; TM2-A1 **Super Star Team**
Row 2: TM3-A1 **Dr. Pepper Team**; TM4-A2 **Brut Team**
Row 3: TM5-A1 (TC8-A1) **7 Up Team**; TM6-A1 (TC9-A1) **Duckhams Team**
Row 4: TMX-A2 & 1 **STP Teams**

115

Row 1: TC10-A1 **Fuji Team**; TC11-A2 **Pirelli Team**
Row 2: TC12-A1 **Tizer Team**; CY111-A1 **Racing Team**
Row 3: TC4-A1 & 2 **Cargo Set**
Row 4: TC4-A2 **Cargo Set**; TC18-A1 **Transport Set**

116

SUPERKINGS
Row 1: K-8E1 **Ferrari F40**; K-15B23, 14 & 15 **London Bus**
Row 2: K-15B16, 17, 18 & 19 **London Bus**
Row 3: K-15B20, 21, 22 & 24 **London Bus**
Row 4: K-15B25 & 26; K-25B3 **Digger & Plow**

Row 1: K-21D12 **Ford Transcontinental**
Row 2: K-170A1 **JCB Excavator**; K-25B6 **Digger & Plow**; K-42C3 **Traxcavator Road Ripper**
Row 3: K-43B2 **Log Transporter**
Row 4: K-69A10 **Volvo Estate & Europa Caravan**; K-70A6 **Porsche Turbo**

Row 1: K-69A9 Volvo Estate & Europa Caravan; K-70A7 Porsche Turbo
Row 2: K-78A17 & 15 Gran Fury Police Car; K-87A2 Massey Ferguson Tractor & Rotary Rake
Row 3: K-88A6 Money Box; K-90A1 & 4 Matra Rancho
Row 4: K-95A6 Audi Quattro; K-98A2, 1 & 3 Porsche 944

Row 1: K-98A4, 5 & 6 Porsche 944; K-99B1 **Dodge Polizei Van**
Row 2: K-100A1, 2, 3 & 4 **Ford Sierra**
Row 3: K-100A5 & 7 **Ford Sierra**; K-101B1 & 2 **Racing Porsche**
Row 4: K-101B4 **Racing Porsche**; K-102B1 **Rally Support Set**

Row 1: K-103B3 **Peterbilt Tanker**
Row 2: K-104B1 **Matra Rescue Set;** K-105B1 **Peterbilt Tipper**
Row 3: K-105B2 **Peterbilt Tipper;** K-106B1 **DAF Aircraft Transporter**
Row 4: K-107B1 **Powerboat Transporter**

Row 1: K-108B2 **Digger Transporter**
Row 2: K-109B2 **Iveco Tanker;** K-110B1 **Fire Engine;** K-111B1 **Peterbilt Refuse Truck**
Row 3: K-112B1 **Fire Spotter Transporter**
Row 4: K-114B1 & 3 **Crane Truck**

Row 1: K-115B1, 2 & 3 Mercedes 190E
Row 2: K-115B4, 5 & 7 Mercedes 190E
Row 3: K-116B1 **Ferrari Racing Transporter**
Row 4: K-117B1 **Scania Bulldozer Transporter**

Row 1: K-118B1 **Road Construction Set**
Row 2: K-119A1 **Fire Set**
Row 3: K-120A1 **Car Transporter**; K-121A1 **Peterbilt Wrecker**
Row 4: K-122A1 **DAF Road Train**

Row 1: K-122A4 **DAF Road Train**
Row 2: K-122A2 **DAF Road Train**
Row 3: K-122A5 **DAF Road Train**
Row 4: K-122A3 **DAF Road Train**

125

Row 1: K-123A2 **Leyland Cement Truck;**
K-124A1 **Mercedes Container Truck**
Row 2: K-124A2 **Mercedes Container Truck**
Row 3: K-126A1 **DAF Helicopter Transporter**
Row 4: K-126A2 **DAF Helicopter Transporter**

Row 1: K-127A1 **Peterbilt Tanker**
Row 2: K-127A2 **Peterbilt Tanker**
Row 3: K-127A3 **Iveco Tanker**
Row 4: K-128A1 **DAF Aircraft Transporter**

127

Row 1: K-129A6 **Powerboat Transporter**

Row 2: K-130A1 **Scania Digger Transporter**

Row 3: K-131A3 & 2 **Iveco Tanker**; K-132A3 **Fire Engine**

Row 4: K-133A1, 3 & 4 **Iveco Refuse Truck**

Row 1: K-135A1 **Garage Transporter**   *Row 3:* K-137A1 **Road Construction Set**
Row 2: K-135A3 **Garage Transporter**   *Row 4:* K-138A1 **Fire Rescue Set**

Row 1: K-139A1, 3 & 4 Iveco Tipper Truck
Row 2: K-140A1 Car Recovery Vehicle; K-141A1 & 2 Skip Truck
Row 3: K-142A1, 2 & 3 BMW 5 Series
Row 4: K-143A1 Bedford Emergency Van; K-144A1 & 2 Land Rover

Row 1: K-144A3 & 4 **Land Rover;** K-146A2 **Jaguar XJ6**
Row 2: K-145A1 **Iveco Twin Tippers**
Row 3: K-146A1 & 4 **Jaguar XJ6;** K-147A2 **BMW 750iL**
Row 4: K-147A3 & 4 **BMW 750iL;** K-148A1 **Crane Truck**

Row 1: K-149A1 **Ferrari Testarossa**; K-151A1 & 4 **Skip Truck**
Row 2: K-150A1 **Leyland Truck (3 pcs.)**; K-153A1 **Jaguar XJ6 Police Car**
Row 3: K-154A1 & 2 **BMW 750iL Police Car**; K-155A1 **Ferrari Testarossa Rally**
Row 4: K-156A1 **Porsche Turbo Rally**; K-157A1 **Porsche 944 Turbo**; K-158A1 **Ford Sierra Rally**

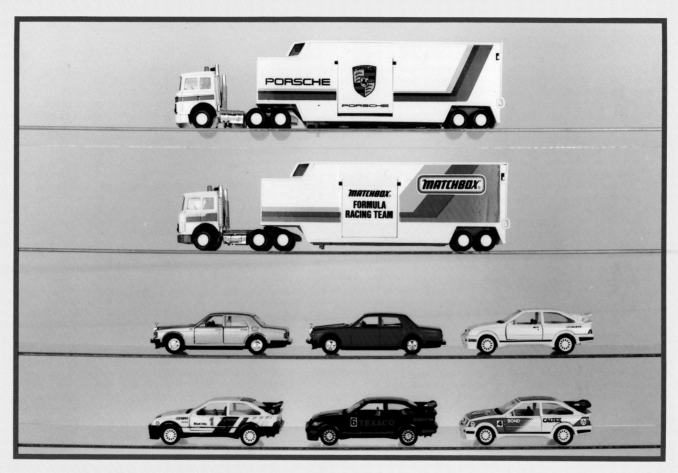

Row 1: K-159A1 **Porsche Transporter**
Row 2: K-160A1 **Matchbox Transporter**
Row 3: K-161A1 & 2 Rolls Royce Silver Spirit; K-162A1 **Ford Sierra Cosworth**
Row 4: K-162A2, 3 & 4 **Ford Sierra Cosworth**

Row 1: K-163A1 **Unimog with Plow**; K-164A1 & 3 **Range Rover**
Row 2: K-164A2 & 4 **Range Rover**; K-165A1 **Range Rover Police**
Row 3: K-165A2 **Range Rover Police**; K-166A1 **Mercedes 190E Taxi**; K-168A1 **Porsche 911 Carrera**
Row 4: K-168A2 **Porsche 911 Carrera**; K-167A2 & 1 **Ford Transit**

Row 1: K-167A3 **Ford Transit**; K-169A1 **Ford Transit Ambulance**; K-171A1 **Toyota Hi-Lux**
Row 2: K-171A2 **Toyota Hi-Lux**; K-172A1 & 2 **Mercedes 500SL**
Row 3: K-173A1 **Lamborghini Diablo**; CS-5A1 **Unimog Tar Sprayer**; FM-3A1 **Massey Ferguson Tractor**
Row 4: FM-5A1 **Muir Hill Tractor & Trailer**; EM-13A1 **Helicopter**

SPECIALS
Row 1: SP1/2A5, 1, 3 & 7 **Kremer Porsche**
Row 2: SP1/2A9 & 10 **Kremer Porsche**; SP3/4A2 & 1 **Ferrari 512BB**
Row 3: SP3/4A3, 5, 7 & 6 **Ferrari 512BB**
Row 4: SP3/4A1 **Ferrari 512BB**; SP5/6A3, 1 & 2 **Lancia Rallye**

Row 1: SP7/8A2, 1, 5 & 7 **Zakspeed Mustang**
Row 2: SP7/8A9 **Zakspeed Mustang**; SP9/10A2, 7 & 1 **Chevy Prostocker**
Row 3: SP11/12A2, 3, 7 & 8 **Chevrolet Camaro**
Row 4: SP13/14A1, 2, 4 & 3 **Porsche 959**

MUSCLE CARS/TURBO SPECIALS/ALARM CARS
Row 1: SP13/14A6 & 7 **Porsche 959**; Muscle Cars SP9/10A9 & 8 **Chevy Prostocker**
Row 2: Muscle Cars SP7/8A10 & 9 **Zakspeed Mustang**; SP11/12A9 & 10 **Chevrolet Camaro**
Row 3: Turbo Specials SP11/12A4 **Chevrolet Camaro**; SP7/8A3 & 6 **Zakspeed Mustang**; SP1/2A6 **Kremer Porsche**
Row 4: Turbo Specials SP9/10A4 & 5 **Chevy Prostocker**, Alarm Car K-8E3 **Ferrari F40**

YESTERYEARS
Row 1: Y-1C7 **1936 Jaguar SS;** Y-2D1, 3 & 5 **1930 Bentley**
Row 2: Y-3D4, 6, 7, 9 & 11 **1912 Model T Tanker**
Row 3: Y-3E1 **1912 Model T Tanker,** Y-4D11, 13 & 15 **1930 Model J Duesenberg**
Row 4: Y-5D19, 21, 23 & 25 **1927 Talbot Van;** Y-5E1 **1929 Leyland Titan**

139

Row 1: Y-5E3 **1929 Leyland Titan**; Y-6D8 **1920 Rolls Royce Fire Engine**; Y-6E1 **1932 Mercedes Benz Lorry**; Y-7D3 **1930 Model A Wrecker**

Row 2: Y-8D10 & 13 **1945 M.G. TD**; Y-8E1, 3 & 4 **1917 Yorkshire Steam Wagon**

Row 3: Y-9C1 **1920 Leyland 3 Ton Lorry**; Y-9D1 **1936 Leyland Cub Fire Engine**; Y-10D1 **1957 Maserati 250F**; Y-10E1 **1931 Diddler Trolley Bus**

Row 4: Y-11C13 **1911 Lagonda Drophead Coupe**; Y-11D1 & 2 **1932 Bugatti Type 51**; Y-12C18 & 12 **1912 Model T Van**

Row 1: Y-12C19, 20, 22, 24 & 26 **1912 Model T Van**
Row 2: Y-12C28, 31 & 30 **1912 Model T Van**; Y-12D1 **1912 Model T Pickup**; Y-12E1 **Stephenson's Rocket**
Row 3: Y-12F1, 3 & 4 **1937 GMC Van**; Y-13C13 **1913 Crossley**
Row 4: Y-13C16 **1913 Crossley**; Y-14C9 & 10 **1934 Stutz Bearcat**; Y-14D1 **1935 E.R.A.**

141

Row 1: Y-14D2 **1935 E.R.A.**; Y-15B14 **1930 Packard Victoria**; Y-15C1 & 3 **Preston Type Tramcar**
Row 2: Y-15C4 & 5 **Preston Type Tramcar**; Y-16B17 & 18 **1928 Mercedes SS**
Row 3: Y-16C1 **1960 Ferrari Dino 246/V12**; Y-16E1 **Scammell 100 Ton Transporter with Class 2-4-0 Locomotive**
Row 4: Y-16D1 **1922 Scania Vabis Postbus**; Y-17A8 **1938 Hispano Suiza**; Y-18A6 **1937 Cord 812**; Y-18B1 **1918 Atkinson**

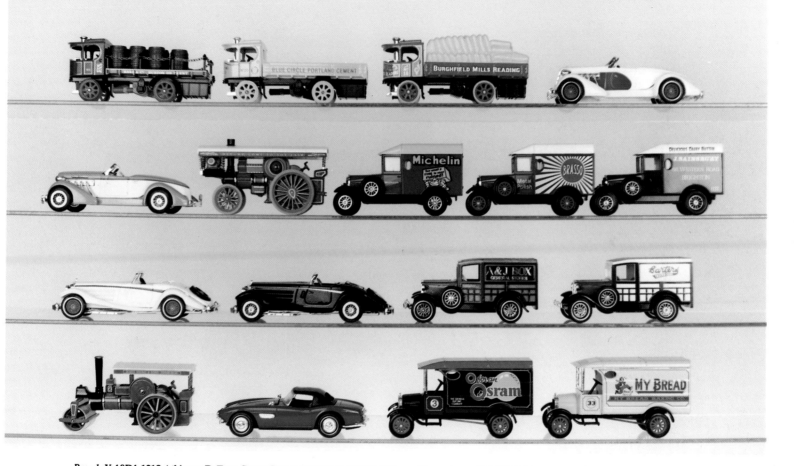

Row 1: Y-18D1 **1918 Atkinson D-Type Steam Lorry;** Y-18C1 & 2 **1918 Atkinson Steam Lorry;** Y-19A7 **1935 Auburn 851 Supercharged Roadster**
Row 2: Y-19A8 **1935 Auburn 851 Supercharged Roadster;** Y-19B1 **Fowler B6 Showman's Engine;** Y-19C2, 1 & 3 **1929 Morris Light Van**
Row 3: Y-20A6 & 9 **1937 Mercedes 540K;** Y-21A5 & 7 **1930 Model A Woody Wagon**
Row 4: Y-21B2 **Aveling-Porter Steam Roller;** Y-21C1 **1957 BMW 507;** Y-21D1 & 3 **Ford Model TT Van**

Row 1: Y-21D4 **Ford Model TT Van;** Y-22A2, 4 & 3 **1930 Model A Ford Van**
Row 2: Y-22A5, 7, 6, 8 & 11 **1930 Model A Ford Van**
Row 3: Y-23A2, 3, 4, 5 & 7 **1922 AEC 'S' Type Bus**
Row 4: Y-23A6 & 8 **1922 AEC 'S' Type Bus;** Y-23B1 & 2 **Mack Petrol Tanker**

Row 1: Y-24A2 & 5 **1928 Bugatti T44**; Y-25A1 & 3 **1910 Renault Type AG Van**
Row 2: Y-25A6, 8, 7, 9 & 10 **1910 Renault Type AG Van**
Row 3: Y-25B1 **1910 Renault Type AG Ambulance**; Y-26A1, 4 & 5 **Crossley Beer Lorry**
Row 4: Y-27A1, 4, 3 & 5 **1922 Foden C Type Steam Lorry**

Row 1: Y-27A6, 7 & 8 **1922 Foden C Type Steam Wagon**; Y-28A1 **1907 Unic Taxi**
Row 2: Y-28A7 & 9 **1907 Unic Taxi**; Y-27B1 **1922 Foden Steam Wagon & Trailer**
Row 3: Y-29A1, 2, 3 & 4 **1919 Walker Electric Van**; Y-30A1 **1920 Model AC Mack Truck**
Row 4: Y-30A3 & 4 **1920 Model AC Mack Truck**; Y-30B1 **1920 Mack Canvasback Truck**; Y-31A1 **1931 Morris Pantechnicon**

Row 1: Y-31A2 **1931 Morris Pantechnicon;** Y-32A1 **1917 Yorkshire Steam Wagon;** Y-33A1 **1920 Mack Truck;** Y-34A1 **1933 Cadillac V-16**
Row 2: Y-34A2 **1933 Cadillac V-16;** Y-35A1 & 2 **1930 Model A Ford Pickup;** Y-36A1 **1926 Rolls Royce Phantom I**
Row 3: Y-36A2 **1926 Rolls Royce Phantom I;** Y-37A1 & 2 **1931 Garrett Steam Wagon;** Y-38A1 **Rolls Royce Armoured Car**
Row 4: Y-39A1 **1820 English Coach;** Y-40A1 **1931 Mercedes Benz Type 770;** Y-41A1 **1932 Mercedes Benz Lorry**.

Row 1: Y-42A1 **1938 Albion CX27**; Y-43A1 **1905 Busch Steam Fire Engine**; Y-44A1 & 2 **1910 Renault Bus**
Row 2: Y-45A1 **1930Bugatti Royale**; Y-46A1 **1868 Merryweather Fire Engine**; Y-47A1 **1928 Morris Van**; Y-61A1 **1933 Cadillac Fire Engine**
Row 3: Y-62A1 **1932 Ford AA Truck**; Y-63A1 **1939 Bedford Truck**; Y-64A1 **1938 Lincoln Zephyr**; Y-901 **1936 Jaguar SS100** in pewter
Row 4: Y-65A1 **Austin 7/BMW/Rosengart** (3 piece set); Y-66A1 **Gold State Coach**

Left to right: Y-15C2 Preston Type Tramcar, Y-8E2 1917 Yorkshire Steam Wagon, Y-5E2 1929 Leyland Titan

Row 1: Y-1C6 **1936 Jaguar SS100**; Y-9B14 **1912 Simplex**; Y-17A5 **1938 Hispano Suiza**

Row 2: Y-20A8 **1937 Mercedes 540K**; Y-24A4 **1928 Bugatti Type 44**; Connoisseur set models—Y-1B9 **1911 Model T Ford**; Y-3B23 **1910 Benz Limousine**; Y-4C9 **1909 Opel Coupe**

Row 3: Connoisseur set models—Y-11B6 **1912 Packard Laundalet**; Y-13B7 **1911 Daimler**; Y-14B6 **1911 Maxwell**; preproductions of Y-12C **Model T Van**; Y-22A **1930 Model A Ford Van**

Row 4: preproductions of Y-12F **1937 GMC Van**; Y-20A **Mercedes Benz 540K**; Y-23A **1922 AEC Type 'S' Bus**; Y-24A **Bugatti Type 44**

DINKY
Row 1: DY-1A1 & 2 **1967 Series E Type Jaguar; DY-2A1 1957 Chevrolet Bel Air Sports Coupe; DY-3A1 1965 MGB GT**
Row 2: DY-3A2 **1965 MGB GT; DY-4A1 & 2 1950 Ford E.83.W Van; DY-5A1 1949 Ford V-8 Pilot**
Row 3: DY-5A2 & 3 **1949 Ford V-8 Pilot; DY-6A1, 3 & 2 1951 Volkswagen Deluxe Sedan**
Row 4: DY-8A1 & 2 **1948 Commer CWT Van; DY-7A1 & 2 1959 Cadillac Coupe De Ville**

Row 1: DY-9A1 & 2 **1949 Land Rover Series**; DY-10A1 **1950 Mercedes Benz Konfernz Type Bus**; DY-11A1 **1948 Tucker Torpedo**
Row 2: DY-11A2 **1948 Tucker Torpedo**; DY-12A1 & 2 **1955 Mercedes 300SL**; DY-13A1 **1955 Bentley Continental**
Row 3: DY-13A2 **1955 Bentley Continental**; DY-14A2 & 1 **1948 Delahaye Chapron**; DY-15A1 **1952 Austin A40 GV4 lO-CWT Van**
Row 4: DY-15A2 **1952 Austin A40 GV4 lO-CWT Van**; DY-16A1 & 2 **1967 Ford Mustang Fastback 2+2**; DY-17A1 **1938 Triumph Dolomite**

Row 1: DY-18A1 **1967 Series E Type Jaguar Convertible**; DY-19A1 **1973 MGB GT V-8**; DY-20A1 **1967 Triumph TR-4A**; DY-21A **1964 Mini Cooper 'S'**; DY-22A1 **1952 Citroen 15CV**

Row 2: DY-22A2 **1952 Citroen 15CV**; DY-23A2 & 1 **1956 Chevrolet Corvette**; DY-24A1 **1973 Ferrari Dino**

Row 3: DY-25A **1958 Porsche 956A**; DY-26A1 **1958 Studebaker Golden Hawk**; DY-27A1 **1957 Chevrolet Bel Air Sports Coupe Convertible** ; DY-28A1 **1973 Triumph Stag**

Row 4: DY-29A1 **1953 Buick Skylark**; DY-30A1 **1956 Austin Healey**; DY-31A1 **1956 Ford Thunderbird**; DY-32A **1957 Citroen 2CV**

Row 1: DY-902 on plinth including DY-25A2 **1958 Porsche 956A**; DY-12A3 **1955 Mercedes 300SL**; DY-24A2 **Ferrari Dino**
Row 2: DY-903 on plinth including DY-20A2 **1965 Triumph TR-4A**; DY-18A2 **1967 Series E Type Jaguar Convertible**; DY-30A2 **1956 Austin Healey**
Row 3: DY-921 **1967 Series E Type Jaguar in pewter**
Row 4: original Dinky releases by Matchbox MB7C24 **V.W. Golf**; MB9D/74E14 **Fiat Abarth**; MB12E/51E18 **Pontiac Firebird SE**; MB14E/69E12 **1983/84 Corvette**; MB39C/60E8 **Toyota Supra**; MB44-E8 **Citroen CV15**

SKYBUSTERS

Row 1: SB 1-A6, 7, 8, 9 & 11 **Lear Jet**

Row 2: SB 1-A12 **Lear Jet**; SB 2-A6 & 7 **Corsair A7D**; SB 4-A5 & 7 **Mirage F1**

Row 3: SB 4-A8 & 10 **Mirage F1**; SB 6A9 & 10 **MIG 21**; SB 7-A4 **Junkers**

Row 4: SB 8-A10 **Spitfire**; SB 9-A13, 14 & 12 **Cessna 402**; SB10-A13 **Boeing 747**

Row 1: SB10-A11, 12, 17, 15 & 14 **Boeing 747**
Row 2: SB10-A18, 19, 22, 24 & 25 **Boeing 747**
Row 3: SB11A-4 & 5 **Alpha Jet**; SB12B2, 3 & 4 **Pitts Special**
Row 4: SB13A-12, 11, 10, 9 & 16 **DC.10**

Row 1: SB13A-18 **DC.10**; SB15A7 **Phantom F4E**; SB19A2 & 3 **Piper Commanche**; SB20-A4 **Helicopter**
Row 2: SB22-A6, 8 & 7 **Tornado**; SB23-A4 **Supersonic Transport**
Row 3: SB23-A5, 6 & 7 **Supersonic Transport**; SB24-A5 **F.16**
Row 4: SB24-A7, 9 & 8 **F.16**; SB25-A6 **Rescue Helicopter**

Row 1: SB25A-5 & 7 **Rescue Helicopter**; SB26-A2 & 3 **Cessna Floatplane**
Row 2: SB26-A4 & 5 **Cessna Floatplane**; SB27-A2, 7 & 3 **Harrier Jet**
Row 3: SB27-A6 & 8 **Harrier Jet**; SB28-A3 & 2 **A.300 Airbus**
Row 4: SB28-A5, 7 & 1 **A.300 Airbus**; SB29-A1 **Lockheed**; SB30-A1 **F.30 Tomcat**

Row 1: SB28-A9 **A.300 Airbus**; SB31-A1, 2, 3 & 4 **Boeing 747-400**

Row 2: SB32-A1 **Fairchild Thunderbolt**; SB33-A1 **Bell Jet Ranger**; SB34-A1 **Hercules**; SB35-A1 **MIL Mi Hind-D**; SB36-A1 **F.117 Stealth**

Row 3: SB36-A2 **F.117 Stealth**; SB37-A1 **Hawk**; SB38-A1 & 2 **BaE 46**

Row 4: SB39-A2 & 3 **Stearman**; SB40-A1 & 2 **Boeing 737**

High Riders

SUPERCHARGERS
Row 1: SC1-A1 **Awesome Kong II**; SC2-A1 **Taurus**; SC3-A1 **USA 1**; SC4-A1 **Big Foot**; SC5-A1 **Mad Dog II**
Row 2: SC6-A1 **Hawk**; SC7-A1 **Flying Hi**; SC8-A1 **Rollin Thunder**; SC9-A1 **So High**; SC12-A1 **Bog Buster**
Row 3: SC12A4 **Bog Buster**; SC11A1, 3 & 4 **Mud Ruler**; SC16Al **Doc Crush**
Row 4: SC18A1 & 2 **'57 Chevy**; SC10A1 & 3 **Toad**; SC13A1 **Hog**

Row 1: SC13-A3 **Hog**; SC15-A3 **Big Pete**; SC17-A2 & 3 **Mud Slinger**
Row 2: pullsleds
Row 3: SC20-A1 **Voo Doo**; SC19-A1 **Drag-On**; SC21-A1 **Hot Stuff**
Row 4: SC24-A1 **12 Pac**; SC22-A1 **Showtime**; SC23-A1 **Checkmate**

HARLEY DAVIDSON
Row 1: MB50C9 & 10 **Harley Davidson Motorcycle**; CY8-A17 **Kenworth Container Truck**
Row 2: K-83A3, 4 & 5 **Harley Davidson Motorcycle**
Row 3: Stunt Cycles

Row 1: Thunderbirds TB-001 **Thunderbirds 1**; TB-002 & 004 **Thunderbirds 2**; TB-003 **Thunderbirds 3**; TB-005 **Penelope's Fab 1**

Row 2: Dodge Las Vegas boxes with **Dodge Caravan & Dodge Daytona**

Row 3: Dodge Las Vegas **Dodge Caravan & Dodge Daytona** on wooden plaque

SUPERFAST MINIS
Row 1: Superfast Minis **2X Jaguar XJ6, 2X Lamborghini Countach, 2X BMW M1**
Row 2: Superfast Minis **2X Chevrolet Camaro, 2X '57 Chevy, 2X Ford Thunderbird**
Row 3: MD-250 **Transporter** with **Porsche Turbo & Ferrari F40**
Row 4: **Porsche Turbo & Chevy Lumina** with MD-250 **Transporter**

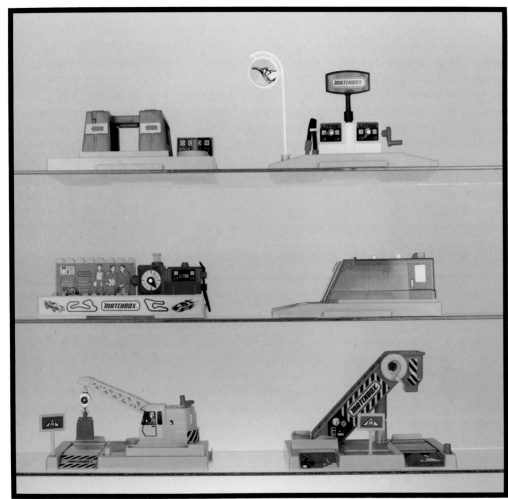

MOTOR CITY
Row 1: MC1-A2 **Car Wash**; MC2-A2 **Service Station**
Row 2: MC3-A2 **Pit Stop**; MC4-A2 **Garage**
Row 3: MC5-A2 **Crane**; MC6-A2 **Gravel Pit**

Row 1: assorted Cap Cars
Row 2: assorted Cap Cars
Row 3: assorted Flash Force 2000
Row 4: assorted Parasites

Row 1: assorted Trickshifters
Row 2: assorted Trickshifters
Row 3: assorted Turbo 2s
Row 4: assorted Turbo 2s

CODE 2
Row 1: CY9-A17 **Kenworth Container Truck**; CY16A14 **Scania Container Truck**
Row 2: CY-16A18 **Scania Container Truck**; CY9-A20 **Kenworth Container Truck**
Row 3: MB44H15 & 14 **1921 Model T Van**; Y-5E4 **1929 Leyland Titan**; MB38E66 **Model A Ford Van**
Row 4: MB17C/28F/51F44 **London Bus**; MB38E66 **Model A Ford Van**; CY27-A2 **Mack Container Truck**

169

CODE 2
Row 1: MB32E/12G13 **Modified Racer**; MB54H12 **Chevy Lumina**; MB65D20 **F.1 Racer**; MB66E/46E14 **Sauber Group C Racer**; MB74H/14H7 **G.P. Racer**
Row 2: MB7F21 **Porsche 959**; MB24H/70F3 **Ferrari F40**; MB52D9 **BMW M1**; MB61E/37H2 **Nissan 300ZX**; MB67F/11F14 **Lamborghini Countach**
Row 3: MB38E79, 122, 195 & 269 **Model A Ford Van**; DY-21A2 **1964 Mini Cooper 'S'**
Row 4: DY-8A3 **1948 Commer Van**; DY-28A2 **1973 Triumph Stag**; CY8-A8 **Kenworth Box Truck**

Row 1: UK Pocket catalogs 1983, 1984, 1985 & 1986
Row 2: UK Pocket catalogs 1987, 1988, 1989, 1990 & 1991

Row 1: 1992 UK Pocket catalog; Japanese pocket catalogs 1984, 1986, 1987 & 1988
Row 2: Australian pocket catalogs 1983, 1984 & 1985
Row 3: Australian pocket catalogs 1985, 1986 & 1987

Assorted Box Types

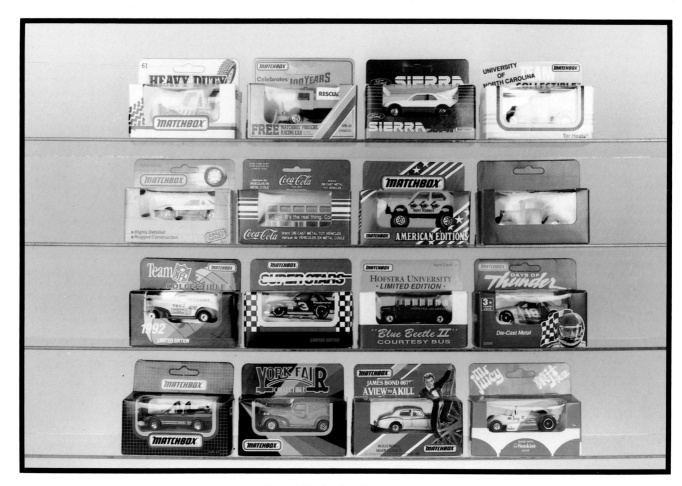

Assorted Window Box Types

Assorted Blisterpack Types

Assorted Pins and Badges

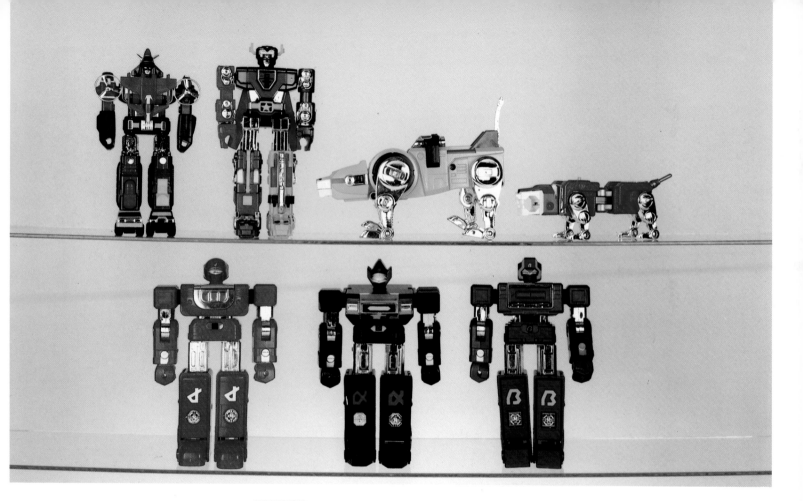

VOLTRON
Row 1: Miniature Voltron I; Miniature Voltron III; Voltron Lions
Row 2: Miniature Voltron II in red, black & blue

ROBOTECH
Row 1: 7352 Bioroid Hovercraft; 7120 Miniature SDF-1; 7131 Miniature Alpha Fighter; 7314 Gladiator
Row 2: 2 Battloids; 850003 Veritech Fighter; 7351 Armoured Cyclone
Row 3: Assorted Battloids

Row 1: Breetai large figure; Rick Hunter Doll; 7354 Tactical Battle Pod

Row 2: Scott Bernard mini figure; 7311 Excalibar MKVI; 7110 MAC II Destroid Cannon

PEE WEE HERMAN TOYS
Row 1: Chairry; Conky; Pterri; Reba; Pee Wee's Scooter
Row 2: Randy & Globey; Pee Wee Herman; Miss Yvonne

BIG MOVERS
Row 1: Stubbies Container Truck, Cement Truck & Crane Truck
Row 2: Stubbies Tanker; Container Truck
Row 3: Stubbies Dump Truck; 4X4 Pickup

Row 1: Screamin' Stocks Chevy Lumina & Ford T-Bird with launchers
Row 2: Hot Foot Racers
Row 3 & 4: Railways

Row 1: Dueling Dragsters
Row 2: Pocket Rockets
Row 3: Colani Cars
Row 4: Robotech vehicles

Row 1: Pocket Rocket Buggies
Row 2, 3 & 4: Assorted Power Lifters

Row 1 & 2: Assorted Key Cars
Row 3 & 4: Assorted Lightning Key Cars

Row 1 & 2: Assorted Speed Riders
Row 3 & 4: Flashbacks

186

Left: Car & Driver Collector Cards
Right: Puffy Stickers

Assorted Connectables

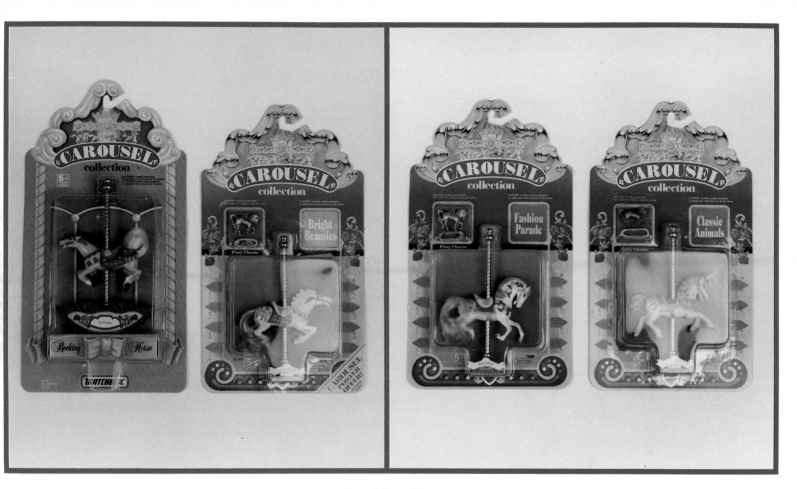

Assorted Carousel Horses

Row 1: 4 Monster In My Pocket "Howlers"; Monster In My Pocket Four Pack
Row 2: Monster In My Pocket Battle Cards; Monster In My Pocket Super Scary

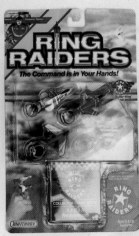

Row 1: World's Smallest Matchbox; Micro Super Chargers
Row 2: Matchbox 2000; Ring Raiders

191

Row 1: Babycise Clutch Ball; Linkits Stridants
Row 2: Oh Jenny Farm Animals; Rubik's Magic Puzzle

Row 1: Live N Learn Helicopter; Plastic Kit
Row 2: Superfast Machine; Popsicle Kids doll

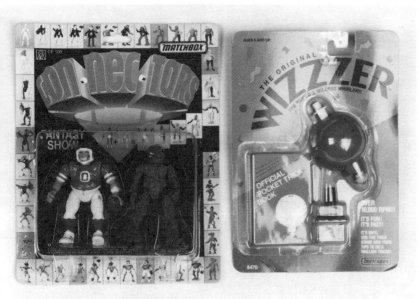

Connectors; Wizzzer Top

My Precious Puffs; Bubble Heads

Row 1: Cobra Radio Control car
Row 2: 2203 Hot Rod RacerFirebird

Twinkletown Freddy Ready Fire Truck

Noid plush doll; Real Model CollectionChristie Brinkley

Assorted Kidco products by Matchbox

Assorted LJN products by Matchbox

COPIES & PIRATE MODELS

Row 1: Big American Container Rig (Convoy pirate); copy of the Y-3A Tram by Typhoo Tea

Row 2: "stolen" Dinky DY-15 & DY-4 in repackaged boxes; copy of the MB17 Ford Escort & look-a-like box

Row 3: copy of MB9D Fiat from Poland; MC Toy brand copies of MB23D Audi Quattro & MB15-D Ford Sierra; Hong Kong copy of MB6-H Ford Supervan II; Majorette look-a-like of MB38-E Model A Van

Row 4: copies of Y-20-A & Y-8-D by Guisval of Spain; Y-12-D Model T Van plastic copy from Poland; MB34-C Chevy Prostocker copy in box from Hungary

MATCHBOX 1-75

Since 1982, the miniature series has diversified itself in such a manner that models in the standard range are modified or issued to become part of a specialized miniatures category. Some of these categories include Superfast, Laser Wheels, Graffic Traffic and Indy. Some of these models were given "new" designations such as "WC" for World Class or "SF" for Superfast. Although these designations will be noted, all models in the miniatures category will be denoted under their original numbers with notes to category listing. The following notations will be listed and described below. These denotations are used for all series.

PROMOTIONALS

UK-United Kingdom	US-United States
AU-Australian	JP-Japanese
GR-German	SW-Swiss
DU-Dutch	HK-Hong Kong
SU-Saudi	BR-Brazil
SP-Spain	CN-Canada
FR-French	IR-Irish
SC-Scottish	WL-Wales
BE-Belgium	AS-Austrian
CH-Chinese	SA-South Africa
GK-Greek	

SUB-CATEGORIES

DM-Dream Machine	KS-King Size
SB-Skybuster	PS-Playset
SF-Superfast	LW-Laser Wheels
LL-Live N Learn	WC-World Class
LS-Lasertronics	LT-Lightning
AP-Action Pack	TH-Triple Heat
NM-Nutmeg Collect.	DT-Days of Thunder
SC-Supercolor Changer	CY-Convoy
	CM-Commando
IN-Indy	RB-Roadblasters
GT-Graffic Traffic	IC-Intercom City
TP-Two Pack Issue	MC-Motor City/Gift
TM-Team Matchbox	TC-Team Convoy
WR-White Rose	OP-On-pack offer
HS-Hot Stocks	HD-Harley Davidson
VG-Vegas models	C2-Code 2 Approved
MP-Multi Pack	JB-James Bond
WP-Whales Project	UC-Ultra Class

SUB-CATEGORY LISTINGS

The name "Superfast" was originally introduced in 1969 by Lesney Products to counter the marketing of Mattel's Hot Wheels. By 1980, the name was gradually phased out of use. The Superfast 1-75 series became known worldwide simply as Matchbox 75 or Miniatures. This is why the prefix MB rather than SF is used when cataloging post 1969 miniatures. The Superfast name was reborn in 1986 with the introduction of 25 selected models from the standard range with new low friction axles and 24 spoke type wheels. These were introduced for the U.S. market only.

In 1987, the U.K. introduced similar fast axles on the same 25 models in revised colors but with "Laser Wheels". The wheel pattern now featured an iridescent design. Additional models were added.

"Roadblasters" were introduced in 1987. These miniatures featured all new colors and decorations, some even included revised castings, especially window ornamentation. Each model featured clip-on armament. The series was divided into the "Turbo Force" (12 good guys) and "Motor Lords" (12 bad guys). The series also included four Superking releases. There was to have been Supercharger and Skybuster Roadblasters but the series had proved unpopular and these were never issued.

In 1989, several new miniature sub-categories emerged. One of these was Super Colorchangers. These miniatures featured models whose color would change when subjected to cold and hot temperature extremes. Although normally water temperature changes the color, extreme room temperature changes also alter the model's color. Super Colorchangers were also offered in the Skybuster series.

The year 1989 also heralded the release of the popular World class series. These highly detailed miniatures featured rubber tires, chrome windshields and fine detailing at 2 to 3 times the cost of a normal miniature. The price was not a factor and the line continues to be upgraded.

Also introduced in 1989, but for the U.S. market only, was the Commando series which also included two Convoys. These miniatures contained military versions of existing as well as older reissued models. The series was divided into "Strike Team" (good guys) and "Dagger Force" (bad guys). The Strike Team featured olive green models and the Dagger Force black and gray models.

The year 1990 saw the release of specially colored miniatures in the Live N Learn/preschool range. These models were colored in primary colors with preschool decorations and featured wheels to match.

Also introduced in 1990 were Lasertronics. Termed Siren Force in the United States, these twelve models (two pairs of six) featured press down front axles which activated working lights and sirens. The series was not very popular, even though in 1991, the U.S. marketed them under a third title of "Rescue 911" to go along with the popular TV series.

The year 1990 saw the release of "Car and Driver." This series flopped because the vehicles were never issued in a special color. The only purchase incentive was the collector card at twice the retail!

To commemorate the release of Paramount Pictures "Days of Thunder", Matchbox introduced a series of miniatures, Convoys and Team Matchbox in 1990.

The year 1991 saw the debut of Lightning. These "fastest cars on the market" featured brightly colored models with specially designed axles to feature extra speed on the track. The wheel hubs feature lightning bolts.

The year 1991 also saw the innovative release of Graffic Traffic. The series extends into the Skybuster, Convoy and Specials ranges. Each model is plain white with usually brightly colored windows. Each model, usually in two or three pack form, is issued with water resistant marking pens so children can decorate their own models. Label sheets are also included for extra decorating. (NOTE: decorating these models reduces the collector value to zero!)

In 1991 Action Packs were released. These recolored miniatures were issued with special added accessories for extra play value.

To commemorate the 75th Anniversary of the Indianapolis 500, Matchbox introduced an Indy range of miniatures, Convoys and Team Matchbox as well as a Superking Transporter. The 76th running of Indy in 1992, saw additional releases including two support vehicle miniatures.

The year 1992 saw additional sub-category miniature releases. These include Hot Stocks-a set of six Chevy Luminas with added plastic signs and racing stuff similar to Action Packs.

Triple Heat was introduced in 1992. The set of six included a three pack release of a miniature, a Superfast Mini and a World's Smallest.

The Harley Davidson series included miniature, Superking, Convoy and large plastic motorcycles. These were 1992 releases.

The final 1992 release was Intercom City. This highly revolutionary idea features electronic bar codes on the bases of the models. When the bar code runs over another bar code on the play environment, it sets up a play in which a spoken message is given and in turn further play is continued as suggested by the message.

OTHER MINIATURE CATEGORIES

There are further miniature categories in this book which included the following: Models manufactured in Bulgaria, Hungary and Brazil, mainly of holder castings. Due to the tremendous number of variations in these three countries alone, only the Brazilian models will be included in the variation text. On Hungarian and Bulgarian combinations alone-body, base, interior, window, etc. there are over 2000 known variations! Some models, like the Hungarian Pizza Van, exist in over 80 varieties each. Only a small sampling are included in this book.

BULGARIAN MODELS

MB 5-A Lotus Europa	MB39-A Clipper
MB 6-B Mercedes Tourer	
MB 9-D Fiat Abarth	MB39-C Toyota Supra
MB16-B Pontiac	MB40-A Vauzhall Guilsdman
MB17-D AMX Prostocker	
	MB41-A Ford GT
MB21-C Renault 5TL	MB45-B BMW 3.0 CSL
MB22-C Blaze Buster	MB46-A Mercedes 300SE

MB24-A Rolls Royce Silver Shadow	
MB25-A Ford Cortina	MB51-B Citroen SM
MB27-A Mercedes 230SL	
	MB51-D Midnight Magic
MB28-C Lincoln Continental	
	MB56-C Mercedes 450SL
MB33-A Lamborghini Miura	
	MB59-C Planet Scout
MB33-E Renault 11	MB66-B Mazda RX500
MB35-C Pontiac T-Roof	
	MB67-C Datsun 260Z
MB37-E Matra Rancho	
	MB74-C Cougar Villager
MB75-C Seasprite Helicopter	
IX-A Flamin Manta	

HUNGARIAN MODELS

MB17-D AMX Prostocker	
	MB39-B Rolls Royce Silver Shadow
MB20-B Police Patrol	MB41-A ford GT
MB23-A Dormobile	MB45-B BMW 3.0 CSL
MB25-A Ford Cortina	MB51-B Citroen SM
MB27-A Mercedes 230SL	
	MB53-C Jeep C16

BRAZILIAN MODELS (MANAUS)

MB 3-C Porsche Turbo	MB40-B Horse Box
MB 6-B Mercedes Tourer	
	MB40-C Corvette T-Roof
MB 8-D Rover 3500	
MB 9-D Fiat Abarth	MB50-E Chevy Blazer
MB11-C Car Transporter	
	MB53-D Flareside Pickup
MB15-E Peugeot 205	MB55-E Racing Porsche 935
MB19-D Peterbilt Mixes	
	MB56-D Peterbilt Tanker
MB21-C Renault 5TL	MB61-C Peterbilt Wrecker
MB23-D Audi Quattro	MB62-E Chevrolet Corvette
MB28-D Formula 5000	MB65-B Airport Coach
MB30-E Peterbilt Quarry Truck	
MB68-E Dodge Caravan	
MB31-E Mazda RX7	MB70-E Ferrari 308GTB

WHITE ROSE COLLECTIBLES/ NUTMEG COLLECTIBLES

White Rose Collectibles had become a driving force in the Matchbox company, contributing as much as 10-15% of the business by 1992. White Rose Collectibles was begun by Ron Slyder and his partners in 1989. The original introductions by White Rose collectibles included four CY-104 Superstar Kenworth Transporter in Nascar liveries. It wasn't until 1990, that the market really began to see the driving promotional force behind White Rose Collectibles. Along with Nascar related Convoys and miniatures, there were also Sports Collectibles including baseball and football sets for 1990, 1991 and 1992 with promotional miniatures and Convoys for college sports teams and the National Hockey League. Other White Rose promotionals include those for the York Fair and the International Fire Chiefs Association. In 1992, White Rose produced a promotional gift set with two Matchbox MB5Y Chevy Luminas and a Winross brand truck in "Goodwrench" liveries. The models were available in cardboard or wooden boxes. These are valued as a set from $100-250.

Nutmeg Collectibles founded by Mark Daddio, covers the Sprint and Modified Racing markets as White Rose does for Nascar and Busch racing. Several Sprint and Modified sets have been issued since 1990.

ON-PACK MODELS

As a category, on-pack offers is unusual; most on-pack offers overlap into the promotional category. The model's promotional offering was through a direct mail-in offer based on sending tokens, coupons, UPC symbols, etc. through the mail, sometimes with a sum of money, to receive these models. Most of the on-pack offers were geared to the United Kingdom although some were for the United States and a few other countries. Some model offer packages are depicted in this book.

VEGAS MODELS

Two models in the miniature series do not fit into any real category. Although introduced later in modified form into the standard line, the original releases of the Dodge Daytona and Dodge Caravan featured sealed castings and special cast bases and are therefore not listed under MB28-E and MB68-E/MB64-E respectively. These models, introduced in 1983, were a 5,000 run each in which the bases read "Expressly for Dodge Las Vegas 1983". One thousand of these models were issued on a wooden plinth and given to certain individuals attending the convention in Las Vegas. The remaining four thousand pieces wre issued in special generic boxes.

MINIATURES 1-75

MB 1-F Jaguar X16 (ROW)	1987
MB 1-F Jaguar X16 Police Car (ROW)	1991
MB 2-E Pontiac Fiero	1985
MB 2-F Rover Sterling (ROW)	1988
MB 2-G 1965 Corvette Grand Sport (USA)	1990
MB 4-E London Taxi (ROW)	1987
MB 5-E Peterbilt Tanker (ROW)	1984
MB 6-C I.M.S.A. Mazda (USA)	1983
MB 6-D F1 Racer (ROW)	1984
MB 6-E Ford Supervan II (USA)	1986
MB 6-F Atlas Excavator (USA reintro)	1990
MB 6-G Alfa Romeo (ROW)	1991

MB 7-E I.M.S.A. Mazda (ROW)	1983
MB 7-F Porsche 959	1986
MB 8-E Greased Lightning (USA)	1983
MB 8-F Scania T142 (USA)	1985
MB 8-G Vauxhall Astra Police Car (ROW)	1987
MB 8-H Mack CH600 Aerodyne (USA)	1990
MB 8-I Airport Tender (USA)	1992
MB 9-E AMX Prostocker (ROW)	1983
MB 9-F Toyota MR2 (ROW)	1986
MB 9-G Caterpillar Bulldozer (ROW reintro)	1986
MB 9-H Faun Dump Truck (USA)	1990
MB10-D Buick Le Sabre	1987
MB11-E I.M.S.A. Mustang (USA)	1983
MB11-F Lamborghini Countach (ROW)	1985
MB12-F Pontiac Firebird Racer (ROW)	1986
MB12-G Modified Racer (ROW)	1988
MB12-H Mercedes 500SL Convertible (USA)	1990
MB12-I Cattle Truck (ROW reintro)	1992
MB14-E 1983/84 Corvette (USA)	1983
MB14-F 4X4 Jeep (ROW)	1984
MB14-G 1987 Corvette (USA)	1987
MB14-H Grand Prix Racer (ROW)	1988
MB15-E Peugeot 205 Turbo (USA)	1988
MB15-F Saab 9000 Turbo (USA)	1988
MB15-G 1965 Corvette Grand Sport (USA)	1990
MB15-H Alfa Romeo (USA)	1991
MB15-I Sunburner (USA)	1992
MB16-D F.1 Racer (USA)	1984
MB16-E Ford LTD Police Car (USA)	1987
MB16-F Land Rover 90 (ROW)	1987
MB17-D AMX Prostocker (USA)	1983
MB17-E Ford Escort Cabriolet (USA)	1985
MB17-F Dodge Dakota (USA)	1987

MB18-C Fire Engine	1984
MB20-D Volvo container Truck (ROW)	1985
MB20-E Volkswagen Transporter (ROW)	1988
MB21-D Corvette Pace Car (USA)	1983
MB21-E Breakdown Van	1985
MB21-F GMC Wrecker (USA)	1987
MB21-G Nissan Prairie (ROW)	1991
MB22-F Saab 9000 Turbo (ROW)	1988
MB22-G Vectra Cavalier Gsi 2000 (USA)	1990
MB22-H Lamborghini Diablo (USA)	1992
MB23-F Honda ATC (USA)	1985
MB23-G Volvo Container Truck (USA)	1987
MB24-E Datsun 280ZX	1983
MB24-F Nissan 300ZX	1986
MB24-G Lincoln Towncar (ROW)	1989
MB24-H Ferrari F40 (USA)	1989
MB24-I Airport Tender (ROW)	1992
MB25-E Ambulance (USA)	1984
MB25-F Audi Quattro (ROW)	1983
MB25-G Peugeot 205 Turbo (Row reintro)	1991
MB26-F Volvo Tilt Truck (ROW)	1984
MB26-G BMW 5 Series (USA)	1989
MB26-H 4 X 4 Chevy Van (USA reintro)	1991
MB27-D Jeep Cherokee	1986
MB27-E Mercedes Tractor (ROW)	1990
MB28-E Dodge Daytona	1984
MB28-F London Bus (USA reintro)	1990
MB28-G 1987 Corvette (ROW)	1990
MB28-H Forklift Truck (USA)	1991
MB28-I BMW Cabriolet (ROW reintro)	1991
MB28-J T-Bird Turbo Coupe (ROW reintro)	1992
MB30-F Mercedes "G" Wagon (ROW)	1985

MB31-E Mazda RX7 1983
MB31-F Rolls Royce Silver Cloud (ROW) 1987
MB31-G Rover Sterling (USA) 1988
MB31-H BMW 5 Series (ROW) 1989
MB31-I Nissan Prairie (USA) 1991
MB32-E Modified Racer (USA) 1988
MB33-D Volkswagen Golf Gti (USA) 1985
MB33-E Renault 11 (ROW) 1987
MB33-F Mercury Sable Wagon (ROW) 1989
MB33-G Utility Truck (USA) 1989
MB33-H Mercedes 500SL Convertible (ROW) 1990
MB34-D Ford RS200 1987
MB34-E Sprint Racer (USA) 1990
MB34-F Dodge Challenger (ROW reintro) 1991
MB35-E 4X4 Mini Pickup (ROW) 1986
MB35-F Land Rover 90 (USA) 1987
MB35-G Ford Bronco II (ROW) 1988
MB37-F Jeep 4 X 4 (ROW) 1984
MB37-G Ford Escort Cabriolet (ROW)1985
MB37-H Nissan 300ZX (ROW) 1990
MB38-F Ford Courier (ROW) 1992
MB38-G Ford Courier (UK) 1992
MB38-H Mercedes 600 (USA) 1992
MB39-C Toyota Supra (ROW) 1983
MB39-D BMW Cabriolet 1985
MB39-E Ford Bronco II (USA) 1987
MB39-F Mack CH600 Aerodyne (ROW) 1990
MB39-G Mercedes 600 (ROW) 1992
MB40-D Rocket Transporter (ROW) 1985
MB40-E Ford Sierra (USA reintro) 1990
MB40-F Road Roller (USA) 1991
MB41-E Racing Porsche (ROW) 1983
MB41-F Jaguar X16 (USA) 1987
MB41-G Vectra Cavalier Gsi 2000 (ROW) 1990

MB41-H Cosmic Blues (USA reintro) 1991
MB41-I Sunburner (ROW) 1992
MB42-E Faun Crane Truck (ROW) 1984
MB42-E Faun Crane Truck (USA) 1987
MB43-E AMG Mercedes 500SEC 1984
MB43-F Renault 11 (USA) 1987
MB43-G Lincoln Town Car (USA) 1989
MB43-H '57 Chevy (ROW reintro) 1990
MB44-E Citroen CV15 (ROW) 1983
MB44-F Datsun 280ZX Police Car (Japan) 1987
MB44-G Skoda LR (ROW) 1987
MB44-H 1921 Model T Ford 1990
MB45-D Ford Cargo Skip Truck (ROW) 1987
MB45-E Highway Maintenance Vehicle (USA) 1990
MB46-E Sauber Group "C" Racer (ROW) 1984
MB46-F Mission Helicopter (USA) 1985
MB47-E School Bus 1987
MB48-E Unimog with Plow (ROW) 1983
MB48-F Vauxhall Astra/Opel Kadette (ROW) 1986
MB49-D Sand Digger 1983
MB49-E Peugeot Quasar 1986
MB49-F Lamborghini Diablo (ROW) 1992
MB49-G Volvo Tilt Truck (USA) 1990
MB50-E Chevy Blazer 1984
MB50-F Dodge Dakota (ROW) 1990
MB50-G Auxiliary Power Truck (ROW) 1991
MB51-F London Bus (USA reintro) 1985
MB51-G Camaro IROC-Z (USA) 1986
MB51-H Ford LTD Police Car (ROW)1987
MB52-D BMW M1 1983
MB52-E Isuzu Amigo 1991
MB53-E Faun Dump Truck (ROW) 1989
MB53-F Breakdown Van (USA reintro) 1990

MB53-G Ford LTD Taxi (USA) 1992
MB54-F Command Vehicle 1984
MB54-G Chevy Lumina 1989
MB54-H Chevy Lumina 1990
MB55-F Racing Porsche (USA) 1983
MB55-G Mercury Sable Wagon (USA)1987
MB55-H Rolls Royce Silver Spirit (USA) 1990
MB55-I Model A Ford (USA reintro) 1991
MB56-E Volkswagen Golf Gti (ROW)1985
MB56-F 4 X 4 Jeep (ROW reintro) 1990
MB56-G Ford LTD Taxi (ROW) 1992
MB57-F Mission Helicopter (ROW) 1985
MB57-G Ford Transit (USA) 1990
MB57-H Auxiliary Power Truck (USA) 1991
MB58-D Ruff Trek 1983
MB58-E Mercedes 300 SE 1986
MB58-F Corvette T-Roof (USA reintro) 1991
MB59-E Porsche 944 (ROW) 1987
MB59-F T-Bird Turbo Coupe (USA) 1987
MB60-E Toyota Supra (ROW) 1983
MB60-F Pontiac Firebird Racer (USA)1985
MB60-F Ford Transit (ROW) 1986
MB60-H Rocket Transporter (USA) 1990
MB61-D T-Bird Turbo Coupe (ROW) 1987
MB61-E Nissan 300ZX (USA) 1990
MB61-F Forklift Truck (ROW) 1991
MB62-F Rolls Royce Silver Cloud (USA) 1986
MB62-G Volvo 760 (ROW) 1986
MB62-H Oldsmobile Aerotech 1989
MB62-I Volvo Container Truck (ROW reintro) 1990
MB63-G Volkswagen Golf Gti (ROW reintro) 1991
MB63-H 0-4-0 Loco (ROW reintro) 1992
MB64-E Dodge Caravan (ROW) 1984
MB64-F Oldsmobile Aerotech (ROW)1989

MB65-D F.1 Racer (USA) 1985
MB65-E Plane Transporter (ROW) 1985
MB65-F Cadillac Allante (ROW) 1987
MB66-E Sauber Group "C" Racer (USA) 1984
MB66-F Rolls Royce Silver Spirit (ROW) 1987
MB67-D Flame Out (USA)1983
MB67-E I.M.S.A. Mustang (ROW) 1983
MB67-F Lamborghini Countach (USA) 1985
MB67-G Ikarus Coach (ROW) 1986
MB68-E Dodge Caravan (USA) 1984
MB68-F Camaro IROC-Z (ROW) 1987
MB68-G TV News Truck (USA) 1989
MB68-H Road Roller (ROW) 1991
MB69-E 1983/84 Corvette (ROW) 1983
MB69-F Volvo 480ES (ROW) 1988
MB69-G Highway Maintenance Vehicle (ROW) 1990
MB70-E Ford Cargo Skip Truck (USA) 1989
MB70-F Ferrari F40 (ROW) 1989
MB71-E Scania T142 (ROW) 1985
MB71-F GMC Wrecker (ROW) 1988
MB71-G Porsche 944 (USA) 1989
MB72-F Sand Racer (USA) 1984
MB72-G Plane Transporter (USA) 1985
MB72-H Ford Supervan II (ROW) 1986
MB72-I Cadillac Allante (USA) 1987
MB72-J Sprint Racer (ROW) 1990
MB72-K Dodge Zoo Truck (ROW) 1992
MB73-D TV News Truck (ROW) 1989
MB73-E Mercedes Tractor (ROW) 1990
MB74-F Mustang GT (USA) 1984
MB74-G Toyota MR2 (ROW) 1986
MB74-H Grand Prix Racer (USA) 1987
MB74-I Utility Truck (ROW) 1989
MB75-E Ferrari Testarossa 1986
MB212 Ford Thunderbird (WR) 1992

MB216 Pontiac Grand Prix (WR)	1992
MB217 D.I.R.T. Modified (WR)	1992
MB245 Chevy Panel Van (WR)	1992
MBXX 1983 Dodge Daytona Turbo Z (Las Vegas)	1983
MBXX 1983 Dodge Caravan (Las Vegas)	1983

From the above list, you will note that some models exist with two, and even three numbers. This is due to the fact that the United States (USA) and the rest of the world (ROW) have two different numbering systems. This began in 1982, in the final Lesney year and continues today. Some models may also be available for the U.S. or ROW only. A third number comes into play when an older model is reintroduced. In most cases, the U.S. number is listed first, except where the U.S. number was introduced more than a year later this is denoted second.

MB 1-D DODGE CHALLENGER, issued 1982
MB34-F DODGE CHALLENGER, reissue 1991 (ROW)(continued)

5. yellow body, silver/gray base, black roof, red interior, 5 arch front wheels, clear windows, "Toyman" tempa, England casting ($2-4)
6. yellow body, pearly silver base, black roof, red interior, 5 arch front wheels, clear windows, "Toyman" tempa, Macau casting ($2-3)
7. yellow body, pearly silver base, black roof, red interior, dot dash front wheels, clear windows, "Toyman" tempa, Macau casting ($2-3)
8. yellow body, pearly silver base, black

roof, red interior, dot dash front wheels, clear windows, "Toyman" tempa, China casting ($1-2)
9. lemon body, pearly silver base, black roof, red interior, dot dash front wheels, clear windows, "Toyman" tempa, China casting ($1-2)
10. orange-yellow body, pearly silver base, black roof, red interior, dot dash front wheels, clear windows, "Toyman" tempa, China casting ($1-2)
11. white body, white base, white roof, blue interior, dot dash front wheels, blue windows, no tempa, China casting ($10-15) (GF)
12. powder blue body, silver/gray base, black roof, white interior, dot dash front wheels, clear windows, "Challenger" tempa, China casting ($8-10) (AP)
13. yellow body, yellow base, black roof, red interior, dot dash front wheels, clear windows no tempa, China casting ($8-12) (GS)(GR)
14. white body, silver/gray base, black roof, red interior, dot dash front wheels, clear windows, "Toyman" tempa, China casting ($8-12) (GS)(US)

MB 1-F JAGUAR XJ6 POLICE CAR, issued 1991 (ROW)

1. white body, tan interior, blue windows & dome lights, 8 dot wheels, "Police" with blue/yellow stripes, Thailand casting ($3-5)
2. white body, tan interior, blue windows & dome lights, 8 dot wheels, "Police" with checkers/stripes, Thailand casting ($3-5)

MB 2-A MERCEDES TRAILER, issued 1970 (continued)

11. yellow body, white canopy, 5 crown wheels, "Alpine Rescue" labels, England casting ($3-5)(TP)
12. yellow body, no canopy, 5 crown wheels, 5 crown wheels, no tempa, England casting ($5-7)(US release in single box)
13. yellow body, white canopy, 5 crown wheels, "Alpine Rescue" tempa, England casting ($3-5)(TP)
14. yellow body, white canopy, 5 crown wheels, "Alpine Rescue" tempa, no origin cast ($3-5)(TP)
15. red body, white canopy, 5 crown wheels, "Unfall Retung" tempa, no origin cast ($3-5)(TP)
16. white body, orange canopy, 5 crown wheels, "C&S" tempa, no origin cast ($3-5)(TP)
17. red body, white canopy, dot dash wheels, "Big Top Circus" tempa, no origin cast ($4-6)(TP)

MB 2-D S.2 JET, issued 1981 (continued)

7. dark blue body, white base, white wings, clear windows, "Viper" tempa, Macau casting ($2-4)
8. dark blue body, white base, white wings, clear windows, "Viper" tempa, Hong Kong casting ($20-25)
9. dark blue body, white base, white wings, clear windows, "Viper" tempa, China casting ($2-4)
10. olive green body, black base, olive green wings, clear windows, "AC152" & tan camouflage tempa, Macau casting ($4-6)(CM)

MB 2-E PONTIAC FIERO, issued 1985

1. white upper body, dark blue lower body, 8 dot silver wheels, "Goodyear 85" tempa, Macau casting ($2-4)
2. white upper body, dark blue lower body, 8 dot gold wheels, "Goodyear 85" tempa, Macau casting ($5-8)
3. white upper body, dark blue lower body, dot dash wheels, "Goodyear 85" tempa, Macau casting ($4-6)
4. yellow upper body, orange lower body, starburst wheels, "Protech 16" tempa, Macau casting ($3-5)(SF)
5. yellow upper body, orange lower body, 8 dot silver wheels, "Protech 16" tempa, Macau casting ($50-75)
6. white upper body, red lower body, 8 dot silver wheels, "GT Fiero" tempa, Macau casting ($2-4)
7. yellow upper body, metallic gold lower body, laser wheels, "Protech 16" tempa, Macau casting ($3-5)(LW)
8. black upper body, red lower body, 8 dot silver wheels, "2 Dog Racing Team" tempa, Macau casting ($8-12)(HK)
9. white body, red lower body, 8 dot silver wheels, "GT Fiero" tempa, China casting ($2-4)
NOTE: above models all have black base insert, clear windows, silver/gray interior.

MB 2-G 1965 CORVETTE GRAND SPORT, issued 1990 (USA)
MB15-G 1965 CORVETTE GRAND SPORT, issued 1990 (ROW)
NOTE: below models have 5 arch front & 5 crown rear wheels unless otherwise noted.
1. met. blue body, clear windows, chrome interior, white stripes & small "15" on

doors tempa, black base, Macau casting ($2-3)

2. met. blue body, clear windows, chrome interior, white stripes & larger "15" on doors tempa, black base, Macau casting ($2-3)

3. met. red body, chrome windows, black interior, white stripes & "63" tempa, black base, gray disc w/rubber tires, Macau casting ($15-20)(WC)

4. met. red body, chrome windows, black interior, white stripes & "63" tempa, chrome base, gray disc w/rubber tires, Macau casting ($5-8)(WC)

5. met. blue body, clear windows, chrome interior, white stripes & larger "15" tempa on doors, Thailand casting ($2-3)

6. met. blue body, clear windows, chrome interior, white stripes & small "15" tempa on doors, China casting ($2-3)

7. met. blue body, clear windows, chrome interior, "Heinz 57" tempa, China casting ($20-25)(US)

8. met. blue body, clear windows, chrome interior, white stripes & "Corvette" on doors tempa, Thailand casting ($2-3)

9. white & blue body, clear windows, black interior, "9" & red stripes tempa, white disc w/rubber tires, Thailand casting ($5-7)(WP)

MB 3-C PORSCHE TURBO, issued 1978 (continued)

28. red body, brown interior, black base, clear windows, "Porsche 90 Turbo" tempa, Macau casting ($2-4)

29. black body, tan interior, black base, amber windows, "Porsche 90 Turbo" tempa, Macau casting ($2-4)

30. red body, tan interior, black base, clear

windows, black edged tempa, Macau casting ($8-12)(JP)

31. white body, tan interior, black base, clear windows, "Boss 14" tempa, Macau casting ($8-12)(JP)

32. dark blue body, white interior, black base, clear windows, "Wrangler 47" tempa, Macau casting ($8-12)(UK)

33. white body, tan interior, black base, clear windows, "Porsche" & "3" tempa, Macau casting ($8-12)

34. black body, tan interior, black base, clear windows, "Porsche" tempa (without side tempa), Macau casting ($8-12)

35. white body, tan interior, black base, clear windows, "Boss 14" with "14" inside square tempa, Macau casting ($3-5)

36. white body, blue interior, black base, clear windows, "Boss 14" with "14" inside square tempa, Macau casting ($3-5)

37. black body, black interior, black base, clear windows, "90 Porsche Turbo" and "BP" tempa, Macau casting ($8-12)(DU)

38. red body, blue interior, black base, clear windows, "8 Dragon Racing Team" tempa, Macau casting ($8-12)(HK)

39. robin egg blue body, blue interior, black base, clear windows, "Boss 14" with "14" in square tempa, Macau casting ($3-5)(MP)

40. dark blue body, tan interior, black base, clear windows, "Porsche" & yellow stripe tempa, Macau casting ($2-3)

41. red body, tan interior, black base, clear windows, "Porsche" & "Porsche 911" tempa, Macau casting ($1-2)

42. black body, tan interior, black base, clear windows, "Porsche" & "Porsche

911" tempa, Macau casting ($2-4)(MP)

43. red body, tan interior, black base, clear windows, "Porsche" & Porsche 911" tempa, Thailand casting ($1-2)

44. dark blue body, tan interior, black base, clear windows, "Porsche" & yellow stripes tempa, Thailand casting ($1-2)

45. yellow body, tan interior, black base, clear windows, "Porsche" logo tempa, Thailand casting ($3-5)(GS)

46. robin egg blue body, blue interior, black base, clear windows, "Boss 14" with "14" in square tempa, Thailand casting ($3-5)

47. black body, tan interior, black base, clear windows, "Porsche 90 Turbo" tempa, Manaus casting ($20-35)(BR)

48. lemon body, black interior, black base, clear windows, spatter design & "Porsche" tempa ($2-3)(TP)

MB 4-D '57 CHEVY, issued 1979
MB43-H '57 CHEVY, reissued 1990 (ROW)(continued)

6. black body, red hood, pearly silver base, flames tempa, Macau casting ($1-2)

7. black body, red hood, pearly silver base, flames tempa, China casting ($1-2)

8. pink body, red hood, pearly silver base, flames tempa, Macau casting ($3-4)(SC)

9. light pea green body, red hood, pearly silver base, flames tempa, Macau casting ($3-4)(SC)

10. light peach body, red hood, pearly silver base, flames tempa, Macau casting ($3-4)(SC)

11. rose red body, red hood, pearly silver base, flames tempa, Macau casting, ($3-4)(SC)

12. dark purple body, red hood, pearly silver base, flames tempa, Macau casting ($3-4)(SC)

13. red body & hood, pearly silver base, "Heinz 57 Chevy" tempa, Macau casting ($15-20)(US)

14. bluish purple body & hood, pearly silver base, "Milky Way" tempa, Macau casting ($15-20)(UK)(OP)

15. bluish purple body & hood, pearly silver base, "Milky Way" tempa, China casting ($15-20)(UK)(OP)

16. black body, dark red hood, pearly silver base, flames tempa, Thailand casting ($1-2)

17. black body, dark red hood, pearly silver base, flames tempa, China casting ($1-2)

18. metallic red body & hood, pearly silver base, silver stripe tempa, chrome windows, gray disc w/rubber tires, China casting ($5-8)(WC)

19. yellow body & hood, pearly silver base, blue/red stripes & "57" tempa, black windows, Thailand casting ($3-5)(TH)

20. black body, dark red hood, black base, flames tempa, Thailand casting ($1-2)

MB 4-E TAXI FX4R, issued 1987 (ROW)

1. black body, unpainted base, gray interior, no tempa, Macau casting ($2-4)

2. black body, unpainted base, gray interior, "Great Taxi Ride London To Sydney" tempa, Macau casting ($5-7)(AU)

3. yellow body, unpainted base, blue interior, "ABC Taxi" tempa, blue wheels with red hubs, Macau casting ($4-6)(LL)

4. black body, unpainted base, gray interior, no tempa, China casting ($2-4)

5. black body, silver/gray base, gray interior, no tempa, China casting ($2-4)

6. black body, unpainted base, gray interior, "London Taxi" & British flag tempa (left side only), China casting ($2-4)

MB 5-D 4 X 4 JEEP, issued 1982
**MB56-F 4 X 4 JEEP, reissued 1990
(ROW) (continued)**

3. brown body, black metal base, "Golden Eagle" tempa, Macau casting ($2-4)
4. red body, black metal base, "Golden Eagle" tempa, Macau casting ($2-4)
5. olive body, black metal base, red/yellow/blue tempa, Macau casting, includes plastic armament ($6-8)(RB)
6. red body, black plastic base, "Golden Eagle" tempa, Macau casting ($1-2)
7. red body, black plastic base, "Golden Eagle" tempa, Thailand casting
8. yellow body, black plastic base, "50th Anniversary Jeep" tempa, Thailand casting ($12-18)(PS)
9. hot pink body, black plastic base, white bolt tempa, white interior, Thailand casting ($2-4)(DM)

**MB 6-B MERCEDES TOURER, issued
1973 (continued)**

25. white body, no roof, silver/gray base, clear windows, translucent red interior, 5 arch wheels, China casting ($8-12)(MP)
26. pale gray body, no roof, black plastic base, clear windows, translucent white interior, 4 arch wheels, Manaus casting ($35-50)(BR)
27. beige body, no roof, black plastic base, clear windows, translucent white interior, 4 arch wheels, Manaus casting ($35-50)(BR)
28. red body, no roof, black plastic base, clear windows, translucent white interior, r arch wheels, Manaus casting ($35-50)(BR)

**MB 6-C I.M.S.A. MAZDA, issued 1983
(USA)**
**MB 7-E I.M.S.A. MAZDA, issued 1983
(ROW)**

1. dark blue body, red interior, 5 arch wheels, white/orange tempa, Macau casting ($2-4)
2. dark blue body, red interior, dot dash wheels, white/orange tempa, Macau casting ($2-4)

**MB 6-E FORD SUPERVAN II, issued
1986 (USA)**
**MB72-H FORD SUPERVAN II, issued
1986 (ROW)**

1. white body, black metal base, dark gray windows, no dome lights, dark blue "Ford Supervan" tempa, Macau casting ($2-4)
2. white body, black metal base, dark gray windows, no dome lights, light blue "Ford Supervan" tempa, Macau casting ($2-4)
3. dark blue body, black metal base, dark gray windows, no dome lights, "Duckhams QXR Engine Oils" tempa, Macau casting ($4-6) (TC)
4. dark gray body, black metal base, dark gray windows, no dome lights, "Danger High Explosive/Heavy Load" tempa, Macau casting, includes plastic armament ($6-8)(RB)
5. white body, black metal base, dark gray windows, no dome lights, "Starfire" tempa, Macau casting ($3-5)(UK)
6. white body, black metal base, dark gray windows, no dome lights, "Fuji Racing Team" tempa, Macau casting ($3-5)(TC)
7. bright yellow body, black metal base, dark gray windows, no dome lights, "Service Car BP Oil" tempa, Macau casting ($10-15)(DU)
8. red body, black metal base, dark gray windows, no dome lights, "Tizer Flavored Soft Drink" tempa, Macau casting ($3-5)(TC)
9. yellow body, black metal base, dark gray windows, no dome lights, "Goodyear Pit Stop" tempa, Macau casting ($3-5)(GS)
10. white body, black plastic base, dark gray windows, no dome lights, light blue "Ford Supervan" tempa, Macau casting ($2-4)
11. white body, black plastic base, dark gray windows, no dome lights, "Fuji Racing Team" tempa, Macau casting ($3-5)(TC)
12. red body, black plastic base, dark gray windows, no dome lights, "Tizer Flavored Soft Drink" tempa, Macau casting ($3-5)(TC)
13. red body, black plastic base, black windows, amber dome lights, "Fire Observer" tempa, Macau casting ($12-15)(SR)
14. red body, black plastic base, black windows, greenish yellow dome lights, "Fire Observer" tempa, Macau casting ($12-15)(SR)
15. white body, black plastic base, black windows, red dome lights, "Ambulance" tempa, Macau casting ($12-15)(SR)
16. dark blue body, black plastic base, black windows, red dome lights, "Police Control Unit" tempa, Macau casting ($12-15)(SR)
17. white body, black plastic base, dark gray windows, no dome lights, "Starfire" tempa, Macau casting ($3-5)(UK)
18. yellow body, black plastic base, dark gray windows, no dome lights, "Goodyear Pit Stop" tempa, Thailand casting ($3-5)(GS)
19. red body, black plastic base, black windows, amber dome lights, "Fire Observer/Rescue 911" tempa, China casting ($12-15)(SR)
20. white body, black plastic base, black windows, red dome lights, "Ambulance/Rescue 911" tempa, China casting ($12-15)(SR)
21. dark blue body, black plastic base, black windows, red dome lights, "Police Control Unit/Rescue 911" tempa, China casting ($12-15)(SR)
22. white body, black plastic base, black windows, no dome lights, no tempa, China casting ($10-15)(GF)
23. white body, black plastic base, dark gray windows, no dome lights, light blue "Ford Supervan" tempa, Thailand casting ($3-5)
24. light gray body, black plastic base, dark gray windows, no dome lights, "Danger High Explosive/Heavy Load" tempa, China casting ($15-20)(Tomy box)(JP)

**MB 7-C V.W.GOLF, issued 1976
(continued)**

22. black body, black base, red interior, clear windows, no rack, red stripe tempa, Macau casting ($5-7)(JP)
23. black body, black base, red interior, clear windows, no rack, red stripe tempa, with tow hook cast, Macau casting ($3-5)(TP)
24. black body, black base, red interior, clear windows, no rack, red & orange stripes with "9" tempa, Macau casting ($8-12)(DY)

MB 7-D ROMPIN RABBIT, issued 1982 (continued)

NOTE: rabbit on hood can face left or right

5. yellow body, blue interior, "Ruff Rabbit" tempa, with tow hook, Macau casting ($3-5)
6. yellow body, blue interior, "Ruff Rabbit" tempa, with tow hook, Hong Kong casting ($3-5)
7. yellow body, blue interior, "Ruff Rabbit" tempa, without tow hook, Hong Kong casting ($3-5)
8. yellow body, blue interior, "Ruff Rabbit" tempa, without tow hook, Macau casting ($3-5)
9. yellow body, red interior, "Ruff Rabbit" tempa, without tow hook, Macau casting ($3-5)

MB 7-F PORSCHE 959, issued 1986

NOTE: all models with red interior, clear windows and black base unless otherwise indicated.

1. pearly silver body, 5 arch silver wheels, "Porsche" on doors tempa, Macau casting ($2-3)
2. pearly gray body, 5 arch silver wheels, "Porsche" on doors tempa, Macau casting ($2-3)
3. pearly silver body, 8 dot silver wheels, "Porsche" on doors tempa, Macau casting ($2-3)
4. white body, 5 arch white wheels, "Porsche" on doors tempa, Macau casting ($3-5)
5. white body, 8 dot silver wheels, "Porsche" on doors tempa, Macau casting ($2-4)

6. white body, 5 arch silver wheels, "Porsche" on doors tempa, Macau casting ($2-4)
7. white body, 8 dot white wheels, "Porsche" on doors tempa, Macau casting ($3-5)
8. white body, 8 dot silver wheels, "Porsche" on doors tempa, China casting ($2-4)
9. white body, 5 arch silver wheels, "Porsche" on doors tempa, China casting ($2-4)
10. charcoal gray body, 5 arch silver wheels, "Porsche 959" tempa, Macau casting ($2-4)
11. white body, 5 arch silver wheels, "Porsche 959" with red/yellow/black stripes tempa, Macau casting ($6-8)(KS)
12. pink body, 5 arch silver wheels, "Porsche 959" tempa, Macau casting ($3-4)(SC)
13. dark purple body, 5 arch silver wheels, "Porsche 959" tempa, Macau casting ($3-4)(SC)
14. white body, 5 arch silver wheels, "Redoxon" tempa, Macau casting ($20-35)(HK)
15. white body, 5 arch silver wheels, "Pace Car/Shell" tempa, Macau casting ($3-5)(GS)
16. white body, 5 arch silver wheels, "313 Pirelli Gripping Stuff" tempa, Macau casting ($3-5)(TC)
17. silver/gray body, gray disc w/rubber tires, chrome windows, "Porsche" tempa, Macau casting ($5-8)(WC)
18. black body, 5 arch silver wheels, "Porsche" logo tempa, Macau casting

($4-6)(MP)
19. white body, 5 arch silver wheels, "Pace Car/Shell" tempa, Thailand casting ($3-5)(GS)
20. charcoal gray body, 5 arch silver wheels, "Porsche 959" tempa, Thailand casting ($2-3)
21. chrome plated body, 5 arch silver wheels, no tempa, Macau casting ($12-18)(C2)
22. white body, 5 arch silver wheels, "Porsche" logo tempa, Thailand casting ($3-5)(GS)
23. silver/gray body, gray disc w/rubber tires, chrome windows, "Porsche" tempa, Thailand casting ($5-8)(WC)
24. white body, 5 arch silver wheels, red windows, no tempa, Thailand casting ($10-15)(GF)
25. silver/gray body, 5 arch silver wheels, black windows, checkers & stripes with "959" tempa, Thailand casting ($3-5)(TH)
26. white body, 5 arch silver wheels, "Lloyds" tempa, Thailand casting ($8-12)(UK)
27. silver/gray body, 5 arch silver wheels, "Porsche 959" (all red) tempa, Thailand casting ($1-2)

MB 8-D ROVER 3500, issued 1982 (continued)

NOTE: all models with dot dash wheels unless otherwise noted.

4. bronze body, white interior, clear windows, small maltese cross wheels, England casting ($5-8)
5. white body, light tan interior, blue windows, blue & yellow "Police" tempa, blue dome lights & silver beacon,

England casting ($2-4)
6. white body, light tan interior, clear windows, blue & yellow "Police" tempa, clear dome lights & silver beacon, England casting ($3-5)
7. white body, light tan interior, blue windows, no tempa, blue dome lights & silver beacon, England casting ($2-4)
8. white body, black interior, blue windows, no tempa, blue dome lights & silver beacon, England casting ($2-4)
9. white body, black interior, blue windows, blue & yellow "Police" tempa, blue dome lights & silver beacon, England casting ($2-4)
10. white body, black interior, blue windows, blue & yellow "Police" tempa, blue dome lights & black beacon, England casting ($2-4)
11. white body, light tan interior, blue windows, blue & yellow "Police" tempa, blue dome lights & silver beacon, England casting ($2-4)
12. white body, black interior, blue windows, blue & yellow "Police" tempa, blue dome lights & silver beacon, England casting ($2-4)
13. white body, light tan interior, blue windows, blue & yellow "Police" tempa, blue dome lights & black beacon, Macau casting ($2-4)
14. white body, light tan interior, blue windows, blue & orange "Police" tempa, blue dome lights & black beacon, Macau casting ($2-4)
15. white body, light tan interior, blue windows, blue & red "Police" tempa, blue dome lights & black beacon, China casting ($3-5)
16. white body, translucent tan interior, blue

windows, blue & red "Police" tempa, blue dome lights & black beacon, China casting ($3-5)

17. white body, orange-yellow interior, blue windows, blue & yellow "Police" tempa, blue dome lights & black beacon, Manaus casting ($25-40)

MB 8-E GREASED LIGHTNING, issued 1983 (USA)

1. red body, white base, black interior, Macau casting ($5-8)
2. red body, white base, black interior, Hong Kong casting ($5-8)

MB 8-F SCANIA T142, issued 1985 (USA)
MB71-E SCANIA T142, issued 1985 (ROW)

NOTE: all models listed have white interior, blue windows & blue dome lights

1. white body, black base, 8 dot white wheels, "Police" with red/blue stripes tempa, Macau casting ($3-5)
2. white & dark green body, black base, 8 dot white wheels, "Polizei" tempa, Macau casting ($3-5)
3. white & dark green body, black base, 8 dot silver wheels, "Polizei" tempa, Macau casting ($3-5)
4. white body, black base, 8 dot silver wheels, "Police" with red/blue stripes tempa, Macau casting ($2-4)
5. white body, black base, 8 dot silver wheels, "Police" with red/blue stripes tempa, China casting ($2-4)
6. white & dark green body, black base, 8 dot silver wheels, "Polizei" tempa, China casting ($3-5)
7. white body, white base, blue wheels with

8 dot yellow, "Police" with orange/blue stripes & face on hood tempa, China casting ($3-5)(LL)
8. white body, white base, blue wheels (plain hubs), "Police" with orange/blue stripes & face on hood tempa, China casting ($3-5)(LL)
9. white body, white base, 8 dot silver wheels, no tempa, China casting ($10-15)(GF)
10. white body, white base, 8 dot silver wheels, "Police" with orange/yellow stripes & orange dot on roof tempa, China casting ($1-2)
11. white body, white base, 8 dot silver wheels, "Police" with orange/yellow stripes but without dot on roof tempa, China casting ($1-2)

MB 8-G VAUXHALL ASTRA POLICE CAR, issued 1987 (ROW)

Note: all models listed have white interior, blue windows & blue dome lights.

1. white body, black base, 8 dot white wheels, "Police" with red/blue stripes tempa, Macau casting ($3-5)
2. white & dark green body, black base, 8 dot white wheels, "Polizei" tempa, Macau casting ($3-5).
3. white & dark green body, black base, 8 dot silver wheels, "Polizei" tempa, Macau casting ($3-5)
4. white body, black base, 8 dot silver wheels, "Police" with red/blue stripes tempa, Macau casting ($2-4)
5. white body, black base, 8 dot silver wheels, "Police" with red/blue stripes tempa, China casting ($2-4)
6. white & dark green body, black base, 8 dot silver wheels, "Polizei" tempa, China

casting ($3-5)
7. white body, white base, blue wheels with 8 dot yellow, "Police" with orange/blue stripes & face on hood tempa, China casting ($3-5)(LL)
8. white body, white base, blue wheels (plain hubs), "Police" with orange/blue stripes & face on hood tempa, China casting ($3-5)(LL)
9. white body, white base, 8 dot silver wheels, no tempa, China casting ($10-15)(GF)
10. white body, white base, 8 dot silver wheels, "Police" with orange/yellow stripes & orange dot on roof tempa, China casting ($1-2)
11. white body, white base, 8 dot silver wheels, "Police" with orange/yellow stripes but without dot on roof tempa, China casting ($1-2)

MB 8-H MACK CH600 AERODYNE, issued 1990 (USA)
MB39-F MACK CH600 AERODYNE, issued 1990 (ROW)

NOTE: this model was used as a component for Convoy models but only models released as singles are listed here.

1. white body, black chassis, gray base, black/red stripes with "CH600" tempa, Macau casting ($1-2)
2. white body, black chassis, gray base, black/red stripes with "CH600" tempa, Thailand casting ($1-2)

MB 9-A BOAT & TRAILER, issued 1970 (continued)

11. milk white deck, blue hull, white interior, orange trailer, "8" tempa, dot dash wheels, Macau casting ($3-5)(TP)

12. white deck, white hull, gray interior, black trailer, "Seaspray" tempa, dot dash wheels, no origin cast ($3-5)(TP)
13. white deck, white hull, gray interior, dark metallic blue trailer, yellow dashes on sides tempa, no origin cast.
14. blue deck, blue hull, gray interior, black trailer, white/orange spatter tempa, no origin cast ($3-5)(TP)

MB 9-D FIAT ABARTH, issued 1982 (USA)
MB74-E FIAT ABARTH, issued 1982 (ROW)

6. white body, black base, red interior, clear windows, red & orange "Matchbox" tempa, Macau casting ($2-4)
7. white body, black base, red interior, very light amber windows, red & orange "Matchbox" tempa, Macau casting ($2-4)
8. white body, black base, red interior, clear windows, purple & orange "Matchbox" tempa, Macau casting ($2-4)
9. white body, black base, red interior, amber windows, purple & orange "Matchbox" tempa, Macau casting " ($2-4)
10. white body, black base, red interior, clear windows, green & red "Alitalia" tempa, Macau casting ($3-5)
11. white body, charcoal base, red interior, clear windows, dark green & red "Alitalia" tempa, Macau casting ($3-5)
12. white body, black base, red interior, amber windows, green & red "Alitalia" tempa, Macau casting ($3-5)
13. white body, black base, red interior, amber windows, dark green & red "Alitalia" tempa, Macau casting ($3-5)
14. white body, black base, red interior,

clear windows, red/orange/yellow stripes tempa, Macau casting ($8-12)(DY)

15. white body, black base, red interior, clear windows, "Matchbox 11" tempa, 4 arch wheels, Manaus tempa ($35-50)

16. white body, black base, red interior, clear windows, "Matchbox 11" tempa, 8 spoke wheels, Manaus tempa ($35-50)

MB 9-F TOYOTA MR2, issued 1986 (USA)

MB74-G TOYOTA MR2, issued 1986 (ROW)

NOTE: all models with white lower body, white base insert, black interior and clear windows.

1. white body, 8 dot wheels, "MR2 Pace Car" tempa, Macau casting ($2-4)

2. dark blue body, starburst wheels, "MR2" & pink stripe tempa, Macau casting ($3-5)(SF)

3. metallic dark blue body, laser wheels, "MR2" & pink stripe tempa, Macau casting ($3-5)(LW)

4. green body, 8 dot wheels, "7 Snake Racing Team" tempa, Macau casting ($8-12)(HK)

MB 9-H FAUN DUMP TRUCK, issued 1990 (USA)

MB53-E FAUN DUMP TRUCK, issued 1990 (ROW)

1. yellow body, pearly silver dump, Maltese cross wheels, orange stripes tempa, Macau casting ($2-3)

2. yellow body, pearly silver dump, Maltese cross wheels, orange stripes tempa, China casting ($2-3)

3. blue body, yellow dump, red Maltese cross wheels (yellow hubs), orange

stripes & tools tempa, China casting ($3-5)(LL)

4. orange-yellow body, orange-yellow dump, Maltese cross wheels, orange stripes tempa, China casting ($2-3)(AP)

5. orange-yellow body, red dump, Maltese cross wheels, no tempa, China casting ($1-2)

6. blue body, yellow dump, red Maltese cross wheels (plain hubs), orange stripes and tools tempa, China casting ($3-5)(LL)

MB10-C GRAN FURY POLICE CAR, issued 1979 (continued)

8. white body, pearly silver base, dark blue windows & dome lights, dot dash wheels, "Metro" tempa, Macau casting ($3-5)

9. white body, light gray base, blue windows & dome lights, dot dash wheels, blue "Police" tempa, Macau casting ($2-4)

10. white body, pearly silver base, blue windows & dome lights, dot dash wheels, blue "Police" tempa, Macau casting ($2-4)

11. white body, pearly silver base, blue windows & dome lights, 8 dot wheels, blue "Police" tempa, Macau casting ($3-5)

12. white body, pearly silver base, dark blue windows & dome lights, dot dash wheels, "Police SFPD" tempa, Macau casting ($2-4)

13. white body, pearly silver base, dark blue windows & dome lights, dot dash wheels, "Sheriff SP-5" tempa, Macau casting ($2-4)(MP)

14. white body, pearly silver base, green windows & dome lights, dot dash wheels,

"Police SFPD" tempa, Macau casting ($75-100)

15. white body, pearly silver base, dark blue windows & dome lights, dot dash wheels, "Police SFPD" tempa, Thailand casting ($2-4)

MB10-D BUICK LESABRE, issued 1987

1. black body, white base, red interior, 8 dot wheels, "4" & "355" CID" tempa, Macau casting ($2-4)

2. grape & white body, white base, gray interior, laser wheels, "Ken Wells/ Quicksilver" tempa, white airscoop cast on hood, Macau casting ($5-8)(LW)

3. light pea green body, white base, red interior, 8 dot wheels, "4" & "355 CID" tempa, Macau casting ($3-4)(SC)

4. light brown body, white base, red interior, 8 dot wheels, "4" & "355 CID" tempa, Macau casting ($3-4)(SC)

5. orange body, white base, red interior, 8 dot wheels, "4" & "355 CID" tempa, Macau casting ($6-8)(SC)

6. rose red body, white base, red interior, 8 dot wheels, "4" & "355 CID" tempa, Macau casting ($6-8)(SC)

7. yellow body, red base, gray interior, 8 dot wheels, "10/Shell/Marshall" tempa, Macau casting ($1-2)

8. bright yellow body, red base, gray interior, 8 dot wheels, "10/Shell/ Marshall" tempa, Thailand casting ($1-2)

9. bright yellow body, red base, gray interior, 8 dot wheels, "10/Shell/ Marshall" tempa, Thailand casting (large rear side windows) ($25-40)

10. white body, red base, gray interior, 8 dot wheels, "10/Shell/Marshall" tempa, Thailand casting ($3-5)

MB11-C CAR TRANSPORTER, issued 1976 (continued)

42. orange body, gray carrier, pearly silver base, blue/red/yellow cars, blue windows, Macau casting ($7-10)

MB11-E I.M.S.A. MUSTANG, issued 1983 (USA)

MB67-E I.M.S.A. MUSTANG, issued 1983 (ROW)

NOTE: all models fitted with amber windows, chrome interior, dot dash rear wheels.

1. black body, black base, 5 spoke front wheels, red/white stripes & "Mustang Ford" tempa, Macau casting ($2-3)

2. black body, black base, 5 arch front wheels, red/white stripes & "Mustang Ford" tempa, Macau casting ($2-3)

3. black body, black base, 5 arch front wheels, yellow/green flames tempa, Macau casting ($1-2)

4. black body, black base, 5 spoke front wheels, yellow/green flames tempa, Macau casting ($1-2)

5. black body, black base, 5 arch front wheels, yellow/green flames tempa, China casting ($1-2)

6. yellow body, black base, 5 arch front wheels, black/red stripes & "47" tempa, China casting ($3-5)(AP)

7. red body, red base, 5 arch front wheels no tempa, China casting ($8-12)(GR)(GS)

8. black body, greenish-black base, 5 arch front wheels, yellow/green stripes tempa, China casting ($1-2)

MB12-D CITROEN CX, issued 1979 (continued)

22. white body, red interior, silver-gray

base, blue windows, "Ambulance" tempa, England casting ($2-4)(TP)

23. white body, red interior, black base, blue windows, "Ambulance" tempa, England casting ($2-4)(TP)

24. white body, red interior, unpainted base, blue windows, "Ambulance" tempa, England casting ($2-4)(TP)

25. white body, red interior, silver/gray base, blue windows, "Marine Division Police" tempa, England casting ($2-4)(TP)

26. white body, red interior, unpainted base, blue windows, "Marine Division Police" tempa, England casting ($2-4)(TP)

27. white body, red interior, pearly silver base, blue windows, "Marine Division Police" tempa, Macau casting ($4-6)(TP)

MB12-E PONTIAC FIREBIRD SE, issued 1982 (USA)
MB51-E PONTIAC FIREBIRD SE, issued 1982 (ROW)(continued)

6. rose red body, pearly silver base, tan interior, clear windows, 5 arch silver wheels, "Firebird" tempa, Macau casting ($4-6)

7. dark red body, pearly silver base, tan interior, clear windows, 5 arch silver wheels, no tempa, Macau casting ($4-6)

8. black body, pearly silver base, tan interior, clear windows, 5 arch silver wheels, "Firebird" tempa, Macau casting ($3-5)

9. black body, pearly silver base, red interior, clear windows, 5 arch silver wheels, "Firebird" tempa, Macau casting ($3-5)

10. black body, black base, gray interior, clear windows, starburst wheels,

"Halley's comet" tempa, Macau casting ($8-10)(MP)(US)

11. black body, pearly silver base, red interior, opaque glow windows, 5 arch gold wheels, "Firebird" tempa, Macau casting ($10-15)(PS)

12. black body, pearly silver base, red interior, clear windows, 5 arch gold wheels, "Firebird" tempa, Macau casting ($4-6)

13. black body, pearly silver base, red interior, opaque glow windows, 5 arch silver wheels, "Firebird" tempa, Macau casting ($10-15)

14. blue body, black base, white interior, blue windows, starburst wheels, red/orange/yellow stripes tempa, Macau casting ($8-12)(SF)

15. blue body, white base, white interior, blue windows, starburst wheels, red/orange/yellow stripes tempa, Macau casting ($3-5)(SF)

16. red body, pearly silver base, red interior, clear windows, 5 arch silver wheels, "Maaco" labels ($7-10)(US)

17. metallic blue body, white base, white interior, blue windows, laser wheels, red/orange/yellow stripes tempa, Macau casting ($3-5)(LW)

18. powder blue body, pearly silver base, red interior, clear windows, green/yellow/white stripes tempa, 5 arch silver wheels, Macau casting ($8-12)(DY)

19. purple body, pearly silver base, red interior, clear windows, 5 arch silver wheels, "Firebird" tempa, Macau casting ($3-4)(SC)

20. black body, dark gray plastic base, red interior, clear windows, 5 arch silver

wheels, "Firebird" tempa, Macau casting ($2-4)

21. blue body, white plastic base, white interior, blue windows, starburst wheels, red/orange/yellow stripes tempa, Macau casting ($4-6)(SF)

22. black body, dark gray plastic base, dark red interior, clear windows, dot dash wheels, "Firebird" tempa, China casting ($3-4)(MP)

23. black body, dark gray plastic base, dark red interior, clear windows, starburst wheels, "Firebird" tempa, China casting ($3-4)(MP)

MB12-H MERCEDES 500SL CONVERTIBLE, issued 1990 (USA)
MB33-H MERCEDES 500SL CONVERTIBLE, issued 1990 (ROW)

1. silver/gray body, silver/gray base, clear windshield, dark blue interior/tonneau, no tempa, 8 dot wheels, Macau casting ($1-2)

2. silver/gray body, silver/gray base, clear windshield, dark blue interior/tonneau, no tempa, 8 dot wheels, China casting ($1-2)

3. black body, metallic charcoal base, chrome windshield/dark gray interior/tonneau, detailed trim tempa, gray disc w/rubber tires, China casting ($5-8)(WC)

4. white body, gray base, clear windshield, brown interior/tonneau, "500SL" tempa, 8 dot wheels, China casting ($1-2)

5. silver/gray body, gray base, chrome windshield, maroon interior/tonneau, detailed trim tempa, gray disc w/rubber tires tempa, China casting ($5-8)(WC)

MB13-D 4 X 4 OPEN BACK TRUCK, issued 1982 (USA)
MB63-F 4 X 4 OPEN BACK TRUCK, issued 1982 (ROW) (continued)

6. yellow body, pearly silver base, red windows, "4X4 Goodyear" tempa, Macau casting ($2-4)

7. yellow body, pearly silver base, red windows, "4X4 Goodrich" tempa, Macau casting ($2-4)

8. white body, pearly silver base, red windows, "Bob Jane T-Mart" tempa, Macau casting ($10-15)(AU)

9. white body, black metal base, red windows, "63" & stripes tempa, Macau casting ($1-2)

10. white body, black plastic base, red windows, "63" & stripes tempa, Macau casting ($1-2)

11. white body, black plastic base, red windows, "63" & stripes tempa, Thailand casting ($1-2)

MB14-D LEYLAND TANKER, issued 1982 (continued)

5. yellow cab, white tank, yellow tank base, "Shell" tempa, England casting ($7-10)(JP)

6. red cab, white tank, red tank base, "Shell" tempa, England casting ($60-85)

7. black cab, black tank, gray tank base, "Gas" tempa, Macau casting ($4-5)(CM)

MB14-E 1983/84 CORVETTE, issued 1983 (USA)
MB69-E 1983/84 CORVETTE, issued 1983 (ROW)

NOTE: early models cast "1983 Corvette" with later models cast "1984 Corvette".

1. pearly silver upper & lower body, red

interior, dot dash wheels, "83 Vette" tempa, Macau casting ($3-5)

2. pearly silver upper & lower body, red interior, 8 dot wheels, "83 Vette" tempa, Macau casting ($3-5)

3. red upper body, light gray lower body, black interior, dot dash wheels, "Vette" & silver/black stripes. Macau casting ($2-4)

4. red upper body, light gray lower body, black interior, 8 dot wheels, "Vette" & silver/black stripes, Macau casting ($2-4)

5. red upper body, light gray lower body, black interior, 5 arch wheels, "Vette" & silver/black stripes, Macau casting ($2-4)

6. red & white upper body, red lower body, black interior, starburst wheels, "350 CID" tempa, Macau casting ($2-4)(SF)

7. red & white upper body, red lower body, black interior, starburst wheels, "350 CID" tempa with "Chef Boyardee" label, Macau casting ($12-18)(US)(OP)

8. red & white upper body, red lower body, black interior, laser wheels, "350 CID" tempa with "Chef Boyardee" label, Macau casting ($75-90)(US)(OP)

9. red & white upper body, red lower body, black interior, laser wheels, "350 CID" tempa, Macau casting ($3-5)(LW)

10. gray upper body, bluish purple lower body, no interior-replaced with light gold plated window armament, starburst wheels, stripes & stars tempa, Macau casting includes plastic armament ($6-8)(RB)

11. gray upper body, bluish purple lower body, no interior-replaced with dark gold plated window armament, starburst wheels, stripes & stars tempa, Macau casting includes plastic armament ($6-8)(RB)

12. red upper body, pearly silver lower body, black interior, 8 dot wheels, blue & silver tempa ($8-12)(DY)

MB14-G 1987 CORVETTE, issued 1987 (USA)
MB28-G 1987 CORVETTE, issued 1990 (ROW)

1. yellow upper & lower body, black interior, clear windshield, 8 dot wheels, "Corvette" & logo tempa, Macau casting ($2-3)

2. white & red upper body, red lower body, black interior, clear windshield, laser wheels, "350 CID" tempa, Macau casting ($3-5)(LW)

3. white & red upper body, red lower body, black interior, clear windshield, starburst wheels, "350 CID" tempa, Macau casting ($3-5)(SF)

4. rose red upper & lower body, black interior, clear windshield, 8 dot wheels, "Corvette" & logo tempa, Macau casting ($3-4)(SC)

5. orange upper & lower body, black interior, clear windshield, 8 dot wheels, "Corvette" & logo tempa, Macau casting ($3-4)(SC)

6. lemon upper & lower body, black interior, clear windshield, 8 dot wheels, "Corvette" & logo tempa, Macau casting ($1-2)

7. metallic blue upper & lower body, black interior with gray seats, chrome windshield gray disc w/rubber tires, "Corvette" tempa, Macau casting ($5-8)(WC)

8. red upper & lower body, black interior, clear windshield, 8 dot wheels, flames tempa, Macau casting ($7-10)(US)(OP)

9. orange-yellow upper & lower body, black interior, clear windshield, 8 dot wheels, "Corvette" & logo tempa, Thailand casting ($1-2)

10. orange-yellow upper & lower body, black interior, clear windshield, 6 spoke ringed wheels, "Corvette" & logo tempa, Thailand casting ($60-85)

11. white upper & lower body, dark pink interior, clear windshield, 8 dot wheels, stripes & zigzags tempa, Thailand casting ($2-4)(DM)

12. lime upper & lower body, black interior, clear windshield, gray disc w/rubber tires, "Rally Official-Joe Bulgin" tempa, Thailand casting ($6-8)(WP)

MB15-D FORD SIERRA, issued 1982 (ROW)
MB55-E FORD SIERRA, issued 1982 (USA)
MB40-E FORD SIERRA, reissued 1990 (USA) (continued)

6. silver/gray body, dark gray lower body, gray base, red interior, clear windows, gray hatch, 8 dot silver wheels, "Ford XRI Sport" tempa, England casting ($3-5)

7. silver/gray body, dark gray lower body, gray base, red interior, clear windows, gray hatch, 5 crown wheels, "Ford XRI Sport" tempa, England casting ($2-3)

8. pearly silver body, dark gray lower body, gray base, red interior, clear windows, gray hatch, dot dash wheels, "Ford XRI Sport" tempa, Macau casting ($2-3)

9. black body, dark gray lower body, gray base, black interior, clear windows, black hatch, starburst wheels, white/green

stripes & "85" tempa, Macau casting ($3-4)(SF)

10. yellow body, black lower body, black base hatch, 8 dot gold wheels, "XR 4X4" tempa, Macau casting ($3-4)

11. yellow body, black lower body, black base, black interior, clear windows, black hatch, 8 dot silver wheels, "XR 4X4" tempa, Macau casting ($2-3)

12. yellow body, dark gray lower body, black base, red interior, clear windows, gray hatch, 8 dot gold wheels, "XR 4X4" tempa, Macau casting ($35-50)

13. yellow body, dark gray lower body, gray base, red interior, clear windows, gray hatch, 8 dot silver wheels, "XR 4X4" tempa, Macau casting ($35-50)

14. yellow body, dark gray lower body, gray base, red interior, clear windows, gray hatch, 8 dot gold wheels, "XR 4X4" tempa, Macau casting ($35-50)

15. yellow body, dark gray lower body, gray base, red interior, clear windows, gray hatch, dot dash wheels, "XR 4X4" tempa, Macau casting ($35-50)

16. cream body, dark gray lower body, gray base, black interior, clear windows, black hatch, 8 dot silver wheels, "55" & black band tempa, Macau casting ($8-12)

17. metallic green body, dark gray lower body, gray base, black interior, clear windows, black hatch, laser wheels, white/gold stripe tempa, Macau casting ($3-4)(LW)

18. dark blue body, black lower body, black base, black interior, clear windows, black hatch, 8 dot silver wheels, "Duckhams Race Team" tempa, Macau casting ($3-5)(TC)

19. white body, red lower body, red base, red interior, clear windows, red hatch, 8 dot silver wheels, "Virgin Atlantic" tempa, Macau casting ($3-5)(GS)

20. yellow body, dark gray lower body, gray base, red interior, amber windows, gray hatch, 8 dot silver wheels, "XR 4X4" tempa, Macau casting ($35-50)

21. black body, black lower body, black base, red interior, clear windows, black hatch, 8 dot silver wheels, "Texaco 6/Pirelli" tempa, Macau casting ($1-2)

22. red body, black lower body, black base, black interior, clear windows, black hatch, 8 dot silver wheels, "Tizer The Appetizer" tempa, Macau casting ($3-5)(TC)

NOTE: above models with metal base, below models with plastic base.

23. red body, black lower body, black base, black interior, clear windows, black hatch, 8 dot silver wheels, "Tizer The Appetizer" tempa, Macau casting ($3-5)(TC)

24. red body, black lower body, black base, no interior, black windows, black hatch, 8 dot silver wheels, amber dome lights, "Fire Dept." tempa, Macau casting ($7-10)(SR)

25. red body, black lower body, black base, no interior, black windows, black hatch, 8 dot silver wheels, greenish yellow dome lights, "Fire Dept" tempa, Macau casting ($7-10)(SR)

26. yellow-orange body, black lower body, black base, no interior, black windows, black hatch, 8 dot silver wheels, red dome lights, "Airport Security" tempa, Macau casting ($7-10)(SR)

27. yellow-orange body, black lower body,

black base, no interior, black windows, black hatch, 8 dot silver wheels, green dome lights, "Airport Security" tempa, Macau casting ($18-25)(SR)

28. white body, white lower body, black base, no interior, black windows, white hatch, 8 dot silver wheels, red dome lights, "Sheriff" tempa, Macau casting ($7-10)(SR)

29. black body, black lower body, black base, red interior, clear windows, black hatch, 8 dot silver wheels, "Texaco 6/Pirelli" tempa, Macau casting ($1-2)

30. white body, red lower body, red base, black interior, clear windows, red hatch, 8 dot silver wheels, "Virgin Atlantic" tempa, Macau casting ($3-5)

31. metallic green body, dark gray lower body, gray base, black interior, clear windows, black hatch, laser wheels, white/gold stripes tempa, Macau casting ($3-5)(LW)

32. metallic green body, dark gray lower body, gray base, black interior, clear windows, black hatch, starburst wheels, white/gold stripes tempa, Macau casting ($8-12)(SF)

33. black body, black lower body, black base, red interior, clear windows, black hatch, 8 dot silver wheels, "Texaco 6/Pirelli" tempa, Thailand casting ($1-2)

34. red body, yellow lower body, black base, blue interior, clear windows, blue hatch, 8 dot red wheels, 3 faces & lion head tempa, Thailand casting ($15-18)(LL)

35. white body, red lower body, red base, black interior, clear windows, red hatch, 8 dot silver wheels, "Virgin Atlantic" tempa, Thailand casting ($2-3)

36. black body, dark gray lower body, gray base, black interior, clear windows, black hatch, starburst wheels, white/green stripes & "85" tempa, Thailand casting ($3-4)(SF)

37. black body, dark gray lower body, gray base, black interior, amber windows, black hatch, starburst wheels, white/green stripes & "85" tempa, Thailand casting ($3-4)(SF)

38. white body, black lower body, black base, red interior, clear windows, black hatch, 8 dot silver wheels, "Gemini/N. Cooper/1" tempa, Thailand casting ($2-3)

MB15-E PEUGEOT 205, issued 1984 (ROW)
MB25-G PEUGEOT 205, reissued 1991 (ROW)

NOTE: all models with clear windows, silver/gray interior & 8 spoke wheels unless noted otherwise.

1. white body, white base, red "205" with stripes tempa, Macau casting ($2-4)

2. white body, white base, black "205" with stripes tempa, Macau casting ($2-4)

3. white body, white base, dark purple "205" with stripes tempa, Macau casting ($2-4)

4. white body, white base, purple "205" with stripes tempa, China casting ($2-4)

5. white body, white base, black "205" & "Matchbox 11" tempa, Manaus casting ($35-40)(BR)

6. white body, white base, black "205" tempa, Manaus casting ($20-30)

7. orange-red body, dark gray base, "48 Michelin/Bilstein" tempa, China casting ($2-3)

8. green body, dark gray base, no tempa,

China casting ($8-12)(GR)(GS)

9. yellow body, dark gray base, "Peugeot 205/48/Bilstein" tempa, China casting ($2-3)

MB15-F SAAB 9000 TURBO, issued 1988 (USA)
MB22-F SAAB 9000 TURBO, issued 1988 (ROW)

1. metallic red body, brown interior, black base, 8 dot wheels, no tempa, Macau casting ($1-2)

2. metallic blue body, gray interior, black base, laser wheels, "Saab Turbo" & stripes tempa, Macau casting ($7-10)(LW)

3. metallic red body, brown interior, black base, 8 dot wheels, no tempa, China casting ($1-2)

4. metallic pink-red body, brown interior, black base, 8 dot wheels, no tempa, China casting ($1-2)

5. white body, brown interior, white base, 8 dot wheels, "Saab 22" & stripes tempa, China casting ($2-3)

6. dark blue body, brown interior, dark blue base, 8 dot wheel, no tempa, China casting ($8-12)(GR)(GS)

7. silver/gray body, brown interior, white base, 8 dot wheels, "Saab 22" & stripes tempa, China casting ($8-10)(GS)

MB15-H ALFA ROMEO, issued 1991 (USA)
MB 6-G ALFA ROMEO, issued 1991 (ROW)

NOTE: all models with black base, tan interior, smoke gray windows & 8 dot wheels.

1. red body, painted black roof, China

casting ($2-3)

2. dull red body, painted black roof, China casting ($2-3)
3. red body, plain roof, China casting ($8-12)(GR)(GS)
4. red body, painted black roof, "Alfa Romeo" tempa, China casting ($1-2)
5. red body, plain roof, "Alfa Romeo" tempa, China casting ($1-2)
6. lime body, plain roof, "Alfa Romeo" tempa, China casting ($4-6)(GS)

MB15-I SUNBURNER, issued 1992 (USA)
MB41-I SUNBURNER, issued 1992 (ROW)
1. fluorescent yellow body, black interior, sun & flames tempa, China casting ($2-4)
2. white body, black interior, sun & flames tempa, China casting ($8-10)(GS)

MB16-B PONTIAC, issued 1979 (continued)
20. white body, pearly silver base, blue hood tempa, dot dash wheels, Macau casting ($4-6)

MB16-D F.1 RACER, issued 1984 (USA)
MB 6-D F.1 RACER, issued 1984 (ROW)
NOTE: In 1985 an identical casting was released. in the U.S.A. both MB16-D & MB65-D were run concurrently and only certain variations appear on either model, so these are listed separately. All English releases are listed here. Chromed/ unchromed refers to the wheel lettering only.
1. red body, black metal base, black driver, black "Pirelli" airfoil, (5) front wheels & (12) chromed rear wheels, chrome

exhausts, "Fiat 3" tempa, Macau casting ($2-4)
2. red body, black metal base, black driver, black "Pirelli" airfoil, (11) front wheels, (12) unchromed rear wheels, chrome exhausts, "Fiat 3" tempa, Macau casting ($2-4)
3. white/orange/green body, black metal base, yellow driver, yellow "Watson's" airfoil, (11) front wheels, (12) chromed rear wheels. "Mr. Juicy/Sunkist" tempa, Macau casting ($20-25)(HK)
4. yellow body, black metal base, dark red driver, dark red "Goodyear" airfoil, (11) front wheels, (12) chromed rear wheels, "Matchbox Racing Team", Macau casting ($2-3)
5. white body, blue metal base, blue driver, blue "Shell" airfoil, (11) front wheels, (12) chromed rear wheels, "Matchbox/ Goodyear" tempa, Macau casting ($5-8)(KS)(GS)
6. white body, blue metal base, blue driver, blue "Shell" airfoil, (11) front wheels, (12) unchromed rear wheels, "Matchbox/ Goodyear" tempa, Macau casting ($5-8)(KS)(GS)
7. red body, black plastic base, black driver, black "Pirelli" airfoil, (11) front wheels, (12) unchromed rear wheels, "Fiat 3" tempa, Macau casting ($2-4)(KS)
8. white body, blue plastic base, blue drive, blue "Shell" air foil, (11) front wheels, (12) unchromed rear wheels, "Matchbox/ Goodyear" tempa, Macau casting ($2-4)(GS)
9. white body, blue plastic base, blue driver, blue "Shell" airfoil, (11) front wheels, (12) unchromed rear wheels, "Matchbox/ Goodyear" tempa, Thailand casting ($2-4)(GS)

MB16-E FORD LTD POLICE CAR, issued 1987 (USA)
MB51-H FORD LTD POLICE CAR, issued 1987 (ROW)
NOTE: all models with white interior, blue windows, red/blue dome lights & 8 dot wheels unless otherwise noted.
1. white body, chrome base, "Police PD-21" tempa, Macau casting ($1-2)
2. dark purple body, chrome base, "Police PD-21" tempa, Macau casting ($3-4)(SC)
3. raspberry red body, chrome base, "Police PD-21" tempa, Macau casting ($3-4)(SC)
4. red body, chrome base, "Fire Dept./Fire Chief" tempa, Macau casting ($2-3)(MC)
5. red body, chrome base, "Fire Dept./Fire Chief" tempa, Thailand casting ($2-3)(MC)
6. white body, chrome base, "Police PD-21" tempa, Thailand casting ($1-2)
7. white body, chrome base, no tempa, Thailand casting ($10-15)(GF)
8. white body, blue base, Police face & cartoon figures tempa, 8 dot red wheels, Thailand casting ($6-8)(LL)
9. dark blue body, chrome base, "Police R-25" tempa, Thailand casting ($3-5)(TH)
10. white body, chrome base, "Police PD-21" but no silver on door shield tempa, Thailand casting ($1-2)
11. white body, black base, "Police PD-21" & "Intercom City" tempa, Thailand casting ($10-15)(IC)

MB17-C LONDON BUS, issued 1982
MB51-F LONDON BUS, reissued 1985 (USA)
MB28-F LONDON BUS, reissued 1990 (USA) (continued)
5. red upper & lower body, white interior,

black base, "Matchbox London Bus" labels, England casting ($3-5)
6. red upper & lower body, white interior, brown base, "Matchbox London Bus" labels, England casting ($10-15)
7. red upper & lower body, white interior, black base, "Nice To Meet You! Japan 1984" labels, England casting ($15-20)(JP)
8. red upper & lower body, white interior, black base, labels, England casting ($15-20)(JP)
9. red upper & lower body, white interior, black base, "York Festival & Mystery Plays" labels, England casting ($10-15)(UK)
10. dark blue upper & lower body, white interior, black base, "Nestle Milkybar" labels, England casting ($10-15)(UK)
11. dark green upper & lower body, white interior, black base, "Rowntree's Fruit Gums" labels, England base ($10-15)(UK)
12. dark blue upper & lower body, white interior, black base, "Keddies No. 1 in Essex" labels, England base ($40-65)(UK)
13. maroon upper & lower body, white interior, black base, "Rapport" labels, England base ($8-12)(UK)
14. white upper body, black lower body, white interior, black base, "Torvale Fisher Engineering Co." labels, Macau casting ($8-12)(UK)
15. white upper body, orange lower body, white interior, black base, "WH Smith Travel" labels, Macau casting ($10-15)(UK)
16. red upper & lower body, white interior, black base, "Matchbox London Bus"

labels, Macau casting ($3-4)

17. red upper & lower body, white interior, black be, "Nestle Milkybar" labels, Macau casting ($10-15)

18. red upper & lower body, white interior, black be, "Rowntree's Fruit Gums" labels, Macau casting ($10-15)

19. red upper & lower body, white interior, black base, "Matchbox No. 1/Montepna" labels, England casting ($150-200)(GK)

20. red upper & lower body, white interior, black base, "You'll ♥ New York" labels, Macau casting ($2-4)

21. blue upper body, white lower body, white interior, black base, "Space For Youth 1985/Staffordshire Police" labels, Macau casting ($8-12)(UK)

22. blue upper & lower body, white interior, black base, "Cityrama" labels, Macau casting ($8-12)(UK)

23. red upper & lower body, white interior, black base, no labels, England casting ($3-5)

24. red upper & lower body, white interior, black base, "You'll ♥ New York" labels, China casting ($2-4)

25. red upper & lower body, white interior, black base, "Nurenburg 1986" labels, Macau casting ($100-150)(GR)

26. red upper & lower body, white interior, black base, "first M.I.C.A. Convention", Macau casting ($250-300)(UK)

27. red upper & lower body, white interior, black base, "First M.I.C.A. Convention", England casting ($250-300)(UK)

28. red upper & lower body, white interior, black base, "First M.I.C.A. Convention", China casting ($250-300)(UK)

29. red upper & lower body, white interior, black base," "Around London Tour Bus"

30. blue upper & lower body, white interior, black be, "National Tramway Museum" labels, China casting ($8-12)(UK)

31. white upper body, dull red lower body, white interior, black base, "Midland Bus Transport Museum", China casting ($8-12)(UK)

32. dull red upper & lower body, white interior, black base "Band-Aid Plasters Playbus", China casting ($8-12)(OP)

33. blue upper & lower body, white interior, black base, "National Girobank", China casting ($8-12)(UK)

34. red upper & lower body, white interior, black base, "Matchbox Niagara Falls" labels, China casting ($6-8)(CN)

35. red upper & lower body, white interior, black base, "Feria Del Juguete Valencia, 12 Febrero 1987" labels, China casting ($125-175)(SP)

36. beige upper body, blue lower body, tan interior, "West Midlands Travel" labels, China casting ($8-12)(UK)

37. white upper & lower body, red interior, "M.I.C.A. Matchbox International Collectors Association" labels, China casting ($8-12)(UK)

38. white upper body, white & dark blue lower body, white interior, "Denney-Happy 1000th Birthday, Dublin" labels ($8-12)(IR)

39. red upper & lower body, orange-yellow interior, orange-yellow wheels with blue hubs, "123 abc" sides tempa and "My First Matchbox Nurenburg 1990" roof tempa, China casting ($10-15)(GR)

40. red upper & lower body, orange-yellow interior, orange-yellow wheels with blue hubs, "123 abc" sides tempa and teddy

bear with "abc" roof tempa, China casting ($6-8)(LL)

41. red upper & lower body, white interior, no labels, China casting ($4-6)

42. yellow upper & lower body, white interior, "It's the Real Thing Coke" labels, China casting ($12-18)(CN)

43. maroon upper & lower body, white interior, "Corning Glass Center" labels, China casting ($8-12)(US)

44. chrome plated upper & lower body, white interior, "Celebrating A Decade of Matchbox Conventions" labels, China casting ($25-40)(C2)

45. red upper & lower body, white interior, "Markfield Project Support Appeal 92", China casting ($15-20)(UK)

46. red upper & lower body, white interior, "London Wide Tour Bus" labels, China casting ($1-2)

47. white upper & lower body, white interior, no labels, China casting ($10-15)(GF)

48. red upper & lower body, white interior, orange-yellow wheels (plain hubs), "123 abc" sides tempa and teddy bear with "abc" roof tempa, China casting ($6-8)(LL)

MB17-D AMX PROSTOCKER, issued 1983 (USA)
MB 9-E AMX PROSTOCKER, issued 1983 (ROW)

1. pearly silver body, black metal base, 5 arch wheels, red interior, red/black stripes & "AMX" tempa, Macau casting ($2-4)

2. pearly silver body, black metal base, dot dash wheels, red interior, red/black stripes & "AMX" tempa, Macau casting ($2-4)

3. maroon body, black metal base, dot dash

wheels, black interior, "Dr. Pepper" tempa, Macau casting ($3-5)

4. maroon body, black plastic base, dot dash wheels, black interior, "Dr. Pepper" tempa, Macau casting ($3-5)

5. silver/gray body, black plastic base, dot dash wheels, black interior, red/black stripes & "AMX" tempa, China casting ($8-10)(MP)

MB17-E FORD ESCORT CABRIOLET, issued 1985 (USA)
MB37-G FORD ESCORT CABRIOLET, issued 1985 (ROW)
NOTE: all models with clear windows, gray interior, black tonneau & bumpers.

1. white body & base, 8 dot silver wheels, no tow hook, "XR3i" tempa, Macau casting ($2-4)

2. white body & base, 8 dot gold wheels, no tow hook, "XR3i" tempa, Macau casting ($4-6)

3. white body & base, starburst wheels, no tow hook, "3" with stripes tempa, Macau basting ($3-5)(SF)

4. white body & base, 5 arch wheels, no tow hook, "XR3i" tempa, Macau casting ($2-4)

5. red body & base, 8 dot silver wheels, no tow hook, "XR3i" & "Ford" tempa, Macau casting ($8-12)

6. white body 7 base, 8 dot silver wheels, with tow hook, "XR3i" tempa, Macau casting ($2-4)(TP)

7. white body & base, 8 dot white wheels, with tow hook, "XR3i" tempa, Macau casting ($3-5)

8. metallic blue body & base, laser wheels, with tow hook, "3" with stripes tempa, Macau casting ($3-5)(LW)

9. white body & base, starburst wheels, with tow hook, "3" with stripes tempa, Macau casting ($3-5)(SF)
10. dark blue body & base, 8 dot silver wheels, with tow hook, "XR3i" tempa, Macau casting ($2-3)(TP)
11. white body & base, 8 dot silver wheels, with tow hook, "XR3i" tempa, Thailand casting ($1-2)(TP)
12. dark blue body & base, 8 dot silver wheels, with tow hook, "XR3i" tempa, Thailand casting ($1-2)(TP)
13. metallic blue body & base, 8 dot silver wheels, with tow hook, white/orange spatter tempa, Thailand casting ($2-3)(TP)

MB17-F DODGE DAKOTA, issued 1987 (USA)
MB50-F DODGE DAKOTA, issued 1990 (ROW)

1. bright red body, chrome base, chrome rollbar, black interior, clear windows, black & white stripes (plain sides) tempa, Macau casting ($3-5)
2. bright red body, chrome base, chrome rollbar, black interior, clear windows, black & white stripes & "Dakota ST" tempa, Macau casting ($1-2)
3. dark red body, chrome base, black rollbar, black interior, clear windows, black & white stripes & "Dakota ST" tempa, China casting ($1-2)
4. metallic green body, chrome base, black rollbar, black interior, clear windows, "MB Construction" & stripes tempa, China casting ($3-5)(AP)
5. white body, chrome base, white rollbar, white interior, blue windows, no tempa, China casting ($10-15)(GF)

6. blue body, chrome base, black rollbar, black interior, clear windows, black & white stripes & "Dakota ST" tempa, China casting ($8-12)(GS)
7. fluorescent orange body, black base, white rollbar, white interior, clear windows, "Fire Chief 1-Intercom City" tempa ($10-15)(IC)

MB18-B HONDARORA, issued 1975 (continued)

18. yellow body, black handlebars, tan driver, silver engine, no label, black seat, mag wheels, Macau casting ($3-5)
19. pearly silver body, black handlebars, red driver, black engine, with tempa, black seat, mag wheels, Macau casting ($8-12)(JP)
20. pearly silver body, black handlebars, red driver, black engine, with tempa, charcoal seat, mag wheels, Macau casting ($8-12)(JP)
21. orange body, black handlebars, tan driver, silver engine, no label, black seat, mag wheels, Macau casting ($3-5)

MB18-C FIRE ENGINE, issued 1984
NOTE: all models with blue windows
1. red body, chrome base, white ladder, black turret, 5 arch wheels, no tempa, Macau casting ($1-2)
2. red body, chrome base, white ladder, black turret, 8 dot wheels, no tempa, Macau casting ($1-2)
3. red body, chrome base, white ladder, black turret, 5 arch wheels, no tempa, China casting ($1-2)
4. red body, chrome base, white ladder, black turret, 5 arch wheels, "Fire Dept." & shield tempa, China casting ($1-2)

5. yellow body, chrome base, white ladder, black turret, 8 dot wheels, no tempa, Macau casting ($2-4)(MP)
6. red body, chrome base, white ladder, black turret, 8 dot wheels, Japanese lettered tempa, Macau casting ($8-12)(JP)(GS)
7. red body, chrome base, white ladder, black turret, 8 dot wheels, "Fire Dept." & shield tempa, Macau casting ($1-2)
8. red body, chrome base, yellow ladder & turret, 8 dot blue wheels with yellow hubs, fireman head tempa, Macau casting ($6-8)(LL)
9. red body, chrome base, white ladder, black turret, 8 dot wheels, "3" & crest tempa, Macau casting ($3-5)(GS)
10. red body, chrome base, white ladder, black turret, 8 dot wheels, "Fire Dept." & shield tempa, Thailand casting ($1-2)
11. red body, chrome base, white ladder, black turret, 5 arch wheels, "Fire Dept." & shield tempa, no origin cast ($8-12)(MP)
12. red body, chrome base, white ladder, black turret, 8 dot wheels, "3" & crest tempa, Thailand casting ($1-2)
13. yellow body, chrome base, white ladder, black turret, 8 dot wheels, no tempa, Thailand casting ($3-5)(MP)
14. red body, blue base, green & blue ladder, yellow turret, 8 dot blue wheels with yellow hubs, fireman head tempa, Macau casting ($6-8)(LL)
15. red body, blue base, green & blue ladder, yellow turret, 8 dot blue wheels with yellow wheels, fireman head tempa, Thailand casting ($6-8)(LL)
16. red body, blue base, green & blue ladder, yellow turret, 8 dot yellow wheels

(plain hubs), fireman head tempa, Thailand casting ($6-8)(LL)
17. fluorescent orange body, chrome base, white ladder, black turret, 8 dot wheels, "4" & checkers tempa ($1-2)
18. white body, white base, white ladder & turret, 8 dot wheels, no tempa, Thailand casting ($10-15)(GF)
19. fluorescent orange body, black base, white ladder, black turret, 8 dot wheels, "5" & "Intercom City" tempa, China casting ($10-15)(IC)

MB19-D PETERBILT CEMENT TRUCK, issued 1982 (continued)

3. pale green body, orange barrel, chrome exhausts, yellow/white "Big Pete" tempa, amber windows, black mounts, Macau casting ($2-4)
4. dark green body, orange barrel, chrome exhaust, yellow/white "Big Pete" tempa, amber windows, black mounts, Macau casting ($2-4)
5. dark green body, orange barrel, chrome exhaust, yellow/white "Big Pete" tempa, amber windows, charcoal mounts, Macau casting ($2-4)
6. dark green body, orange barrel, chrome exhausts, yellow/white "Big Pete" tempa, clear windows, black mounts, Macau casting ($2-4)
7. blue body, lemon barrel, chrome exhaust, "Kwik Set Cement" tempa, clear windows, black mounts, Macau casting ($1-2)
8. blue body, orange barrel, chrome exhausts, "Kwik Set Cement" (roof only) tempa, clear windows, black mounts, Macau casting ($8-12)
9. yellow body, orange barrel, chrome

exhausts, "Dirty Dumper" tempa, clear windows, black mounts, Macau casting ($60-80)

10. yellow body, gray barrel, chrome exhausts, "Pace Construction" tempa, clear windows, black mounts, Macau casting ($2-3)(MP)

11. blue body, lemon barrel, gray exhausts, "Kwik Set Cement" tempa, clear windows, black mounts, Macau casting ($1-2)

12. yellow body, gray barrel, gray exhausts, "Pace Construction" tempa, clear windows, black mounts, Macau casting ($2-3)(MP)

13. red body, lime barrel, lime exhausts, face & stripe tempa, clear windows, yellow mounts, blue base, yellow wheels with blue hubs, Macau casting ($5-7)(LL)

14. light pink body, white barrel, gray exhausts, "Readymix" tempa, clear windows, light pink mounts, black base, Thailand casting ($8-12)(AU)

15. blue body, lemon barrel, gray exhausts, "Kwik Set Cement" tempa, clear windows, black mounts, Thailand casting ($1-2)

16. yellow body, dark gray barrel, gray exhausts, "Pace Construction" tempa, clear windows, black mounts, Thailand casting ($3-4)(MC)

17. red body, lime barrel, lime exhausts, face & stripe tempa, clear windows, yellow mounts, blue base, yellow wheels with blue hubs, Thailand casting ($5-7)(LL)

18. red body, lime barrel, lime exhausts, face & stripe tempa, clear windows, yellow mounts, blue base, yellow wheels (plain hubs), Thailand casting ($5-7)(LL)

19. orange-yellow body, red barrel, gray exhausts, "Pace Construction" tempa, clear windows, black mounts, Thailand casting ($1-2)

20. red body, orange barrel, gray exhausts, man tempa, clear windows, black mounts, gray base, Manaus casting ($35-50)(BR)

21. red body, orange barrel, gray exhausts, man tempa, clear windows, dark gray mounts, gray base, Manaus casting ($35-50)(BR)

MB20-B POLICE PATROL, issued 1975 (continued)

43. beige body, gloss black base, frosted windows, red dome/spinner, "Paris Dakar 83" tempa, Maltese cross wheels, England casting (Int'l)($6-8)

44. beige body, unpainted base, frosted windows, red dome/spinner, "Paris Dakar 83" tempa, maltese cross wheels, England casting (Int'l)($4-6)

45. white and black body, black base, frosted windows, red dome/spinner, Japanese lettered tempa, maltese cross wheels, England casting (Int'l) ($8-12)(JP)

MB20-C 4 X 4 JEEP, issued 1982 (USA)
MB14-F 4 X 4 JEEP, issued 1984 (ROW) (continued)

2. black body, black metal base, red interior, white canopy, "Laredo" tempa, Macau casting ($1-2)

3. black body, black metal base, red interior, white canopy, "Laredo" tempa, Hong Kong casting ($18-25)

4. dark tan body, black metal base, black interior, red canopy, "Golden Eagle" tempa, England casting ($18-25)

5. red body, black metal base, gray interior, white canopy, "Golden Eagle" tempa, Macau casting ($4-5)

6. olive body, black metal base, black interior, tan canopy, black/tan camouflage tempa, Macau casting ($4-6)(CM)

7. olive body, black plastic base, black interior, tan canopy, black/tan camouflage tempa, Macau casting ($4-6)(CM)

8. black body, black plastic base, red interior, white canopy, "Laredo" tempa, Macau casting ($1-2)

9. black body, black plastic base, blue interior, lime canopy, caricature tempa, red wheels with green hubs, Macau casting ($7-10)(LL)

10. black body, black plastic base, red interior, white canopy, "Laredo" tempa, Thailand casting ($1-2)

11. red body, greenish black metal base, gray interior, white canopy, "Golden Eagle" tempa, Macau casting ($4-5)

12. yellow body, black plastic base, blue interior, lime canopy, caricature tempa, red wheels (plain hubs), Thailand casting ($6-8)(LL)

13. black body, black plastic base, red interior, red canopy, "Laredo" tempa, Macau tempa ($8-12)

MB20-D VOLVO CONTAINER TRUCK, issued 1985 (ROW)
MB23-G VOLVO CONTAINER TRUCK, issued 1987 (USA)
MB62-1 VOLVO CONTAINER TRUCK, reissued 1990 (ROW)

NOTE: all models with black base, amber windows, 5 arch wheels.

1. blue body, white container, "Coldfresh" labels, Macau casting ($3-5)

2. white body, white container, "Scotch Corner" labels, Macau casting ($8-12)(SC)

3. gray body, gray container, "Supersaver Drugstores" labels, Macau casting ($8-12)(UK)

4. blue body, white container, "MB1-75 #1 in Volume Sales" labels, Macau casting ($35-50)(UK)

5. blue body, white container, "MB1-75 #1 in Volume Sales" labels, China casting ($35-50)(UK)

6. blue body, white container, "Coldfresh" labels, China casting ($3-5)

7. white body, white container, "Federal Express" labels, China casting ($3-5)

8. blue body, white container, "Unic" (towards front or rear) labels, China casting ($8-12)(BE)

9. blue body, blue container, "Crooke's Healthcare" labels, China casting ($6-8)(UK)

10. white body, white container, "Kellogg's/ Milch-Lait-Latte" labels, China casting ($35-50)(GR)(SW)

11. white body, white container, "TNT Ipec" labels, china casting ($4-6)(TC)

12. green body, gray container, "Hikkoshi Semmon Center" (in Japanese) labels, China casting ($10-15)(UK)

13. blue body, blue container, "Alders" labels, China casting ($6-8)(UK)

14. white body, white container, "XP Parcels" labels, China casting ($4-6)(MP)

15. blue body, blue container, "Comma Performance Motor Oils" labels, China casting ($8-12)(UK)(OP)

16. blue body, white container, "Kelloggs/ Milch-Lait-Latte" labels, Macau casting ($35-50)(GR)(SW)

17. white body, white container, "Federal Express" on container labels with "XP Parcels" on doors tempa, China casting ($15-20)
18. red body, brown container, "Merkur Kaffee" labels, China casting ($8-12)(SW)
19. green body, white container, "M" & green stripe labels, China casting ($8-12)(SW)
20. red body, white container, "Denner" labels, China casting ($8-12)(SW)
21. white body, white container, "Family Trust" labels, China casting ($8-12)(CN)
22. blue body, red container, "Christiansen" labels, China casting ($8-12)(BE)
23. white body, dark cream container, "Federal Express" labels, China casting ($2-3)
24. white body, white container, "Kit Kat" labels, China casting ($25-40)(UK)
25. white body, white container, "Yorkie" labels, China casting ($25-40)(UK)
26. red body, white container, "Big Top Circus" labels, china casting ($1-2)
27. white body, powder blue container, "Co Op-People Who Care" labels, China casting ($6-8)(UK)
28. blue body, white container, "Big Top Circus" labels, China casting ($4-6)(GS)
29. white body, powder blue container, "99 Tea" labels, China casting ($7-10)(OP)

MB20-E VOLKSWAGEN TRANSPORTER, issued 1988 (ROW)
NOTE: all models with blue windows & dome lights, gray interior.
1. white body, black base, 8 dot silver wheels, orange stripe & cross tempa, Macau casting ($2-4)

2. black body, black base, 8 dot black wheels, green cross & "LS2081" tempa, Macau casting ($4-5)(CM)
3. white body, black base, 8 dot silver wheels, red stripe & cross tempa, Macau casting ($2-4)
4. white body, black base, 8 dot silver wheels, red stripe & cross tempa, China casting ($2-4)
5. black body, black base, 8 dot black wheels, green cross & LS2081 tempa, China casting ($4-5)(CM)
6. white body, white base, 8 dot silver wheels, no tempa, China casting ($10-15)(GF)
7. white body, black base, 8 dot silver wheels, red stripe/cross & "Ambulance" tempa, China casting ($1-2)

MB21-C RENAULT 5TL, issued 1978 (continued)
43. white body, tan interior, black base, white tailgate, clear windows, "Roloil" tempa, Macau casting ($3-4)(TP)
44. pearly silver body, black interior, black base, dark gray tailgate, clear windows, "Scrambler" tempa, Macau casting ($5-8)(TP)

MB21-D CORVETTE PACE CAR, issued 1983 (USA)
1. pearly silver body, red interior & base, without "Pace" on trunk, Macau casting ($5-8)
2. pearly silver body, red interior & base, with "Pace" on trunk, Macau casting ($5-8)

MB21-E BREAKDOWN VAN, issued 1985
MB53-F BREAKDOWN VAN, reissued 1990 (USA)
NOTE: all models with blue windows, black tow hook, 8 spoke wheels unless otherwise denoted.
1. red body, pearly silver base, silver/gray interior, white boom, "24 Hour Service" tempa, Macau casting ($1-2)
2. red body, pearly gray base, silver/gray interior, white boom, "24 Hour Service" tempa, China casting ($1-2)
3. red body, pearly gray base, silver/gray interior, white boom, "24 Hours" with white stripes tempa, China casting ($2-3)(MP)
4. yellow body, black base, black interior, black boom, "Auto Relay 24 Hr Tow" tempa, China casting ($1-2)
5. black body, gray base, gray interior, gray boom, yellow stripes tempa, black hubs, China casting ($4-6)(CM)
6. black body, gray base, gray interior, gray boom, yellow stripes tempa, silver hubs, China casting ($5-7)(CM)
7. red body, yellow base, silver/gray interior, green boom, cartoon towed car tempa, blue wheels with yellow hubs, China casting ($6-8)(LL)
8. dark orange body, black base, orange-yellow interior, orange-yellow boom, "Auto Relay 24 Hr Tow" tempa, China casting ($3-5)(AP)
9. white body, black base, black interior, black boom, "Auto Rescue" tempa, China casting ($2-4)(EM)
10. white body, white base, red interior, white boom, no tempa, red windows, China casting ($4-6)(GF)

11. orange-yellow body, greenish black base, black interior, black boom, "Auto Relay 24 Hr Tow" tempa, China casting ($1-2)
12. red body, yellow base, silver/gray interior, green boom, cartoon towed car tempa, all blue wheels, China casting ($6-8)(LL)
13. fluorescent orange body, black base, black interior, black boom, "Intercom City Auto Service 7" tempa, China casting ($10-15)(IC)

MB21-F GMC WRECKER, issued 1987 (USA)
MB710F GMC WRECKER, issued 1987 (ROW)
NOTE: all models with white rear bed, white boom, cloudy amber windows, 8 spoke wheels.
1. white body, unpainted metal base, "Frank's Getty" with phone no. tempa, Macau casting ($1-2)
2. white body, unpainted metal base, "Accessory Wholesalers Inc." tempa, Macau casting ($18-25)(US)
3. white body, chrome plastic base, "Frank's Getty" with phone no. tempa, Macau casting ($1-2)
4. white body, chrome plastic base, "Frank's Getty" with phone no. tempa, China casting ($1-2)
5. white body, chrome plastic base, "Frank's Getty" with phone tempa, Thailand casting ($1-2)
6. white body, chrome plastic base, "Frank's Getty" with phone no. tempa, no origin cast ($3-5)
7. white body, chrome plastic base, "Frank's Getty" without phone no. tempa,

Thailand casting ($1-2)

8. white body, chrome plastic base, "Frank's Getty" without phone no. and solid color print tempa, Thailand casting ($1-2)

MB22-D 4X4 MINI PICKUP, issued 1982 (USA)
MB35-E 4X4 MINI PICKUP, issued 1986 (ROW)

6. pearly silver body, black metal base, white flat roof, "Big Foot" tempa, Hong Kong casting ($5-8)

7. pearly silver body, black metal base, white flat roof, "Big Foot" tempa, Macau casting ($4-6)

8. red body, black metal base, white flat roof, "Aspen Ski Holidays" tempa, Macau casting ($2-4)

9. white body, black metal base, white flat roof, "SLD Pump Service" tempa, Macau casting ($10-15)(UK)

10. red body, black plastic base white flat roof, "Aspen Ski Holidays" tempa, Macau casting ($2-3)

11. red body, black plastic base, white flat roof, "Aspen Ski Holidays" tempa, Thailand casting ($2-3)

MB22-E JAGUAR XK120, issued 1982 (ROW) (continued)

2. dark green body, red interior, unpainted base, clear windows, no tempa, England casting ($2-4)

3. dark green body, red interior, pearly silver base, clear windows, no tempa, Macau casting ($2-4)

4. dark green body, red interior, pearly silver base, amber windows, no tempa, Macau castings ($2-4)

5. cream body, red interior, pearly silver base, clear windows, "414" tempa,

Macau casting ($1-2)

6. white body, dark maroon interior, pearly silver base, chrome windows, gray disc w/rubber tires, no tempa ($5-8)(WC)

7. dark cream body, dark red body, pearly silver base, clear windows, "414" tempa, Macau casting ($1-2)

8. cream body, red interior, pearly silver base, clear windows, "414" tempa, Thailand casting ($1-2)

9. white body, purplish black interior, pearly silver base, chrome windows, gray disc w/rubber tires, no tempa, Thailand casting ($5-8)(WC)

10. white body, orange interior, pearly silver base, amber windows, blue & orange flash tempa, Thailand casting ($3-5)(DM)

MB22-G VECTRA CAVALIER GSi 2000, issued 1990 (USA)
MB41-G VECTRA CAVALIER GSi 2000, issued 1990 (ROW)

NOTE: all models with black base, clear windows, gray interior, 8 dot wheels, no tempa.

1. metallic red body, Macau casting ($1-2)

2. metallic pink-red body, Macau casting ($1-2)

3. metallic pink-red body, China casting ($1-2)

4. green body, China casting ($8-12)(GR)(GS)

MB22-H LAMBORGHINI DIABLO, issued 1992 (USA)
MB49-F LAMBORGHINI DIABLO, issued 1992 (ROW)

1. yellow body, yellow base, black interior, clear windows, 8 dot wheels, "Diablo" tempa ($1-2)

2. red body, red base, black interior, chrome windows, gray disc w/rubber tires, detailed trim tempa ($5-8)(WC)

MB23-B ATLAS, issued 1975 (continued)

12. mid blue body, yellow dump, black plastic base, clear windows, gray interior, maltese cross wheels ($7-10)(GS)

MB23-D AUDI QUATTRO, issued 1982 (USA)
MB25-F AUDI QUATTRO, issued 1982 (ROW) (continued)

NOTE: all models fitted with black interior.

4. white body, black base, clear windows, 5 arch wheels, "Audi 20" tempa, Macau casting ($2-4)

5. white body, black base, light amber windows, 5 arch wheels, "Audi 20" tempa, Macau casting ($2-4)

6. white body, black base, clear windows, 5 arch wheels, "Duckhams/Pirelli" tempa, Macau casting ($3-5)

7. white body, black base, clear windows, 5 arch wheels, "Duckhams/Pirelli" (plain hood) tempa, Macau casting ($3-5)

8. white body, black base, clear windows, 5 arch wheels, "Duckhams" without "Pirelli" tempa, Macau casting ($3-5)

9. white body, black base, amber windows, 5 arch wheels, "Duckhams/Pirelli" tempa, Macau casting ($3-5)

10. plum body, black base, clear windows, 5 arch wheels, "Quattro 0000" tempa, Macau casting ($2-4)

11. dark blue body, black base, clear windows, 8 dot wheels, "Quattro 0000" tempa, Macau casting ($2-4)

12. dark blue body, black base, clear windows, 8 dot wheels, "Quattro 0000"

& "Audi" tempa, Macau casting ($8-12)

13. dark blue body, black base, amber windows, 8 dot wheels, "Quattro 0000" tempa, Macau casting ($8-12)

14. dark plum body, black base, clear windows, 5 arch wheels, "Quattro 0000" tempa, China casting ($2-4)

15. dark gray body, black base, clear windows, 5 arch wheels, "0000 Quattro" and "Audi" tempa, China casting ($1-2)

16. dark blue body, black base, clear windows, 5 arch wheels, "Audi 2584584" & "FT" tempa ($35-50)(CH)

17. white body, black base, clear windows, 8 dot wheels, "Audi 20" tempa, Manaus casting ($35-50)(BR)

MB23-F HONDA ATC, issued 1985 (USA)

1. red body with blue painted seat, red handlebars, yellow wheels, Macau casting ($8-10)

2. red body with blue painted seat, red handlebars, yellow wheels, China casting ($8-10)

3. fluorescent green body with black painted seat, fluorescent orange handlebars, white wheels, China casting ($4-6)(KS)

MB24-C SHUNTER, issued 1979 (continued)

12. dark yellow body, red plastic undercarriage, black base, no panel, "D1496-RF" tempa, Macau casting ($2-4)(MC)

13. dark yellow body, red plastic undercarriage, black base, no panel, "D1496-RF" tempa, China casting ($2-4)(MC)

14. orange-yellow body, red plastic undercarriage, black base, no panel, "D1496-RF" tempa, China casting ($2-4)(TP)

MB24-E DATSUN 280ZX, issued 1983
NOTE: all models with black base, tan interior & clear windows unless otherwise noted.
1. black body, 5 arch silver wheels, gold pin stripe tempa, Macau casting ($2-4)
2. black body, 5 arch silver wheels, "Turbo ZX" tempa, Macau casting ($2-4)
3. white body, 5 arch silver wheels, red/blue "Turbo 33" tempa, Macau casting ($8-12)(JP)
4. black body, small starburst wheels, orange/yellow/white "Turbo" tempa, Macau casting ($3-5)(SF)
5. black body, large starburst wheels, orange/yellow/white "Turbo" tempa, Macau casting ($3-5)(SF)
6. black body, 5 arch gold wheels, "Turbo ZX" tempa, Macau casting ($7-10)
7. black body, 8 dot gold wheels, "Turbo ZX" tempa, Macau casting ($10-15)
8. black body, laser wheels, orange/yellow/white tempa, Macau casting ($3-5)(LW)
9. charcoal gray body, laser wheels, orange/yellow/white tempa, Macau casting ($6-8)(LW)
10. red body, starburst wheels, black interior, black modified windows, black & orange tempa, Macau casting, includes plastic armament ($6-8)(RB)

MB24-F NISSAN 300ZX, issued 1986
1. pearly silver body, lt. brown interior, 8 dot wheels, black base, gold stripe with "Turbo" tempa, Macau casting ($2-4)

2. pearly silver body, lt. brown interior, 5 arch wheels, black base, gold stripe with "Turbo" tempa, Macau casting ($2-4)
3. white body, red interior, 8 dot wheels, green base, "Fujicolor" tempa, Macau casting ($2-4)
4. red body, lt. brown interior, starburst wheels, black base, red & orange stripes tempa, Macau casting ($3-5)(SF)
5. metallic red body, lt. brown interior, laser wheels, black base, red & orange stripes tempa, Macau casting ($3-5)(LW)
6. white body, red interior, 8 dot wheels, black base, "96/BP Racing Team" tempa, Macau casting ($8-12)(DU)
7. yellow body, red interior, 8 dot wheels, green base, "4 Monkey Racing Team" tempa, Macau casting ($8-12)(HK)

MB24-H FERRARI F40, issued 1989 (USA)
MB70-F FERRARI F40, issued 1989 (ROW)
1. red body, clear windows, black interior, 8 dot wheels, "Ferrari" logo tempa, Macau casting ($1-2)
2. red body, chrome windows, black interior, gray disc w/rubber tires, "Ferrari" logo tempa, Macau casting ($5-8)(WC)
3. chrome plated body, clear windows, black interior, 8 dot wheels, no tempa, Macau casting ($12-18)(C2)
4. dull red body, clear windows, black interior, 8 dot wheels, "Ferrari" logo tempa, Macau casting ($1-2)
5. red body, clear windows, black interior, 8 dot wheels, "Ferrari" logo tempa, Thailand casting ($1-2)
6. dull red body, clear windows, black

interior, 8 dot wheels, "Ferrari" log tempa, China casting ($1-2)
7. red body, chrome/black windows, no interior, silver/yellow lightning wheels, lemon stripe & "40" tempa, China casting ($2-4)(LT)
8. lemon body, blue-chrome/black windows, no interior, pink/yellow lightning wheels, blue with "F40" tempa, China casting ($2-4)(LT)
9. lemon body, blue-chrome/black windows, no interior, peach/yellow lightning wheels, blue with "F40" tempa, China casting ($2-4)(LT)
10. lemon body, blue-chrome/black windows, no interior, silver/red lightning wheels, blue with "F40" tempa, China casting ($2-4)(LT)
11. white body, blue-chrome/black windows, no interior, peach/silver lightning wheels, "40" tempa, China casting ($3-5)(LT)
12. black body, black/chrome windows, no interior, pink/silver lightning wheels, "F40" tempa, China casting ($3-5)(LT)
13. red body, black windows, black interior, 8 dot wheels, yellow/white "F40" tempa, Thailand casting ($3-5)(TH)
14. yellow body, chrome windows, black interior, gray disc w/rubber tires, detailed trim tempa, China casting ($3-4)(WC)
15. red body, clear windows, black interior, 8 dot wheels, "Ferrari" logo without multicolor design tempa, Thailand casting ($1-2)

MB25-C FLAT CAR & CONTAINER, issued 1979 (continued)
18. black flatbed, white container, cast doors, no labels, China casting

($10-15)(PS)
19. black flatbed, yellow container, cast doors, no labels, China casting ($10-15)(PS)
NOTE: playset comes with label sheet that supplies labels for containers!

MB25-E AMBULANCE, issued 1983 (USA)
MB25-E AMBULANCE, issued 1990 (ROW)
1. white body, pearly silver base, "Pacific Ambulance" tempa, Macau casting ($4-6)
2. white body, unpainted base, "Pacific Ambulance" tempa, Macau casting ($4-6)
3. white body, pearly silver base, "Paramedics E11" tempa, Macau casting ($1-2)
4. white body, dark gray base, "Paramedics E11" tempa, Macau casting ($1-2)
5. white body, pearly silver base, "Paramedics E11" tempa, China casting ($1-2)
6. white body, pearly silver base, "EMT Ambulance" tempa, China casting ($3-4)(AP)
7. white body, white base, no tempa, China casting ($10-15)(GF)
8. yellow body, pearly silver base, "Paramedics E11" tempa, China casting ($8-12)(GS)
9. fluorescent orange body, black base, "Ambulance 7/Intercom City" tempa, China casting ($10-15)(IC)

MB26-D COSMIC BLUES, issued 1982 (USA)
MB41-H COSMIC BLUES, reissued 1991 (USA) (continued)
3. white body, 5 crown front & maltese

cross rear wheels, chrome exhausts, Macau casting ($1-2)
4. white body, 5 crown front & rear wheels, chrome exhausts, Macau casting ($1-2)
5. white body, 5 arch front & 5 crown rear wheels, chrome exhausts, China casting ($1-2)
6. blue body, 5 arch front & 5 crown rear wheels, chrome exhausts, China casting ($1-2)
7. blue body, 5 arch front & 5 crown rear wheels, black exhausts, China casting ($1-2)

MB26-F VOLVO TILT TRUCK, issued 1984 (ROW)
MB49-G VOLVO TILT TRUCK, issued 1990 (USA)
1. metallic blue body, yellow canopy, no tempa, England casting ($25-35)
2. metallic blue body, yellow canopy, "Fresh Fruit Co." tempa, Macau casting ($2-4)
3. yellow body, yellow canopy, "Ferrymasters Groupage" tempa, China casting ($2-4)
4. yellow body, yellow canopy, "Ferrymasters Groupage" tempa, China casting ($2-4)
5. white body, white canopy, "Federal Express" tempa, China casting ($1-2)
6. dark blue body, yellow canopy, "Michelin" tempa, dark gray base, China casting ($1-2)
7. olive body, tan canopy, "LS2020" tempa, black hubs, China casting ($15-20)
8. black body, dark gray canopy, "LS1506" tempa, black hubs, China casting ($15-20)
9. blue body, blue canopy, "Henneiz" tempa, China casting ($8-12)(SW)

10. red body, green canopy, cartoon face & forklift tempa, yellow wheels with red hubs, China casting ($6-8)(LL)
11. red body, no canopy, "123" on doors tempa, yellow wheels with blue hubs, China casting ($6-8)(LL)
12. red body, green canopy, cartoon face & forklift tempa, yellow wheels with blue hubs, China casting ($6-8)(LL)
13. white body, white canopy, "Pirelli Gripping Stuff" & gray tread pattern tempa, China casting ($1-2)
14. white body, white canopy, "Pirelli Gripping Stuff" & black tread pattern tempa, China casting ($1-2)
15. red body, green canopy, cartoon face & forklift tempa, all yellow wheels, China casting ($6-8)(LL)

MB26-G BMW 5 SERIES, issued 1989 (USA)
MB31-H BMW 5 SERIES, issued 1989 (ROW)
1. dark charcoal body, black base, no tempa, Macau casting ($2-4)
2. dark charcoal body, black base, no tempa, China casting ($2-4)
3. white body, white base, "Fina 31/BMW Team/Sachs/Bilstein" tempa, China casting ($2-4)
4. white body, white base, "Fina 31/BMW Team" without "Sachs/Bilstein" tempa, China casting ($2-4)

MB27-C SWEPT WING JET, issued 1981 (continued)
5. dark red body, white base, white wings, red windows, black/red tempa, Macau casting ($2-4)
6. rose red body, white base, white wings,

red windows, black/red tempa, Macau casting ($2-4)
7. dark red body, white base, light gray wings, red windows, black/red tempa, Macau casting ($4-6)
8. dark red body, white base, white wings, red windows, black/red tempa, China casting ($2-4)
9. black body, gray base, dark gray wings, red windows, yellow/black camouflage tempa, Macau casting ($3-5)(CM)

MB27-D JEEP CHEROKEE, issued 1986
NOTE: all models with clear windows & 8 spoke wheels. When interior is noted this also includes the color of the side molding.
1. white body, black interior, black metal base, "Quadtrak" tempa, Macau casting ($3-5)
2. beige body, black interior, black metal base, "Holiday Club" tempa, Macau casting ($2-4)
3. yellow body, dark gray interior, black metal base, "Forest Ranger County Park" tempa, Macau casting ($7-10)(UK)
4. yellow body, dark gray interior, black metal base, "Mr. Fixer" tempa, Macau casting ($1-2)
5. bright yellow body, dark gray interior, black metal base, "Mr. Fixer" tempa, Macau casting ($1-2)
6. orange-yellow body, dark gray interior, black metal base, "Mr. Fixer" tempa, Macau casting ($1-2)
7. light pea green body, dark gray interior, black metal base, "Mr. Fixer" tempa, Macau casting ($3-4)(SC)
8. light brown body, dark gray interior, black metal base, "Mr. Fixer" tempa,

Macau casting ($3-4)(SC)
9. dark brown body, dark gray interior, black metal base, "Mr. Fixer" tempa, Macau casting ($3-4)(SC)
10. beige body, dark gray interior, black metal base, "Holiday Club" tempa, Macau casting ($2-4)(TP)
11. yellow body, dark gray interior, black plastic base, "Mr. Fixer" tempa, Macau casting ($1-2)
12. reddish brown body, dark gray interior, black plastic base, Mr. Fixer" tempa, Macau casting ($3-4)(SC)
13. dark brown body, dark gray interior, black plastic base, "Mr. Fixer" tempa, Macau casting ($3-4)(SC)
14. white body, dark gray interior, black plastic base, "National Ski Patrol" tempa, Macau casting ($8-12)(US)
15. beige body, dark gray interior, black plastic base, "Holiday Club" tempa, Macau casting ($2-4)(MP)
16. orange-yellow body, dark gray interior, black plastic base, "Mr. Fixer" tempa, Thailand casting ($1-2)
17. silver/gray body, black interior, black plastic base, "Sport" & "Jeep" tempa, Thailand casting ($2-4)(US)

MB28-D FORMULA 5000, issued 1982 (continued)
5. tan body, black metal base, white driver, chrome engine, 5 arch front & dot dash rear wheels, Macau casting ($3-5)
6. tan body, black plastic base, white driver, chrome engine, 4 spoke front & maltese cross rear wheels, Manaus casting ($35-50)(BR)
7. rust red body, black plastic base, cream driver, gray engine, dot dash front & 8

spoke rear wheels, Manaus casting ($35-50)(BR)

MB28-E DODGE DAYTONA, issued 1984

NOTE: all models with clear windows & black interior. Model that has "Las Vegas" base is cataloged at the end of this listing.

1. maroon body, silver/gray base with black insert, 8 dot silver wheels, no tempa, England casting ($3-5)
2. maroon body, silver/gray base with dark gray insert, 5 arch wheels, no tempa, England casting ($3-5)
3. maroon body, silver/gray base with dark gray insert, 8 dot silver wheels, no tempa, England casting ($3-5)
4. maroon body, silver/gray base with no insert, 8 dot silver wheels, no tempa, Macau casting ($50-75)
5. pearly silver body, black base, 8 dot silver wheels, red/black stripes tempa, Macau casting ($2-4)
6. pearly silver base, large 5 crown wheels, red/black stripes tempa, Macau casting ($5-8)
7. pearly silver body, black base, dot dash wheels, red/black stripes tempa, Macau casting ($5-8)
8. pearly silver body, black base, 8 dot gold wheels, red/black stripes tempa, Macau casting ($4-6)
9. white body, blue base, starburst wheels, red/blue stripes & "8" tempa, Macau casting ($3-5)(SF)
10. white body, blue base, laser wheels, red/blue stripes & "8" tempa, Macau casting ($3-5)(LW)
11. plum body, gold base, starburst wheels,

red/yellow/dark blue tempa, gray redesigned windows, Macau casting, includes plastic armament ($6-8)(RB)
12. dark blue body, black base, 8 dot silver wheels, "5 Goat Racing Team" tempa, Macau casting ($8-12)(HK)
13. red body, red base, 8 dot silver wheels, yellow/blue "turbo Z" tempa, Macau casting ($2-3)
14. red body, red base, 8 dot silver wheels, yellow/blue "Turbo Z" tempa, hood cast shut, Macau casting ($2-3)
15. red body, red base, starburst wheels, yellow/blue "Turbo Z" tempa, hood cast shut, China casting ($5-7)(GS)

MB28-H FORKLIFT TRUCK, issued 1991 (USA)
MB61-F FORKLIFT TRUCK, issued 1991 (ROW)

NOTE: all models with black forks & arms, black base & 4 spoke wheels.

1. lime green body, red/white stripes tempa, Thailand casting ($1-2)
2. white body, red stripes tempa, Thailand casting ($2-4)(TC)
3. fluorescent green body, red/white stripes tempa, Thailand casting ($1-2)
4. orange-yellow body, red stripes tempa, Thailand casting ($2-4)(CS)
5. bright lime body, red/white stripes tempa, Thailand casting ($1-2)

MB29-C TRACTOR SHOVEL, issued 1976 (continued)

23. purple body, black shovel, black base, black motor, lime/orange tempa, Macau casting, includes plastic armament ($6-8)(RB)
24. dark yellow body, red shovel, black

base, black motor, stripes tempa, Macau casting ($2-3)(TC)
25. dark yellow body, black shovel, black base, black motor, "Thomas Mucosolvan" tempa, Macau casting ($15-25)(GR)
26. light orange body, black shovel, black base, black motor, "Thomas Mucosolvan" tempa, Macau casting ($20-30)(GR)
27. blue body, red shovel, black base, red motor, yellow wheels with orange hubs, green stripes tempa, Macau casting ($6-8)(LL)
28. light yellow body, black shovel, black base, black motor, "Thomas Mucosolvan" tempa, Macau casting ($6-8)(GR)
29. light yellow body, black shovel, black base, black motor, "Thomas Mucosolvan" tempa, Thailand casting ($18-25)(GR)
30. dark yellow body, black shovel, black base black motor, black stripes tempa, Thailand casting ($1-2)
31. dark yellow body, red shovel, black base, black motor, black stripes tempa, Thailand casting ($1-2)
32. light yellow body, black shove, black base, black motor, "Thomas MucosolvanS" tempa, Thailand casting ($20-25)(GR)
33. blue body, black shovel, black base, black motor, "Spasmo Mucosolvan" tempa, Thailand casting ($10-15)(GR)

MB30C SWAMP RAT, issued 1976 (continued)

5. olive deck, tan hull, tan & black camouflage tempa, tan drive, Macau casting ($4-5)(CM)
6. olive deck, tan hull, tan & black camouflage tempa, black driver, Macau casting ($12-15)(CM)

MB30-D LEYLAND ARTICULATED TRUCK, issued 1982 (continued)

13. bright blue cab, lemon cab base, yellow dump with cast panel, amber windows, "International" tempa, Macau casting ($3-5)

MB30-E PETERBILT QUARRY TRUCK, issued 1982 (USA)
MB23-E PETERBILT QUARRY TRUCK, issued 1982 (ROW) (continued)

5. yellow cab, dark gray dump, amber windows, chrome exhausts, "Dirty Dumper" tempa, Macau casting ($1-2)
6. yellow-orange cab, dark gray dump, amber windows, chrome exhausts, "Dirty Dumper" tempa, Macau casting ($1-2)
7. yellow-orange cab, dark gray dump, clear windows, chrome exhausts, "Dirty Dumper" tempa, Macau casting ($1-2)
8. light yellow body, dark gray dump, "Pace" tempa, Macau casting ($1-2)
9. dark yellow body, dark gray dump, clear windows, chrome exhausts, "Pace" tempa, Macau casting ($1-2)
10. dark yellow body, dark gray dump, clear windows, gray exhausts, "Pace" tempa, Macau casting ($1-2)
11. orange body, dark gray dump, clear windows, chrome exhausts, "Losinger" tempa, Macau casting ($8-12)(SW)
12. dark yellow body, dark gray dump, clear windows, gray exhausts, "Pace" tempa, Thailand casting ($1-2)
13. dark yellow body, red dump, clear windows, gray exhausts, "Pace" tempa, Thailand casting ($1-2)
14. white body, silver/gray dump, gray base, clear windows, gray exhausts, "Cement Company" tempa, Manaus casting ($35-50)(BR)

15. orange-yellow body, red dump, black base, clear windows, gray exhausts, "Pace" & "Intercom City" tempa, China casting ($10-15)(IC)

MB30-F MERCEDES G WAGON, issued 1985 (ROW)

NOTE: all models with silver/gray interior except lasertronic models which have no interior fitted. All models with 8 spoke wheels.
1. red body, white roof, black base, blue windows & dome lights, white "Rescue Unit" & checkers tempa, Macau casting ($2-3)
2. orange body, white roof, black base, blue windows & dome lights, "Lufthansa" tempa, Macau casting ($3-5)(GS)
3. white body, white roof, black base, blue windows & dome lights, "Polizei" & checkers tempa, Macau casting ($2-4)(TP)
4. white body, white roof, black base, blue windows & dome lights, "Polizei" with green doors/roof tempa, Macau casting ($2-3)
5. red body, white roof, black base, blue windows & dome lights, yellow "Rescue Unit" & checkers tempa, Macau casting ($2-3)(MC)
6. olive body, tan roof, black base, blue windows & dome lights, red cross in circle & "LS 2014" tempa, black hubs, Macau casting ($4-5)(CM)
7. white body, orange roof, orange base, blue windows & dome lights, "Ambulance" & checkers tempa, Macau casting ($3-5)(MC)
8. white body, white roof, black base, black windows & green dome lights, "Auto

Rescue 24 Hrs. Towing" tempa, Macau casting ($7-10)(SR)
9. red body, red roof, black base, black windows & red dome lights, "Fire Metro Airport" tempa, Macau casting ($7-10)(SR)
10. navy blue body, navy blue roof, black base, black windows & red dome lights, "Swat Unit Team Support" tempa, Macau casting ($7-10)(SR)
11. red body, white roof, black base, blue windows & dome lights, yellow "Rescue Unit" & checkers tempa, Thailand casting ($2-3)(MC)
12. white body, white roof, black base, blue windows & dome lights, "Polizei" with green roof/doors tempa, Thailand casting ($1-2)
13. white body, dark orange roof, orange base, blue windows & dome lights, "Ambulance" & checkers tempa, Thailand casting ($2-3)(MC)
14. white body, white roof, black base, blue windows & dome lights, "Lufthansa" tempa, Thailand casting ($1-2)(MP)
15. fluorescent orange body, white roof, black base, blue windows & dome lights, "Auto Rescue 24 Hrs. Towing", Thailand casting ($1-2)
16. white body, fluorescent orange roof & base, blue windows & dome lights, "Marine Rescue" & checkers tempa, Thailand casting ($2-3)(EM)

MB31-C CARAVAN, issued 1977 (continued)
17. white body, pearly silver base, chocolate door, lemon interior, amber windows, "Mobile 500" tempa, Macau casting ($2-4)(TP)

18. white body, black base, chocolate door, cream interior, amber windows, "Mobile 500" tempa, no origin cast ($2-4)(TP)
19. beige body, black base, chocolate door, cream interior, amber windows, "Mobile 500" tempa, no origin cast ($1-2)(TP)
20. gray body, black base, gray door, cream interior, amber windows, blue/red stripes tempa, no origin cast ($1-2)(TP)
21. white body, black base, red door, cream interior, clear windows, yellow/orange/red stripes tempa, no origin cast ($2-3)(MC)
22. white body, black base, white door, cream interior, blue windows, blue/maroon/orange design tempa, no origin cast ($1-2)(TP)

MB31-E MAZDA RX7, issued 1983
1. black body, red interior, 5 arch wheels, gold stripe tempa, Macau casting ($2-4)
2. white body, red interior, 5 arch wheels, "7" with stripes tempa, Macau casting ($2-4)
3. black body, red interior, 5 arch wheels, "RX7" & "Mazda" tempa, Manaus tempa ($35-50)(BR)
4. black body, red interior, dot dash wheels, "RX7" & "Mazda" tempa, Manaus casting ($35-50)(BR)

MB31-G STERLING, issued 1988 (USA)
MB 2-F STERLING , issued 1988 (ROW)
NOTE: all models with smoke gray windows, tan interior and 8 dot wheels unless noted otherwise.
1. metallic red body, charcoal base, no tempa, Macau casting ($1-2)
2. pearly silver body, metallic blue base, laser wheels, blue/white/red stripes

tempa, Macau casting ($4-6)(LW)
3. metallic red body, charcoal base, no tempa, China casting ($1-2)
4. blue body, yellow base, blue wheels with yellow hubs, exposed engine design & tools tempa, China casting ($6-8)(LL)
5. blue body, orange-yellow base, blue wheels with yellow hubs, exposed engine & tools design tempa, China casting ($6-8)(LL)
6. silver/gray body, metallic blue base, black door posts tempa, China casting ($1-2)
7. yellow body, yellow base, no tempa, China casting ($8-12)(GR)(GS)
8. silver/gray body, metallic blue base, black door posts & "Rover Sterling" tempa, China casting ($1-2)
9. silver/gray body, metallic blue base, "Rover Sterling" tempa, China casting ($1-2)
10. white body, white base, white interior, no tempa, China casting ($10-15)(GF)

MB31-I NISSAN PRAIRIE, issued 1991
MB21-G NISSAN PRAIRIE, issued 1991
1. metallic blue, black base, clear windows, blue interior, silver sides tempa, China casting ($5-7)
2. silver/gray body, black base, clear windows, blue interior, "Nissan" tempa, China casting ($2-3)
3. white body, white base, green windows, white interior, no tempa, China casting ($10-15)(GF)
4. red body, black base, clear windows, blue interior, "Nissan" tempa, China casting ($8-12)(GS)

MB32-D ATLAS EXCAVATOR, issued 1981

MB 6-F ATLAS EXCAVATOR, reissued 1990 (USA) (continued)

8. yellow body, black scoop, black platform, black case, stripes without "C" tempa, Macau casting ($1-2)
9. yellow-orange body, black scoop, black platform, black base, stripes without "C" tempa, Macau casting ($1-2)
10. light yellow body, black scoop, light yellow platform & base, "JCB" tempa, Macau casting ($7-10)(GS)
11. yellow body, black scoop, black platform, black base, stripes without "C" tempa, Thailand casting ($1-2)
12. yellow body, red scoop, black platform, black base, red stripes tempa, Thailand casting ($1-2)

MB32-E MODIFIED RACE, issued 1988

MB12-G MODIFIED RACER, issued 1988

NOTE: all models with chrome base & racing special slicks.

1. orange body, black interior, chrome exhausts, "12" tempa, Macau casting ($1-2)
2. rose red body, black interior, chrome exhausts, "12" tempa, Macau casting ($3-4)(SC)
3. orange body, black interior, chrome exhausts, "12" tempa, Macau casting ($3-4)(SC)
4. orange body, black interior, black exhaust, "12" tempa, Macau casting ($1-2)
5. red body, black interior, black exhausts, "12" tempa, Macau casting ($3-4)(SC)
6. orange body, black interior, black

exhausts, "12" tempa, Macau casting ($3-4)(SC)
7. orange body, black interior, black exhausts, "12" tempa, China casting ($1-2)
8. red body, black interior, chrome exhausts, "Mike 15" tempa, China casting ($4-6)(NM)
9. yellow body, green interior, chrome exhausts, "44 Reggie/Magnum Oils" tempa, China casting ($4-6)(NM)
10. white body, red interior, chrome exhausts, "U2 Jamie" tempa, China casting ($4-6)(NM)
11. white body, black interior, chrome exhausts, "1 Tony/Universal Joint Sales" tempa, China casting ($4-6)(NM)
12. dark purple body, black interior, black exhausts, "12" tempa, China casting ($3-4)(AP)
13. chrome plated body, black interior, black exhausts, no tempa, China casting ($12-18)(C2)
14. red body, red interior, black exhausts, "36" & stripes tempa, China casting ($4-6)(NM)
15. red body, orange-yellow interior, black exhausts, "12" & stripes tempa, China casting ($4-6)(NM)
16. white & blue body, blue interior, black exhausts, "ADAP 15" tempa, China casting ($4-6)(NM)
17. white body, translucent blue interior, black exhausts, "41" & stripes tempa, China casting ($4-6)(NM)
18. white body, lavender interior, black exhausts, no tempa, China casting ($6-8)(GF)
19. red body, red interior, chrome exhausts, "38 Jerry Cook" tempa, China casting

($4-6)(NM)
20. white body, orange interior, chrome exhausts, "Maynard Troyer" tempa, China casting ($4-6)(NM)
21. dark blue body, black interior, chrome exhausts, "3 Ron Bouchard" tempa, China casting ($4-6)(NM)
22. orange-yellow body, red interior, chrome exhausts, "4 Bugs" tempa, China casting ($4-6)(NM)
23. red body, red interior, chrome exhausts, "42 Jamie Tomaino" tempa, China casting ($4-6)(NM)
24. orange-yellow body, red interior, chrome exhausts, "4 Satch Wirley" tempa, China casting ($4-6)(NM)
25. dark blue body, blue interior, chrome exhausts, "3 Doug Heveron" tempa, China casting ($4-6)(NM)
26. black body, black interior, chrome exhausts, "21 George Kent" tempa, China casting ($4-6)(NM)
27. dark blue body, black interior, black exhausts, "12" tempa, China casting ($4-6)(GS)
28. blue body, black interior, chrome exhausts, "3 Mike McLaughlin" tempa, China casting ($6-8)(NM)

MB33-C POLICE MOTORCYCLE, issued 1977 (continued)

21. white body, "Police" label, black seat, shiny blue rider, silver motor, mag wheels, Macau casting ($2-4)
22. white body, "Honda" (tank) & "Police" (seat) tempa, white seat, bright blue rider, silver motor, mag wheels, Macau casting ($2-4)
23. black body, "Police" tempa, white seat, dark blue (painted) driver, silver motor,

mag wheels, Macau casting ($2-4)
24. white body, Japanese lettered tempa, white seat, dark blue driver, silver motor, mag wheels, Macau casting ($8-12)

MB33-D VOLKSWAGEN GOLF GTi, issued 1985 (USA)

MB56-E VOLKSWAGEN GOLF GTi, issued 1985 (ROW)

MB63-G VOLKSWAGEN GOLF GTi, reissued 1991 (ROW)

NOTE: all models with black interior, grille & bumpers, clear windows.

1. red body & base, 8 dot silver wheels, black & silver "Golf GTi" tempa, Macau casting ($2-4)
2. red body & base, 8 dot silver wheels, white "GTi" tempa, Macau casting ($2-4)
3. red body & base, 8 dot gold wheels, white "GTi" tempa, Macau casting ($3-5)
4. red body & base, dot dash wheels, white "GTi" tempa, Macau casting ($3-5)
5. white body & base, 8 dot silver wheels, "Federal Express" tempa, Macau casting ($3-5)(GS)
6. white body & base, 8 dot silver wheels, "Quantam" tempa, Macau casting ($6-8)(UK)
7. dark gray body & base, 8 dot silver wheels, silver sides tempa, Macau casting ($2-4)(TP)
8. yellow body & base, 8 dot silver wheels, "PT" tempa, Macau casting ($8-12)(SW)
9. dark gray body & base, 8 dot silver wheels, silver sides tempa, Thailand casting ($2-4)(TP)
10. white body & base, 8 dot silver wheels, "Abstract" tempa, Thailand casting ($3-5)

11. white body & base, 8 dot silver wheels, "Lippische Landes-Zeitung" tempa, Thailand casting ($15-20)(GR)

MB33-F RENAULT 11, issued 1987 (ROW)
MB43-F RENAULT 11, issued 1987 (USA)
NOTE: all models with gray base and clear windows.
1. metallic blue body, gray interior, 8 dot wheels gray side stripe tempa, "Taxi Parisian" roof sign, England casting ($10-15)(JB)
2. black body, gray interior, dot dash wheels, "Turbo" & silver stripes tempa, England casting ($2-4)
3. black body, gray interior, 8 dot wheels, no tempa, England casting ($4-6)
4. black body, gray interior, 8 dot wheels, "Turbo" & silver stripes tempa, England casting ($2-4)
5. black body, tan interior, dot dash wheels, "Turbo" & silver stripes tempa, England casting ($2-4)
6. black body, tan interior, dot dash wheels, "Turbo" & silver stripes tempa, China casting ($2-4)
7. black body, tan interior, dot dash wheels, "Turbo" & silver stripes tempa, China casting ($2-4)
8. black body, tan interior, 8 dot wheels, "Turbo" & silver stripes tempa, China casting ($2-4)

MB33-G UTILITY TRUCK, issued 1989 (USA)
MB74-I UTILITY TRUCK, issued 1989 (ROW)
NOTE: all models with blue windows & 8 spoke wheels.
1. gray body, black base, white boom, turret & bucket, "Energy Inc." tempa, Macau casting ($1-2)
2. red body, yellow base, blue turret, no boom or bucket, "53"/circle/bolt tempa, yellow heels with blue hubs, China casting ($6-8)(LL)
3. red body, yellow base, yellow turret, no boom or bucket, "53"/circle/bolt tempa, yellow wheels with blue hubs, China casting ($6-8)(LL)
4. red body, yellow base, blue & yellow boom, blue turret, lime bucket "53"/circle/bolt tempa, yellow wheels with blue hubs, China casting ($6-8)(LL)
5. red body, yellow base, blue & lime boom, yellow turret, blue bucket, "53"/circle/bolt tempa, yellow wheels with blue hubs, China casting ($6-8)(LL)
6. gray body, black base, white boom, turret & bucket, "Energy Inc." tempa, China casting ($1-2)
7. orange-yellow body, black base, white boom, turret & bucket, "Energy Inc." with red cab front tempa, China casting ($3-4)(AP)
8. dark yellow body, black base, white boom, turret & bucket, "Telephone Co. Unit 4" & checkers tempa ($1-2)

MB34-C CHEVY PROSTOCKER, issued 1981 (continued)
13. yellow-orange body, red interior, black base, "4" & stripes tempa, 5 arch front & 5 crown rear wheels, Macau casting ($2-4)
14. yellow-orange body, red interior, black base, "4" & stripes tempa, 5 spoke front & 5 crown rear wheels, Macau casting ($2-4)
15. yellow-orange body, red interior, black base, "4" & stripes tempa, 5 spoke front & rear wheels, Macau casting ($3-5)
16. white body, red interior, black base, "Pepsi 14" tempa, amber windows, 5 spoke front & 5 crown rear wheels, Macau casting ($2-4)
17. white body, red interior, black base, "Pepsi 14" tempa, amber windows, 5 arch front & 5 crown rear wheels, Macau casting ($2-4)
18. white body, black interior, black base, "Pepsi 14" tempa, 5 arch front & 5 crown rear wheels, Macau casting ($65-90)(TM)
19. white body, black interior, black base, "Pepsi 14" tempa, 5 spoke front & 5 crown rear wheels, Macau casting ($65-90)(TM)
20. white body, red interior, black base, "Superstar 217" tempa, 5 arch front & 5 crown rear wheels, Macau casting ($3-5)(TM)
21. white body, red interior, black base, "Halley's Comet" tempa, starburst wheels, Macau casting ($8-12)(MP)
22. white & orange body, red interior, black base, "21 355 CID" tempa, starburst wheels, Macau casting ($4-6)(SF)

23. white & orange body, red interior, black base, "21 355 CID" tempa, laser wheels, Macau casting ($4-6)(LW)
24. white body, red interior, black base, "7 Up" tempa, 5 arch front & 5 crown rear wheels, Macau casting ($3-5)(TM)
25. white body, black interior, black base, "7 Up" tempa, 5 arch front & 5 crown rear wheels, Macau casting ($75-100)(TM)
26. white body, red interior, black base, "Pepsi 14" tempa, maltese cross front & 5 crown rear wheels, Macau casting ($3-5)
27. white body, red interior, black base, "Pepsi 14" tempa, 5 arch front & 5 crown rear wheels, Thailand casting ($3-5)(GS)
28. blue & white body, black interior, silver/gray base, "70 Bailey Excavating" tempa, Goodyear slicks wheels, Thailand casting ($4-6)(WR)

MB34-D FORD RS200, issued 1987
1. white body, black base, clear windows, silver/gray interior, blue with "7" tempa, Macau casting ($1-2)
2. blue body, black base, clear windows, silver/gray interior, white with "2" tempa, Macau casting ($1-2)
3. mid blue body, black base, clear windows, silver/gray interior, white with "2" tempa, China casting ($1-2)
4. white body, black base, clear windows, silver/gray interior, red with "7" tempa, China casting ($1-2)
5. white body, white base, purple windows, white interior, no tempa, China casting ($10-15)(GF)

6. dark blue body, dark blue base, clear windows, silver/gray interior, no tempa, China casting ($8-12)(GR)(GS)
7. white body, black base, clear windows, silver/gray interior, red & blue with "7" tempa, China casting ($1-2)
8. white body, greenish black base, clear windows, silver/gray interior, red & blue with "7" tempa, China casting ($1-2)

MB34-E SPRINT RACER, issued 1990 (USA)
MB72-J SPRINT RACER, issued 1990 (ROW)
NOTE: all models with black base, chrome airfoils, Goodyear slicks front wheels & racing special rear wheels.
1. red body, white driver, "2 Rollin Thunder" tempa, China casting ($1-2)
2. red body, white driver, "Williams 5M" (blue letters) tempa, China casting ($4-6)(NM)
3. red body, white driver, "Williams 5M" (white letters) tempa, China casting ($75-100)(NM)
4. black body, white driver, "TMC 1" tempa, China casting ($4-6)(NM)
5. white body, red driver, "Maxim 11" tempa, China casting ($4-6)(NM)
6. white body, red driver, "Schnee 8D" tempa, China casting ($4-6)(NM)
7. yellow body, white driver, "Ben Cook & Sons 33x" tempa, China casting ($4-6)(NM)
8. blue body, white driver, "Ben Allen la" tempa, China casting ($4-6)(NM)
9. metallic blue body, white driver, "Lucky 7" tempa, China casting ($3-5)(AP)
10. red body, white driver, "7 Joe Gaerte" tempa, China casting ($4-6)(NM)
11. red body, white driver, "4 Gambler" tempa, China casting ($4-6)(NM)
12. yellow body, white driver, "17 F&G Classics Eash" tempa, China casting ($4-6)(NM)
13. yellow body, white driver, "7c Vivarin-D. Blaney" tempa, China casting ($4-6)(NM)
14. powder blue body, white driver, "69 Schnee-D. Krietz" tempa, China casting ($4-6)(NM)
15. black body, white driver, "49 Doug Wolfgang" tempa, China casting ($4-6)(NM)
16. metallic blue body, white driver, "2 Rollin Thunder" tempa, China casting ($4-6)(GS)

MB35-C PONTIAC T-ROOF, issued 1982 (USA)
MB16-C PONTIAC T-ROOF, issued 1982 (ROW) (continued)
3. black body, pearly silver base, red interior, white eagle & "Trans Am" tempa, Macau casting ($2-4)
4. black body, pearly silver base, red interior, yellow eagle & "Trans Am" tempa, Macau casting ($2-4)
5. black body, pearly silver base, red interior, orange tiger stripes tempa, Macau casting ($12-15)(UK)(OP)
6. pearly silver body, black base, red interior, red/orange/yellow eagle & red/orange stripes tempa, Macau casting ($2-3)
7. red body, black base, red interior, "3 Rooster Racing Team" tempa, Macau casting ($8-12)(HK)

MB35-F LAND ROVER 90, issued 1987 (USA)
MB16-F LAND ROVER 90, issued 1987 (ROW)
NOTE: all models with clear windows, black interior/grille and dot dash wheels unless otherwise noted.
1. blue body, pearly gray base, white roof, yellow & orange stripes tempa, Macau casting ($2-3)
2. green body, pearly gray base, white roof, yellow & orange stripes tempa, Macau casting ($1-2)(MC)
3. dark blue body, pearly gray base, white roof, "RN Royal Navy" tempa, Macau casting ($2-3)(GS)
4. red body, pearly gray base, white roof, blue & gray stripes with "Country" tempa, Macau casting ($1-2)
5. black body, dark gray base, dark gray roof & interior, black hubs, gray & yellow camouflage tempa, Macau casting ($4-5)(CM)
6. red body, pearly gray base, white roof, black & red stripes with "Country" tempa, Macau casting ($1-2)(TP)
7. white body, pearly gray base, white roof, black & red stripes with "Country" tempa, Thailand casting ($1-2)(TP)
8. green body, pearly gray base, white roof, yellow & orange stripes tempa, Thailand casting ($1-2)(MC)(TP)
9. red body, pearly gray base, white roof, blue & gray stripes with "Country" tempa, Thailand casting ($1-2)
10. dark blue body, pearly gray base, white roof, "RN Royal Navy" tempa, Thailand casting ($2-3)(GS)
11. light gray/dark navy body, dark navy base, light gray roof, red stripe tempa, Thailand casting ($3-4)(MC)
12. yellow body, pearly silver base, white roof, large "Park Ranger" tempa, Thailand casting ($1-2)
13. white body, white base, blue roof, "KLM" tempa, Thailand casting ($3-5)(MP)
14. white body, white base, blue roof, "SAS" tempa, Thailand casting ($3-5)(MP)
15. white body, white base, blue roof, "Alitalia" tempa, Thailand casting ($3-5)(MP)
16. white body, pearly silver base, white roof, "Bacardi Rum" tempa, Thailand casting ($25-40)(TP)(OP)
17. yellow body, yellow base, white roof, small "Park Ranger" tempa, Thailand casting ($1-2)
18. red body, black base, white roof, "Red Arrows/Royal Air Force" tempa, Thailand casting ($3-5)(MC)
19. white body, white base, white roof, pink windows, neon red interior, no tempa, Thailand casting ($10-15)(GF)
20. white body, pearly silver base, white roof, "Rescue Police" & checkers tempa, Thailand casting ($1-2)(EM)
21. white body, red base, white roof, "Circus Circus" tempa, Thailand casting ($2-4)(MC)
22. white body, pearly silver base, green roof, "Garden Festival Wales" tempa, Thailand casting ($6-8)(WL)(GS)
23. yellow body, pearly silver base, white roof, small "Park Ranger" tempa, Thailand casting ($1-2)

MB36-D REFUSE TRUCK, issued 1980 (continued)

11. blue body, orange dump & hatch, pearly silver base, "Metro" label, red trigger, Macau casting ($2-4)
12. white body, blue dump & hatch, light metallic gray base, "Metro" label, blue trigger, Macau casting ($2-3)
13. white body, blue dump & hatch, pearly silver base, Japanese lettered tempa, blue trigger, Macau casting ($8-12)(JP)
14. white body, blue dump & hatch, light metallic gray base, no label, blue trigger, Macau casting ($2-3)
15. white body, blue dump & hatch, pearly silver base, "Metro" label, red trigger, China casting ($2-3)
16. green body, yellow dump & hatch, pearly gray base, "State City" tempa, red trigger, China casting ($1-2)
17. green body, lemon dump & hatch, pearly gray base, "State City" tempa, red trigger, China casting ($1-2)
18. orange body, gray dump & hatch, black base, "Refuse Disposal" tempa, red trigger, China casting ($1-2)
19. green body, light yellow dump & hatch, black base, "Refuse Disposal" tempa, red trigger, China casting ($1-2)
20. red body, light yellow dump & hatch, black base, "Refuse Disposal" tempa, red trigger, China casting ($8-12)(GS)

MB37-D SUNBURNER, issued 1982 (USA) (continued)

3. black body, black base, red interior, yellow & red flames tempa, Macau casting ($3-5)

4. black body, black base, red interior, yellow & red flames tempa, Hong Kong casting ($3-5)

MB37-E MATRA RANCHO, issued 1982 (ROW) (continued)

6. dark blue body, black base, no tempa, blue tailgate, 5 arch silver wheels, England casting ($7-10)
7. dark blue body, black base, no tempa, blue tailgate, 5 arch gold wheels, England casting ($15-18)
8. dark blue body, yellow base, no tempa, yellow tailgate, 5 arch gold wheels, England casting ($18-25)(TP)
9. dark blue body, yellow base, no tempa, yellow tailgate, 5 arch silver wheels, England casting ($15-18)(TP)
10. dark blue body & base, no tempa, blue tailgate, 5 arch gold wheels, England casting ($15-18)(TP)
11. yellow body & base, no tempa, yellow tailgate, 5 arch silver wheels, England casting ($6-8)
12. yellow body & base, red stripe tempa, yellow tailgate, 5 arch gold wheels, England casting ($15-18)
13. yellow body & base, red stripe tempa, yellow tailgate, 5 arch gold wheels, England casting ($3-5)
14. navy blue body, white base, "Surf Rescue" tempa, blue tailgate, 5 arch silver wheels, England casting ($2-4)(TP)
15. navy blue body, white base, "Surf Rescue" tempa, black tailgate, 5 arch silver wheels, England casting ($3-5)(TP)
16. navy blue body, yellow base, "Surf Rescue" tempa, black tailgate, 5 arch silver wheels, England casting ($12-15)(TP)

17. black body, white base, "Surf Rescue" tempa, black tailgate, 5 arch silver wheels, Macau casting ($2-3)(TP)
18. orange body, black base, "Surf 2" tempa, orange tailgate, 5 arch silver wheels, Macau casting ($2-3)(TP)
19. fluorescent yellow body, yellow base, "Marine Rescue" tempa, fluorescent yellow tailgate, 5 arch silver wheels, China casting ($1-2)(EM)
20. white body, white base, white interior, white tailgate, amber windows, no tempa, 5 arch silver wheels, China casting ($10-15)(GF)

MB38-E MODEL A FORD VAN, issued 1982 (continued)

NOTE: all models with clear windows, 5 crown front wheels and dot dash rear wheels unless noted otherwise.

3. blue body, black chassis, white roof, "Champion" labels, Macau casting ($4-6)
4. blue body, black chassis, white roof, "Kelloggs" labels, Macau casting ($7-10)(UK)(OP)
5. blue body, black chassis, white roof, "Matchbox On The Move in '84" labels, Macau casting ($20-40)(US)
6. blue body, black chassis, white roof, "Matchbox On The Move in '84" labels with "Toy Fair 84" roof label, Macau casting ($20-40)(US)
7. white body, blue chassis, red roof, "Pepsi" with "Come Alive" tempa, Macau casting ($7-10)
8. white body, blue chassis, red roof, "Pepsi" without "Come Alive" tempa, Macau casting ($4-6)
9. white body, blue chassis, red roof, "Matchbox U.S.A." labels, Macau casting ($20-35)(US)
10. white body, blue chassis, red roof, "Ben Franklin" labels, Macau casting ($500-700)(US)
11. red body, black chassis, black roof, "Arnott's Biscuits" tempa, Macau casting ($7-10)(AU)
12. gray body, red chassis, red roof, "Tittensor First School" tempa, Macau casting ($5-8)(UK)
13. beige body, brown chassis, brown roof, "Larkland Motor Museum" tempa, Macau casting ($5-8)(UK)
14. dark blue body, black chassis, red roof, "Bass Museum" tempa, Macau casting ($5-8)(UK)
15. yellow body, green chassis, green roof, "Toy Collectors Pocket Guide" tempa, Macau casting ($6-8)(UK)
16. white body, black chassis, black roof, "The Australian" tempa, Macau casting ($6-8)(AU)
17. olive green body, black chassis, black roof, "BBC 1925" tempa, Macau casting ($6-8)(UK)
18. dark blue body, black chassis, black roof, "Matchbox Speedshop" with yellow pinstriping tempa, 5 arch front wheels, racing special rear wheels (chromed letters), Macau casting ($75-100)
19. dark blue body, black chassis, black roof, "Matchbox Speedshop" with yellow pinstriping tempa, 5 arch front wheels, racing special rear wheels (unchromed letters), Macau casting ($1-2)
20. dark blue body, black chassis, black roof, "Matchbox Speedshop" with orange pinstriping tempa, 5 arch front wheels, racing special rear wheels (unchromed letters), Macau casting ($1-2)

21. cream body, black chassis, green roof, "H.H. Brain" tempa, Macau casting ($18-25)(UK)(OP)
22. powder blue body, dark blue chassis, red roof, "Isle of Man TT86" tempa, Macau casting ($5-7)(UK)
23. dark green body, black chassis, dark green roof, "Weetabix/Sanitarium Food" tempa, Macau casting ($7-10)(AU)(OP)
24. dark blue body, red chassis, white roof, "Smith's Potato Crisps" tempa, Macau casting ($7-10)(AU)(OP)
25. black body, red chassis, red roof, "Isle of Man TT87" tempa, Macau casting ($5-7)(UK)
26. dark yellow body, blue chassis, blue roof, "Junior Matchbox Club" tempa, Macau casting ($6-8)(CN)
27. black body, black chassis, black roof, "2nd M.I.C.A. Convention" decals, Macau casting ($375-450)(UK)
28. white body, black chassis, black roof, "Chesty Bonds" tempa, Macau casting ($5-7)(AU)
29. red body, black chassis, black roof, "I.O.M. Post Office" tempa, Macau casting ($6-8)(UK)
30. blue body, black chassis, black roof, "Silvo" tempa, Macau casting ($7-10)(UK)
31. red body, black chassis, black roof, "W.H. Smith & Sons" tempa, Macau casting ($18-25)(UK)(OP)
32. red body, black chassis, black roof, "Dewhurst Master Butcher" tempa, Macau casting ($8-12)(UK)
33. dark olive body, red chassis, red roof, "John West Salmon" tempa, Macau casting ($8-12)(UK)
34. light blue body, black chassis, white

roof, "Kelloggs Rice Krispies" labels, Macau casting ($8-12)(UK)(OP)
35. dark blue body, black chassis, red roof, "Matchbox-This Van Delivers" with phone # on doors tempa, Macau casting ($60-75)(HK)
36. dark blue body, black chassis, red roof, "Matchbox-This Van Delivers" without phone # on doors ($12-18)(UK)
37. red body, black chassis, black roof, gold grille, "Royal Mail" tempa, Macau casting ($6-8)(MP)
38. orange body, black chassis, black roof, "North America M.I.C.A. Convention 1988", Macau casting ($6-8)(US)
39. yellow body, black chassis, red roof, "Isle of Man TT88" tempa, Macau casting ($5-7)(UK)
40. yellow body, yellow chassis, red roof, "3rd M.I.C.A. Convention" tempa, Macau casting ($8-12)(UK)
41. yellow body, dark blue chassis, white roof, "James Neale & Sons" tempa, Macau casting ($8-12)(UK)
42. red body, black chassis, black roof, "I.O.M. Post Office" with phone # on rear doors ($6-8)(UK)
43. red body, black chassis, black roof, "Manx Cattery" labels, Macau casting ($12-15)(UK)
44. red body, black chassis, black roof, "Mervyn Wynne" labels, Macau casting ($12-15)(UK)
45. dark orange body, black chassis, black roof, "P.M.G." tempa, Macau casting ($7-10)(AU)
46. red body, black chassis, black roof, "Alex Munro Master Butcher" tempa, Macau casting ($8-12)(UK)

47. red body, black chassis, black roof, "Rayner's Crusha" tempa, Macau casting ($8-12)(UK)(OP)
48. red body, greenish black chassis, black roof, "Rayner's Crusha" tempa, Macau casting ($8-12)(UK)(OP)
49. powder blue body, dark gray chassis, dark gray roof, "Chester Heraldry Centre" tempa, Macau casting ($6-8)(UK)
50. yellow body, black chassis, red roof, "Barratt Sherbert" tempa, Macau casting ($10-15)(UK)(OP)
51. blue body, black chassis, blue roof, "Guernsey Post" tempa, Macau casting ($5-7)(UK)
52. dark green body, black chassis, black roof, "Historical Collection/Powerhouse" tempa ($6-8)(AU)
53. dark green body, greenish black chassis, black roof, "Historical Collection/ Powerhouse" tempa ($6-8)(AU)
54. brown body, black chassis, black roof, "Cobb of Knightbridge" tempa, Macau casting ($7-10)(UK)
55. red body, black chassis, black roof, "Big Sister" tempa, Macau casting ($6-8)(AU)(OP)
56. powder blue body, dark gray chassis, dark gray roof, "Chester Toy Museum" labels ($6-8)(UK)
57. green body, black chassis, yellow roof, "Rowntree's Table Jelly" tempa, Macau casting ($8-12)(UK)(OP)
58. blue body, silver/gray chassis, gray roof, "Nat West Action Bank" tempa, Macau casting ($6-8)(UK)
59. black body, light maroon chassis, red roof, "Uniroyal Royal Care" tempa, Macau casting ($18-25)(CN)

60. yellow body, black chassis, black roof, "Matchbox Series Model A Ford Van" tempa, Macau casting ($2-3)
61. yellow body, black chassis, black roof, 5 arch front wheels, "Matchbox Series Model A Ford Van" tempa, Macau casting ($2-3)
62. light yellow body, black chassis, black roof, "W.H. Smith & Sons" tempa, Macau casting ($12-15)(UK)
63. dark green body, dark green chassis & roof, "Green's Sponge Mixture" tempa, Macau casting ($8-12)(UK)(OP)
64. olive green body, red/brown chassis & roof, black grille & hubs, "Aldershot-4th M.I.C.A. Convention" tempa, Macau casting ($8-10)(UK)
65. black body, orange chassis, orange roof, "2nd M.I.C.A. NA Convention 1989" tempa, Macau casting ($6-8)(US)
66. yellow body, blue chassis, blue roof, "Welcome Ye To Chester" labels, Macau casting ($15-20)(C2)
67. dark blue body, black chassis, black roof, "Cheeses of England & Wales" tempa, Macau casting ($10-15)(UK)(OP)
68. cream body, red chassis, red roof, "Isle of Man TT89" tempa, Macau casting ($5-7)(UK)
69. cream body, black chassis, black roof, "Jordan's" tempa, Macau casting ($10-15)(UK)(OP)
70. dark blue body, black chassis, black roof, "Ribena" tempa, Macau casting ($8-12)(UK)
71. yellow body, red chassis, red roof, "Barratt's Sherbet" tempa, Macau casting ($8-12)(UK)
72. yellow body, black chassis, black roof, "Barratt's Sherbert" tempa, Macau

casting ($8-12)(UK)

73. light blue body, blue chassis, blue roof, "Junior Matchbox Club-The Gang '89" tempa, Macau casting ($6-8)(CN)

74. dark blue body, cream chassis, cream roof, "Chasewater Light Railway" tempa, Macau casting ($6-8)(UK)

75. red body, blue chassis, blue roof, "Tandy Electronics" tempa, Macau casting ($12-15)(AU)

76. light gray body, dark gray chassis, dark gray roof, "Moorland Centre" tempa, Macau casting ($6-8)(UK)

77. black body, orange chassis, orange roof, "Baltimore Orioles 1989" tempa, Macau casting ($6-8)(WR)

78. yellow body, dark green chassis, dark green roof, "York Fair 1989" tempa, Macau casting ($6-8)(WR)

79. silver body, silver chassis, black roof, "Matchbox Collectors Club 1989" labels, Macau casting ($125-150)(C2)

80. white body, red chassis, blue roof, "Ten Years Lion" tempa, Macau casting ($7-10)(UK)

81. white body, black chassis, black roof, "Jacky Maeder" tempa, Macau casting ($8-12)(SW)

82. cream body, brown chassis, brown roof, "Johnson's Seeds" tempa, Macau casting ($8-12)(UK)(OP)

83. red body, red chassis, red roof, "Asda Baked Beans" tempa, Macau casting ($8-12)(UK)(OP)

84. dark blue body, dark blue chassis, dark blue roof, "Matchbox 40th Anniversary 1990" tempa, Macau casting ($7-10)(US)

85. white body, blue chassis, white roof, "New York Yankees 1990" tempa, Macau casting ($3-5)(WR)

86. white body, blue chassis, white roof, "Los Angeles Dodgers 1990" tempa, Macau casting ($3-5)(WR)

87. white body, blue chassis, white roof, "Texas Rangers 1990" tempa, Macau casting ($3-5)(WR)

88. white body, blue chassis, white roof, "Chicago White Sox 1990" tempa, Macau casting ($3-5)(WR)

89. white body, blue chassis, white roof, "Houston Astros 1990" tempa, Macau casting ($3-5)(WR)

90. white body, blue chassis, white roof, "Atlanta Braves 1990" tempa, Macau casting ($3-5)(WR)

91. white body, blue chassis, white roof, "Cleveland Indians 1990" tempa, Macau casting ($3-5)(WR)

92. white body, blue chassis, white roof, "New York Mets 1990" tempa, Macau casting ($3-5)(WR)

93. white body, blue chassis, white roof, "Minnesota Twins 1900" tempa, Macau casting ($3-5)(WR)

94. white body, blue chassis, white roof, "Toronto Blue Jays 1990" tempa, Macau casting ($3-5)(WR)

95. white body, mid blue chassis, white roof, "Milwaukee Brewers 1990" tempa, Macau casting ($3-5)(WR)

96. white body, mid blue chassis, red roof, "Chicago Cubs 1990" tempa, Macau casting ($3-5)(WR)

97. white body, blue chassis, orange roof, "Detroit Tigers 1990" tempa, Macau casting ($3-5)(WR)

98. white body, blue chassis, yellow roof, "Mariners 1990" tempa, Macau casting ($3-5)(WR)

99. white body, red chassis, white roof, "Boston Red Sox 1990" tempa, Macau casting ($3-5)(WR)

100. white body, red chassis, white roof, "Cincinnati Reds 1990" tempa, Macau casting ($3-5)(WR)

101. white body, red chassis, white roof, "St. Louis Cardinals 1990" tempa, Macau casting ($3-5)(WR)

102. white body, red chassis, white roof, "Montreal Expos 1990" tempa, Macau casting ($3-5)(WR)

103. white body, red chassis, yellow roof, "California Angels 1990" tempa, Macau casting ($3-5)(WR)

104. white body, maroon chassis, white roof, "Philadelphia Phillies 1990" tempa, Macau casting ($3-5)(WR)

105. white body, black chassis, white roof, "San Francisco Giants 1990" tempa, Macau casting ($3-5)(WR)

106. white body, black chassis, black roof, "Pittsburgh Pirates 1990" tempa, Macau casting ($3-5)(WR)

107. white body, brown chassis, orange roof, "San Diego Padres 1990" tempa, Macau casting ($3-5)(WR)

108. powder blue body, blue chassis, white roof, "Kansas City Royals 1990" tempa, Macau casting ($3-5)(WR)

109. black body, orange chassis, black roof, "Baltimore Orioles 1990" tempa, Macau casting ($3-5)(WR)

110. yellow body, green chassis, white roof, "Oakland A's 1990" tempa, Macau casting ($3-5)(WR)

111. yellow body, red chassis, red roof, "Pava" tempa, Macau casting ($8-12)(DU)

112. dark blue body, black chassis, black roof, "Swarfega" tempa, Macau casting ($18-25)(UK)(OP)

113. dark blue body, black chassis, black roof, "Kellogg's Rice Krispies" tempa, Macau casting ($4-6)(US)(OP)

114. dark green body, dark green chassis, dark green roof, "Carmelle" tempa, Macau casting ($15-20)(SU)

115. dark blue body, black chassis, black roof, "Lyceum Theater" tempa, Macau casting ($6-8)(UK)

116. light blue body, black chassis, white roof, "Fresh Dairy Cream" tempa, Macau casting ($8-12)(UK)(OP)

117. yellow body, red chassis, red roof, "Drink Coca Cola" with brown bottles tempa, Macau casting ($7-10)(CN)

118. yellow body, red chassis, red roof, "Drink Coca Cola" with red bottles tempa, Macau casting ($8-12)(SW)

119. yellow body, dark green chassis, dark green roof, "Drink Coca Cola" tempa, Macau casting ($7-10)(CN)

120. white body, maroon chassis, white roof, "Matchbox U.S.A. 9th Annual Convention 1990" tempa, Macau casting ($8-10)(US)

121. blue body, red chassis, yellow roof, "Isle of Man TT90" tempa, Macau casting ($5-7)(UK)

122. fluorescent green body, fluorescent orange chassis, black roof, "Matchbox Collectors Club 1990" labels, Macau casting ($85-110)(C2)

123. dark blue body, white chassis, white roof, "Penn State 1990" tempa, Macau casting ($6-8)(WR)

124. white body, dark blue chassis, dark blue roof, "Penn State 1990" tempa, Macau casting ($6-8)(WR)

125. cream body, red chassis, red roof,

"York Fair 1990-225 Years" tempa, Macau casting ($5-7)(WR)

126. white body, navy blue chassis, orange roof, "Bears 1990" tempa, Macau casting ($3-5)(WR)

127. white body, blue chassis, blue roof, "Colts 1990" tempa, Macau casting ($3-5)(WR)

128. white body, green chassis, green roof, "Jets 1990" tempa, Macau casting ($3-5)(WR)

129. white body, red-brown chassis, maroon roof, "Cardinals 1990" tempa, Macau casting ($3-5)(WR)

130. white body, red chassis, lemon roof, "KC Chiefs 1990" tempa, Macau casting ($3-5)(WR)

131. navy blue body, yellow chassis, white roof, "Chargers 1990" tempa, Macau casting ($3-5)(WR)

132. blue body, white chassis, white roof, "Giants 1990" tempa, Macau casting ($3-5)(WR)

133. blue body, red chassis, white roof, "Bills 1990" tempa, Macau casting ($3-5)(WR)

134. bright blue body, red chassis, bright blue roof, "Oilers 1990" tempa, Macau casting ($3-5)(WR)

135. silver/gray body, blue chassis, white roof, "Seahawks 1990" tempa, Macau casting ($3-5)(WR)

136. silver/gray body, light blue chassis, white roof, "Lions 1990" tempa, Macau casting ($3-5)(WR)

137. silver/gray body, navy blue chassis, navy blue roof, "Cowboys 1990" tempa, Macau casting ($3-5)(WR)

138. silver/gray body, black chassis, black roof, "Raiders 1990" tempa, Macau

casting ($3-5)(WR)

139. silver/gray body, green chassis, white roof, "Eagles 1990" tempa, Macau casting ($3-5)(WR)

140. green-gold body, black chassis, white roof, "Saints 1990" tempa, Macau casting ($3-5)(WR)

141. gold body, red chassis, white roof, "49ers 1990" tempa, Macau casting ($3-5)(WR)

142. red body, black chassis, light gray roof, "Falcons 1990" tempa, Macau casting ($3-5)(WR)

143. red body, white chassis, blue roof, "Patriots 1990" tempa, Macau casting ($3-5)(WR)

144. red/brown body, lemon chassis, white roof, "Redskins 1990" tempa, Macau casting ($3-5)(WR)

145. dark orange body, brown chassis, white roof, "Browns 1990" tempa, Macau casting ($3-5)(WR)

146. dark orange body, blue chassis, orange roof, "Broncos 1990" tempa, Macau casting ($3-5)(WR)

147. light orange body, black chassis, black roof, "Buccaneers 1990" tempa, Macau casting ($3-5)(WR)

148. blue-green body, orange chassis, white roof, "Dolphins 1990" tempa, Macau casting ($3-5)(WR)

149. olive body, lemon chassis, white roof, "Packers 1990" tempa, Macau casting ($3-5)(WR)

150. yellow body, blue chassis, yellow roof, "Rams 1990" tempa, Macau casting ($3-5)(WR)

151. yellow body, black chassis, yellow roof, "Steelers 1990" tempa, Macau casting ($3-5)(WR)

152. black body, dark orange chassis, white roof, "Bengals 1990" tempa, Macau casting ($3-5)(WR)

153. purple body, purple chassis, yellow roof, "Vikings 1990" tempa, Macau casting ($3-5)(WR)

154. metallic green body & chassis, green roof, "Canada Dry" tempa, Macau casting ($8-12)(CN)

155. dark blue body, black chassis, black roof, "Kelloggs Rice Krispies" tempa, Thailand casting ($4-6)(US)(OP)

156. cream body, green chassis, green roof, "William Lusty" tempa, Macau casting ($6-8)(UK)

157. dark plum body, black chassis, black roof, "Johnnie Walker" tempa, Macau casting ($18-25)(UK)(JP)

158. blue body, black chassis, blue roof, "Mitre 10" tempa, Macau casting ($7-10)(AU)

159. orange-red body, black chassis, black roof, "Tyne Brand" tempa, Macau casting ($12-15)(UK)(OP)

160. dark blue body, black chassis, black roof, "Tyne Brand" tempa, Macau casting ($12-15)(UK)(OP)

161. cream body, green chassis, red roof, "PG Tips" tempa, Macau casting ($12-15)(UK)(OP)

162. powder blue body, dark gray chassis, dark gray roof, "Open Every Day" tempa on doors/plain main panel, Macau casting ($12-15)(MP)

163. dark blue body, black chassis, black roof, "Kelloggs Rice Krispies" tempa, China casting ($4-6)(US)(OP)

164. white body, black chassis, black roof, "Rutter Bros. Dairy" tempa, Macau casting ($6-8)(WR)

165. dark blue body, black chassis, white roof, "Lyon's Tea" tempa, Macau casting ($6-8)(UK)

166. yellow body, black chassis, black roof, black grille, "Matchbox Dinky Toy Convention 1991", China casting ($8-12)(UK)

167. white body, red chassis, dark blue roof, "Indians 1991" tempa, China casting ($3-5)(WR)

168. white body, red chassis, dark blue roof, "Yankees 1991" tempa, China casting ($3-5)(WR)

169. white body, red chassis, white roof, "Cardinals 1991" tempa, China casting ($3-5)(WR)

170. white body, yellow chassis, white roof, "Brewers 1991" tempa, China casting ($3-5)(WR)

171. white body, black chassis, dark gray roof, "White Sox 1991" tempa, China casting ($3-5)(WR)

172. lemon body, blue body, mid blue roof, "Mariners 1991" tempa, China casting ($3-5)(WR)

173. orange-yellow body, red chassis, dark blue roof, "Angels 1991" tempa, China casting ($3-5)(WR)

174. orange body, black chassis, white roof, "Orioles 1991" tempa, China casting ($3-5)(WR)

175. orange body, blue chassis, white roof, "Mets 1991" tempa, China casting ($3-5)(WR)

176. orange body, dark blue chassis, orange-yellow roof, "Astros 1991" tempa, China casting ($3-5)(WR)

177. red body, blue chassis, white roof, "Rangers 1991" tempa, China casting ($3-5)(WR)

178. red body, white chassis, white roof, "Cubs 1991" tempa, China casting ($3-5)(WR)

179. red body, red chassis, white roof, "Reds 1991" tempa, China casting ($3-5)(WR)

180. dark maroon body, dark maroon chassis, white roof, "Phillies 1991" tempa ($3-5)(WR)

181. dark blue body, red chassis, white roof, "Red Sox 1991" tempa, China casting ($3-5)(WR)

182. dark blue body, red chassis, white roof, "Twins 1991" tempa, China casting ($3-5)(WR)

183. blue body, red chassis, white roof, "Braves 1991" tempa, China casting ($3-5)(WR)

184. blue body, baby blue chassis, white roof, "Royals 1991" tempa, China casting ($3-5)(WR)

185. med. blue body, blue chassis, white roof, "Blue Jays 1991" tempa, China casting ($3-5)(WR)

186. baby blue body, red chassis, white roof, "Expos 1991" tempa, China casting ($3-5)(WR)

187. pearly silver body, blue chassis, orange roof, "Dodgers 1991" tempa, China casting ($3-5)(WR)

188. pearly silver body, dark blue chassis, orange roof, "Padres 1991' tempa, China casting ($3-5)(WR)

189. silver/gray body, dark blue chassis, orange roof, "Tigers 1991" tempa, China casting ($3-5)(WR)

190. lavender gray body, black chassis, orange roof, "Giants 1991" tempa, China casting ($3-5)(WR)

191. black body, yellow chassis, white roof, "Pirates 1991" tempa, China casting ($3-5)(WR)

192. green body, yellow chassis, white roof, "Athletics 1991" tempa, China casting ($3-5)(WR)

193. powder blue body, blue chassis, red roof, "Isle of Man TT91" tempa, China casting ($5-7)(UK)

194. orange-yellow body, black chassis, black roof, "Celebrating A Decade of Matchbox Conventions 1991", China casting ($8-10)(US)

195. white body, dark pink chassis, black roof, "Matchbox Collectors Club 1991" labels, Macau casting ($60-75)(C2)

196. chrome plated body, red chassis, red roof, "15th Anniversary Matchbox U.S.A. 1991" tempa, Macau casting ($35-50)(C2)

197. light yellow body, light blue chassis, light blue roof, "Dairylea" tempa, China casting ($15-20)(UK)(OP)

198. white body, white chassis & sub-base, white roof, white grille, no tempa, China casting ($10-15)(GF)

199. yellow body, green chassis, white roof, "Notre Dame 1991" tempa, China casting ($5-7)(WR)

200. silver/gray body, red chassis, white roof, "UNLV 1991" tempa, China casting ($5-7)(WR)

201. white body, blue chassis, blue roof, "Rugby Child Development Centre" tempa, China casting ($7-10)(UK)

202. dark plum body, black chassis, black roof, "Johnnie Walker" tempa, Thailand casting ($18-25)(UK)

203. red body, black chassis, black roof, "Mervyn Wynne" with black island on door labels, Macau casting ($18-25)(UK)

204. silver/gray body, turquoise chassis & roof, "York Fair 1991" tempa, China casting ($6-8)(WR)

NOTE: Football set models for 1991 have roof blades cast with labels applied.

205. lemon body, maroon chassis, maroon roof, "Redskins 1991" tempa, China casting ($3-5)(WR)

206. lemon body, red chassis, white roof, "Chiefs 1991" tempa, China casting ($3-5)(WR)

207. yellow body, navy blue chassis, navy blue roof, "Chargers 1991" tempa, China casting ($3-5)(WR)

208. yellow body, black chassis, black roof, "Steelers 1991" tempa, China casting ($3-5)(WR)

209. yellow body, dark blue chassis, dark blue roof, "Rams 1991" tempa, China casting ($3-5)(WR)

210. lemon body, purple chassis, purple roof, "Vikings 1991" tempa, China casting ($3-5)(WR)

211. orange body, blue-green chassis & roof, "Dolphins 1991" tempa, China casting ($3-5)(WR)

212. dark orange body, dark blue chassis, dark blue roof, "Broncos 1991" tempa, China casting ($3-5)(WR)

213. dark orange body, black chassis, black roof, "Bengals 1991" tempa, China casting ($3-5)(WR)

214. light orange body, red chassis, white roof, "Buccaneers 1991" tempa, China casting ($3-5)(WR)

215. white body, dark orange chassis, navy blue roof, "Bears 1991" tempa, China casting ($3-5)(WR)

216. white body, dull green chassis & roof, "Jets 1991" tempa, China casting ($3-5)(WR)

217. white body, dark blue chassis, red roof, "Bills 1991" tempa, China casting ($3-5)(WR)

218. white body, dark blue chassis & roof, "Giants 1991" tempa, China casting ($3-5)(WR)

219. white body, green-gold chassis, black roof, "Saints 1991" tempa, China casting ($3-5)(WR)

220. blue body, silver/gray chassis, white roof, "Lions 1991" tempa, China casting ($3-5)(WR)

221. bright blue body, red chassis, white roof, "Oilers 1991" tempa, China casting ($3-5)(WR)

222. red body, black chassis, black roof, "Falcons 1991" tempa, China casting ($3-5)(WR)

223. red body, gold chassis, gold roof, "49ers 1991" tempa, China casting ($3-5)(WR)

224. maroon body, white chassis, white roof, "Cardinals 1991" tempa, China casting ($3-5)(WR)

225. brown body, brown chassis, orange roof, "Browns 1991" tempa, China casting ($3-5)(WR)

226. black body, silver/gray chassis, white roof, "Raiders 1991" tempa, China casting ($3-5)(WR)

227. dark blue body, green chassis, white roof, "Seahawks 1991" tempa, China casting ($3-5)(WR)

228. dark blue body, white chassis, white roof, "Colts 1991" tempa, China casting ($3-5)(WR)

229. dark blue body, red chassis, red roof, "Patriots 1991" tempa, China casting ($3-5)(WR)

230. navy blue body, silver/gray chassis, purple-blue roof, "Cowboys 1991" tempa, China casting ($3-5)(WR)

231. olive body, yellow chassis, yellow roof, "Packers 1991" tempa, China casting ($3-5)(WR)

232. silver/gray body, green chassis, green roof, "Eagles 1991" tempa, China casting ($3-5)(WR)

233. orange-yellow body, purple chassis, white roof, "University of Washington 1992" tempa, China casting ($4-6)(WR)

234. dark yellow body, black chassis, white roof, "University of Colorado 1992" tempa, China casting ($4-6)(WR)

235. yellow body, navy blue chassis, yellow roof, "University of Michigan 1992" tempa, China casting ($4-6)(WR)

236. dark orange body, blue chassis, white roof, "Clemson University 1992" tempa, China casting ($4-6)(WR)

237. dark orange body, blue chassis, white roof, "Syracuse University 1992" tempa, China casting ($4-6)(WR)

238. white body, dark blue chassis, dark blue roof, "Penn State 1991" tempa, China casting ($4-6)(WR)

239. red body, blue chassis, white roof, "Canadiennes (logo) 1917" tempa, Thailand casting ($5-7)(WR)

240. white body, red chassis, white roof, "Candiennes (logo) 1992" tempa, Thailand casting ($5-7)(WR)

241. black body, red chassis, white roof, "Chicago Black Hawks 1917" tempa, Thailand casting ($5-7)(WR)

242. white body, black chassis, white roof, Indian's head (Blackhawk logo) "1992" tempa, Thailand casting ($5-7)(WR)

243. tan body, brown chassis, white roof, "Boston Bruins 1917" tempa, Thailand casting ($5-7)(WR)

244. black body, yellow chassis, white roof, Bruins logo & "1992" tempa, Thailand casting ($5-7)(WR)

245. red body, white chassis, white roof, Redwings logo & "1917" tempa, Thailand casting ($5-7)(WR)

246. white body, red chassis, white roof, Redwings logo & "1992" tempa, Thailand casting ($5-7)(WR)

247. bright blue body, red chassis, white roof, "Rangers 1917" tempa, Thailand casting ($5-7)(WR)

248. white body, bright blue chassis, white roof, "New York Rangers 1992" tempa, Thailand casting ($5-7)(WR)

249. navy blue body, white chassis, white roof, "Toronto Maple Leafs 1917" tempa, Thailand casting ($5-7)(WR)

250. white body, blue chassis, white roof, "Toronto Maple Leafs 1992" tempa, Thailand casting ($5-7)(WR)

251. lime body, black chassis, black roof, "Milo" tempa, China casting ($6-8)(AU)(GS)

252. red body, black chassis, black roof, "Uncle Toby's" tempa, China casting ($6-8)(AU)(GS)

253. dark blue body, black chassis, black roof, "IXL" tempa, China casting ($6-8)(AU)(GS)

254. orange body, black chassis, black roof, "Billy Tea" tempa, China casting ($6-8)(AU)(GS)

255. beige body, black chassis, black roof, "Aeroplane Jelly" tempa, China casting ($6-8)(AU)(GS)

256. dark purple body, black chassis, black roof, "Violet Crumble" tempa, China casting ($6-8)(AU)(GS)

257. powder blue body & chassis, white roof, "North Carolina 1992" tempa, China casting ($4-6)(WR)

258. white body, orange chassis, white roof, "Tennessee Vols. 1992" tempa, China casting ($4-6)(WR)

259. white body, blue chassis, blue roof, "Camperdown Cumberland" tempa, China casting ($10-15)(AU)

260. white body, white chassis, white roof, amber windows, no tempa, Thailand casting ($10-15)(GF)

261. brown body, black chassis, black roof, "Cobb of Knightbridge" tempa, Thailand casting ($8-12)(UK)

262. white body, black chassis, red roof, "Big Ben" tempa, China casting ($6-8)(AU)

263. white body, white chassis, white roof, "Matthew Walker" tempa, China casting ($8-12)(UK)

264. red body, black chassis, black roof, "McVities Digestive" tempa, China casting ($12-15)(UK)(OP)

265. orange-yellow body, green chassis, green roof, "1st M.I.C.A. Australia Convention 1992", China casting ($8-12)(AU)

266. red body, black chassis, black roof, "Pritt Stick" tempa, China casting ($7-10)(UK)(OP)

267. black body, black chassis, black roof, "City Ford" tempa, China casting ($7-10)(AU)

268. orange body, black chassis, black roof, "P.M.G." tempa, China casting ($8-12)(AU)

269. brass plated body, black chassis, black roof, "Matchbox Collectors Club 1992" tempa, Macau casting ($40-65)(C2)

270. pale gray body, brown chassis, brown roof, "Yardley" tempa, China casting ($8-12)(UK)

271. white body, white chassis, white roof, no tempa, China casting ($10-15)(GF)

272. orange body, black chassis, black roof, "Philadelphia Flyers 1993" tempa, Thailand casting ($6-8)(WR)

273. black body, silver/gray chassis, silver/gray roof, "LA Kings 1993" tempa, Thailand casting ($6-8)(WR)

274. white body, red chassis, blue roof, "Washington Capitals 1993" tempa, Thailand casting ($6-8)(WR)

275. white body, metallic turquoise chassis, black roof, "San Jose Sharks 1993" tempa, Thailand casting ($6-8)(WR)

276. purple body, black chassis, black roof, "Tyrrells Dry Red" tempa, Thailand casting ($6-8)(AU)(GS)

277. red body, black chassis, black roof, "McWilliams Cream Sherry" tempa, Thailand casting ($6-8)(AU(GS)

278. brown body, black chassis, black roof, "Yalumba Port" tempa, Thailand casting ($6-8)(AU)(GS)

279. blue body, black roof, "Houghton White Burgundy" tempa, Thailand casting ($6-8)(AU)(GS)

280. green body, black chassis, black roof, "Penfold's tempa, Thailand casting ($6-8)(AU)(GS)

281. black body, black chassis, black roof, "Hardy's Black Bottle" tempa, Thailand casting ($6-8)(AU)(GS)

282. dark green body, dark green chassis, black roof, "Selfridge & Co." tempa, China casting ($5-7)(UK)

283. green body, green chassis, green roof,

"William Lusty" tempa, China casting ($6-8)(UK)

284. yellow body, greenish black chassis, black roof, "Matchbox Series Model A Ford Van" tempa, Thailand casting ($2-3)

285. yellow body, black chassis, black roof, "Matchbox Series Model A Ford Van" tempa, Thailand casting ($2-3)

286. red body, white chassis, white roof, "Vileda" tempa, China casting ($8-12)(UK)(OP)

MB38-F FORD COURIER, issued 1992 (ROW)

1. white body, white metal base, gray interior, "Courier" tempa, China casting ($12-15)(UK)(LE)
2. lilac body, lilac plastic base, gray interior, "Milka" tempa, China casting ($2-3)
3. dark blue body, black plastic base, gray interior, "Matchbox-The Ideal Premium" tempa ($20-25)(GR)

MB38-G FORD COURIER, issued 1992 (ENGLAND)

1. red body, red metal base, gray interior, no tempa, China casting ($12-15)(LE)

NOTE: MB38-G features side cast windows, whereas MB38-F features no side windows.

MB38-H MERCEDES 600 SEL, issued 1992 (USA)
MB39-G MERCEDES 600 SEL, issued 1992 (ROW)

1. silver/gray body, gray base, clear windows, light gray interior, 8 dot wheels, gray sides tempa, China casting ($1-2)

MB39-C TOYOTA SUPRA, issued 1983 (USA)
MB60-E TOYOTA SUPRA, issued 1983 (ROW)

NOTE: all models feature red interior & black rear hatch.

1. white body, black base, clear windows, 5 crown wheels, "41" tempa, Macau casting ($2-4)
2. white body, black base, light amber windows, 5 crown wheels, "41" tempa, Macau casting ($2-4)
3. white body, charcoal base, light amber windows, 5 crown wheels, "41" tempa, Macau casting ($2-4)
4. red body, black base, clear windows, 5 crown wheels, "Twin Cam 24", tempa, Macau casting ($7-10)(JP)
5. dark red body, black base, clear windows, 8 dot wheels, "Twin Cam 24" tempa, Macau casting ($7-10)(JP)
6. white body, black base, clear windows, 5 crown wheels, "Supra" & pinstripes tempa, Macau casting ($2-4)
7. white body, black base, clear windows, 8 dot wheels, "Supra" & pinstripes tempa, Macau casting ($2-4)
8. white body, black base, clear windows, 8 dot wheels, red/blue/yellow design tempa, Macau casting ($8-12)(DY)

MB39-D BMW CABRIOLET, issued 1985
MB28-I BMW CABRIOLET, reissued 1991 (ROW)

1. metallic silver blue body & base, red-brown interior, 8 dot silver wheels, no tow hook, "323i" tempa, Macau casting ($2-3)
2. metallic silver blue body & base, red-brown interior, dot dash wheels, no tow hook, "323i" tempa, Macau casting ($2-3)
3. red body & base, red-brown interior, 8 dot gold wheels, no tow hook, "323i" tempa, Macau casting ($1-2)
4. red body & base, red-brown interior, 8 dot silver wheels, no tow hook, "323i" tempa, Macau casting ($1-2)
5. red body & base, red-brown interior, 5 arch wheels, no tow hook, "323i" tempa, Macau casting ($2-3)
6. white body & base, red-brown interior, star burst wheels, no tow hook, "Alpina" tempa, Macau casting ($3-5)(SF)
7. white body & base, red-brown interior, 8 dot silver wheels, no tow hook, "BMW" & "323i" tempa, Macau casting ($8-12)
8. white body & base, red-brown interior, 8 dot silver wheels, with tow hook, "BMW" & 323i" tempa, Macau casting ($8-12)
9. red body & base, red -brown interior, 8 dot silver wheels, with tow hook, "Gliding Club" tempa, Macau casting ($3-4)(TP)
10. white body & base, red-brown interior, starburst wheels, with tow hook, "Alpina" tempa, Macau casting ($3-5)(SF)
11. white body & base, red-brown interior, laser wheels, with tow hook, "Alpina" tempa, Macau casting ($3-5)(LW)
12. dark blue body, black base, black interior, 8 dot silver wheels, with tow hook, "323i" & "BP" tempa, Macau casting ($8-12)(DU)
13. silver blue body & base, gray interior, 8 dot silver wheels with tow hook, dark blue stripe tempa, Macau casting ($2-4)(TP)
14. light silver blue body & base, gray interior, 8 dot silver wheels, with tow hook, dark blue stripe tempa, Thailand casting ($2-4)(TP)
15. red body & base, red-brown interior, 8 dot silver wheels, wit tow hook, "323i" tempa, Thailand casting ($1-2)
16. white body & base, gray interior, 8 dot silver wheels, with tow hook, blue & red with "323i" tempa, Thailand casting ($2-3)(TP)
17. white body & base, maroon interior, 8 dot silver wheels, with tow hook, purple/orange/blue tempa, Thailand casting ($1-2)(TP)

MB39-E FORD BRONCO II, issued 1987 (USA)
MB35-G FORD BRONCO II, issued 1988 (ROW)

NOTE: all models with clear windows, black tire carrier, maltese cross wheels unless noted.

1. white body, chrome base, red interior, "Bronco" & stripes tempa, Macau casting ($1-2)
2. dark brown body, chrome base, red interior, "Bronco" & stripes tempa, Macau casting ($3-4)(SC)
3. orange body, chrome base, red interior, "Bronco" & stripes tempa, Macau casting ($3-4)(SC)
4. white body, chrome base, red interior, "Coast Guard Beach Patrol" tempa, Macau casting ($1-2)
5. white body, chrome base, red interior, "Coast Guard Beach Patrol" tempa, Thailand casting ($1-2)
6. bright yellow body, chrome base, red interior, flames & "4x4" tempa, Thailand

casting ($6-8)(US)(OP)

7. red body, blue base, orange-yellow interior, map & compass tempa, all yellow wheels, no spare tire carrier, Thailand casting ($6-8)(LL)

8. metallic blue body, chrome base, silver/gray interior, white splash marks & orange "4x4" Bronco tempa with window tempa, Thailand casting ($1-2)

9. metallic blue body, chrome base, silver/gray interior, white splash marks & orange "4x4" Bronco tempa, plain window, Thailand casting ($1-2)

10. white body, chrome base, red interior, "Police PD-22" tempa, Thailand casting ($1-3)(EM)

MB40-B HORSE BOX, issued 1977 (continued)

36. orange body, chocolate box, white door, black base, green windows, no tempa, Macau casting ($4-6)

37. blue body, yellow box, lime door, red base, green windows, caricature tempa, lime wheels with red hubs, Macau casting ($6-8)(LL)

38. blue body, yellow box, lime door, red base, green windows, caricature tempa, lime wheels with blue hubs, Thailand casting ($15-20)(LL)

39. red body, dark tan box, dark brown door, black base, green windows, no tempa, Manaus casting ($35-50)(BR)

40. red body, dark tan box, light brown door, black base, green windows, no tempa, Manaus casting ($35-50)(BR)

41. red body, dark tan box, chocolate door, black base, green windows, no tempa, Manaus casting ($35-50)(BR)

42. white body, white box, white door, red

base, blue windows, "Circus Circus" tempa, Thailand casting ($3-5)(MC)

MB40-C CORVETTE T-ROOF, issued 1982 (USA)
MB62-E CORVETTE T-ROOF, issued 1982 (ROW)
MB58-F CORVETTE T-ROOF, reissued 1991 (USA) (continued)

NOTE: all models with clear windows unless noted otherwise.

6. white body, red interior, red metal base, stripes tempa, 5 arch front & 5 crown rear wheels, Macau casting ($2-3)

7. white body, red interior, red metal base, stripes tempa, 5 crown front & 5 arch rear wheels, Macau casting ($2-3)

8. white body, red interior, red metal base, stripes tempa, 5 crown front & rear wheels, Macau casting ($2-3)

9. dark blue body, red interior, pearly silver metal base, flames tempa, 5 arch front & 5 crown rear silver wheels, Macau casting ($2-3)

10. dark blue body, red interior, pearly silver metal base, flames tempa, 5 arch front & 5 crown rear gold wheels, Macau casting ($3-4)

11. dark blue body, red interior, pearly silver metal base, flames tempa, starburst wheels, Macau casting ($75-100)

12. yellow body, silver/gray interior, black metal base, "Corvette" tempa, 5 arch front & 5 crown rear wheels, Macau casting ($1-2)

13. yellow body, silver/gray interior, black plastic base, "Corvette" tempa, 5 arch front & 5 crown rear wheels, Macau casting ($1-2)

14. yellow body, silver/gray interior, black

plastic base, "Corvette" tempa, 5 arch front & 5 crown rear wheels, China casting ($1-2)

15. dark blue body, red interior, black plastic base, yellow/red stripes tempa, 4 arch wheels, Manaus casting ($35-50)(BR)

16. dark blue body, red interior, black plastic base, yellow/red stripes tempa, 8 spoke wheels, Manaus casting ($35-50)(BR)

17. dark orange body, black interior, gray plastic base, detailed trim tempa, gray disc w/rubber tires, chrome windshield, China casting ($3-4)(WC)

18. red body, white interior, gray plastic base, "Vette" & Chevy logo tempa, 5 arch front & 5 crown rear wheels, China casting ($1-2)(US)

19. metallic blue body, dark gray interior, silver/gray plastic base, detailed trim tempa, gray disc w/rubber tires, chrome windshield, China casting ($5-8)(WC)

20. black body, black interior, silver/gray plastic base, "22" & stripes tempa, black disc w/rubber tires, China casting ($5-7)(WP)

MB40-D NASA ROCKET TRANS-PORTER, issued 1985 (ROW)
MB60-H NASA ROCKET TRANS-PORTER, issued 1990 (USA)

1. white body, blue windows, white rockets, "NASA" with US flag tempa, Macau casting ($1-2)

2. white body, blue windows, white rockets, "NASA" with US flag tempa, China casting ($1-2)

3. black body, blue windows, dark gray rockets, yellow/gray camouflage tempa,

black hubs, China casting ($25-35)(CM)

4. white body, blue windows, white rockets, "NASA" with checkers tempa, China casting ($1-2)

MB40-F ROAD ROLLER, issued 1991 (USA)
MB68-H ROAD ROLLER, issued 1991 (ROW)

1. dark orange body & base, dark gray interior, 5 crown rear wheels, blue stripes tempa, Thailand casting ($1-2)

2. orange-yellow body & base, dark gray interior, 5 crown rear wheels, red stripes & design tempa, Thailand casting ($1-2)

MB41-D KENWORTH AERODYNE, issued 1982 (continued)

NOTE: other variations on this model exist as cabs to Convoy models but these are note listed as these were not issued as singles. English based models 1, 2 & 3 have Lesney bases whereas 4 onward English bases have Matchbox International cast.

4. red body, clear windows, black & white curved stripes tempa, England casting ($3-5)

5. red body, clear windows, black & white curved stripes tempa, England casting ($3-5)

6. black body, clear windows, orange/yellow/white stripes tempa, Macau casting ($4-6)

7. pearly silver body, clear windows, red & blue stripes tempa, Macau casting ($3-5)

8. blue body, clear windows, no tempa, Macau casting ($4-6)(MP)

NOTE: this model normally associated with CY-9 Mitre 10 but found as a release in Multipacks!

MB41-F JAGUAR XJ6, issued 1987 (USA)

MB 1-E JAGUAR XJ6, issued 1987 (ROW)

NOTE: all models listed with black base, clear windows, 8 dot wheels.

1. metallic red body, tan interior, no tempa, Macau casting ($1-2)
2. black body, tan interior, "W&M" & crest tempa, Macau casting ($25-35)(UK)(OP)
3. green body, maroon interior, "Redoxon/ Jaguar" tempa, Macau casting ($18-25)(HK)
4. white body, black interior, no tempa, Macau casting ($7-10)(KS)
5. metallic red body, tan interior, no tempa, Thailand casting ($1-2)

MB42-D 1957 FORD THUNDERBIRD, issued 1982 (continued)

4. cream & red body, pearly silver base, red interior, clear windows, with tempa, 5 arch wheels, Macau casting ($3-5)
5. cream & red body, pearly silver base, red interior, light amber windows, with tempa, 5 arch wheels, Macau casting ($3-5)
6. cream & red body, pearly silver base, red interior, clear windows, with tempa, 5 arch wheels, China casting ($3-5)
7. black body, pearly silver base, red interior, clear windows, no tempa, 5 arch wheels, China casting ($3-5)
8. black body, pearly silver base, red interior, clear windows, dot dash front & 5 arch rear wheels ($3-5)

MB42-E FAUN CRANE TRUCK, issued 1984 (ROW)

MB42-E FAUN CRANE TRUCK, issued 1987 (USA)

NOTE: all models with black boom, red (shades) hook, black base.

1. yellow body, yellow metal crane cab, "Reynolds Crane Hire" tempa, England casting ($2-3)
2. yellow body, yellow metal crane cab, "Reynolds Crane Hire" tempa, Macau casting ($1-2)
3. yellow body, yellow metal crane cab, "Reynolds Crane Hire" tempa, China casting ($1-2)
4. yellow body, yellow metal crane cab, no tempa, China casting ($2-3)(MC)
5. yellow body, yellow plastic crane cab, "Reynolds Crane Hire" tempa, Thailand casting ($1-2)
6. yellow body, yellow plastic crane cab, no tempa, Thailand casting ($2-3)(MC)
7. yellow body, red plastic crane cab, bridge & road design tempa, Thailand casting ($1-2)
8. yellow body, fluorescent orange plastic crane cab, checkers & "IC" logo tempa, China casting ($10-15)(IC)

MB43-A PONY TRAILER, issued 1970 (continued)

NOTE: all models with clear windows, dot dash wheels.

13. beige body, black base, "Silver Shoes" tempa, white tailgate, Macau casting ($2-3)(TP)
14. beige body, black base, "Silver Shoes" tempa, white tailgate, no origin cast ($2-3)(TP)
15. white body, black base, "Polizei" &

checkers tempa, blue-gray tailgate, no origin cast ($2-3)(TP)
16. white body, black base, horse silhouette tempa, lime tailgate, no origin cast ($1-2)(TP)
17. green body with white roof, black base, "Polizei" tempa, green tailgate, no origin cast ($1-2)(TP)
18. white body with red roof, black base, black/red dashed stripes tempa, white tailgate, no origin cast ($1-2)(TP)

MB43-C 0-4-0 STEAM LOCO, issued 1978

MB63-H 0-4-0 STEAM LOCO, reissued 1992 (ROW)

8. green body, black boiler, black base, black wheels, "British Railways" tempa, Macau casting ($6-8)(UK)
9. red body, black boiler, black base, black wheels, "4345" tempa, Macau casting ($2-4)(MC)
10. green body, black boiler, black base, black wheels, "4345" tempa, Macau casting ($2-4)(MC)
11. green body, black boiler, black base, black wheels, "West Somerset Railway" tempa ($6-8)(UK)
12. red body, black boiler, black base, black wheels, "North Yorkshire Moors Railway" (all white), Macau casting ($6-8)(UK)
13. red body, black boiler, black base, black wheels, "North Yorkshire Moors Railway" (white & black), Macau casting ($15-20)(SW)
14. yellow body, red boiler, blue base, blue wheels, "123/efg" tempa, Macau casting ($6-8)(LL)
15. blue body, black boiler, black base,

black wheels, "Hutchinson" tempa, Macau casting ($6-8)(UK)
16. green body, black boiler, black base, black wheels, white emblem tempa, Macau casting ($15-20)(UK)(OP)
17. matt olive body, black boiler, black base, black wheels, "GWR" tempa, Macau casting ($6-8)(UK)
18. matt black body, black boiler, black base, black wheels, "British Railways" tempa ($6-8)(UK)
19. yellow body, lime boiler, blue base, blue wheels, "123/456" tempa, Macau casting ($6-8)(LL)
20. blue body, black boiler, black base, black wheels, red coachline tempa, Macau casting ($6-8)(UK)
21. yellow body, lime boiler, blue base, blue wheels, "123/456" tempa, China casting ($6-8)(LL)
22. green body, black boiler, black base, black wheels, "British Railways" (red/ black/white), China casting ($1-2)
23. red body, black boiler, black base, black wheels, "4345" tempa, China casting ($2-3)(MP)
24. green body, black boiler, black base, black wheels, white Kelloggs rooster head tempa, China casting ($35-50)(FR)(DU)
25. white body, white boiler, white base, black wheels, no tempa, China casting ($10-15)(GF)
26. green body, black boiler, black base, black wheels, "British Railways" (red/ white) tempa, China casting ($1-2)

MB43-D PETERBILT CONVEN-TIONAL, issued 1982 (USA) (continued)

NOTE: other variations exist as components to Convoy models. Only single releases are listed here. England bases are cast "Matchbox International" as opposed to "Lesney".

7. black body, amber windows, chrome exhausts, white/red "Ace" tempa, England casting ($2-4)
8. black body, clear windows, chrome exhausts, white/red "Ace" tempa, England casting ($2-4)
9. black body, clear windows, chrome exhausts, white /red "Z" pattern tempa, England casting ($3-5)
10. black body, amber windows, chrome exhausts, white/red "Z" pattern tempa, Macau casting ($2-4)
11. white body, clear windows, gray exhausts, "NASA" & rocket tempa, Macau casting ($3-5)(MP)

NOTE: this model normally associated with CY-15, but some models found as singles in multipacks.

MB43-E MERCEDES 500 SEC, issued 1984

1. white body, black metal base, clear windows, blue interior, 8 dot silver wheels, "AMG" tempa, Macau casting ($2-4)
2. red body, black metal base, clear windows, black interior, 8 dot silver wheels, "AMG" tempa, Macau casting ($2-3)
3. red body, black metal base, clear windows, black interior, starburst wheels, "AMG" with stripes tempa, Macau casting ($3-5)(SF)

4. white body, black metal base, clear windows, black interior, 8 dot gold wheels, red/blue with "7" tempa, Macau casting ($2-4)
5. white body, black metal base, clear windows, black interior, 8 dot silver wheels, red/blue with "7" tempa, Macau casting ($2-4)
6. white body, black metal base, clear windows, black interior, starburst wheels, red/blue interior, starburst wheels, red/blue with "7" tempa, Macau casting ($75-100)
7. white body, black metal base, clear windows, black interior, 8 dot silver wheels, "AMG" on hood & sides tempa, Macau casting ($6-8)
8. black body, black metal base, clear windows, brown interior, 8 dot silver wheels, "500SEC" & stripe tempa, Macau casting ($2-3)
9. metallic red body, black metal base, clear windows, black interior, 8 dot silver wheels, "AMG" & stripes tempa, Macau casting ($3-5)(LW)
10. red body, pearly silver metal base, clear windows, brown interior, 8 dot silver wheels, green/yellow stripes tempa, Macau casting ($8-12)(DU)
11. white body, pearly silver metal base, clear windows, brown interior, 8 dot silver wheels, "1 Pig Racing Team" tempa, Macau casting ($8-12)(HK)
12. black body, pearly silver metal base, clear windows, brown interior, 8 dot silver wheels, "500 SEC" & stripe tempa, Macau casting ($2-3)
13. metallic red body, pearly silver metal base, clear windows, black interior, laser wheels, "AMG" & stripes tempa, Macau

casting ($3-5)(LW)
14. white body, pearly silver metal base, clear windows, brown interior, 8 dot silver wheels, "500SEC" & silver stripe tempa, Macau casting ($18-25)(SU)
15. black body, black metal base, clear windows, brown interior, 8 dot silver wheels, "Redoxon/500SEC AMG" tempa, Macau casting ($18-25)(HK)
16. white body, black metal base, chrome windows, no interior, gray disc w/rubber tires, detailed trim tempa, Macau casting ($5-8)(WC)
17. black body, black metal base, navy blue windows, no interior, 8 dot silver wheels, "Pace Car Heuer" tempa, amber dome lights, Macau casting ($6-8)(SR)
18. black body, black plastic base, navy blue windows, no interior, 8 dot silver wheels, "Pace Car Heuer" tempa, amber dome lights, Macau casting ($6-8)(SR)
19. white body, black metal base, black windows, no interior, 8 dot silver wheels, "Police" tempa, red dome lights, Macau casting ($6-8)(SR)
20. white body, black plastic base, black windows, no interior, 8 dot silver wheels, "Police" tempa, red dome lights, Macau casting ($6-8)(SR)
21. cream body, black metal base, black windows, no interior, 8 dot silver wheels, "Emergency Doctor" tempa, green dome lights, Macau casting ($6-8)(SR)
22. cream body, black plastic base, black windows, no interior, 8 dot silver wheels, "Emergency Doctor" tempa, green dome lights, Macau casting ($6-8)(SR)
23. cream body, black metal base, black windows, no interior, 8 dot silver wheels, "Emergency Doctor/Rescue 911" tempa,

green dome lights, China casting ($7-10)(SR)
24. white body, black metal base, black windows, no interior, 8 dot silver wheels, "Police/Rescue 911" tempa, red dome lights, China casting ($7-10)(SR)
25. black body, black metal base, black windows, no interior, 8 dot silver wheels, "Pace Car Heuer/Rescue 911" tempa, green dome lights, China casting ($7-10)(SR)
26. red body, silver/gray metal base, clear windows, brown interior, starburst wheels, "AMG" with stripes tempa, Macau casting ($3-5)(SF)

MB43-G LINCOLN TOWN CAR, issued 1989 (USA)
MB24-G LINCOLN TOWN CAR, issued 1989 (ROW)

NOTE: all models with black base, smoke gray windows, red interior unless noted otherwise.

1. white body, metal base, dot dash wheels, no tempa, Macau casting ($1-2)
2. black body, plastic base, gray disc w/ rubber tires, no tempa, Macau casting ($5-8)(WC)
3. white body, plastic base, dot dash wheels, no tempa Thailand casting ($1-2)
4. silver/gray body, plastic base, dot dash wheels, pink/yellow design tempa, Thailand casting ($2-3)(DM)
5. orange-yellow body, plastic base, dot dash blue wheels, face on hood tempa, Thailand casting ($10-15)(LL)
6. metallic maroon body, plastic base, dot dash wheels, brown rear half of roof tempa, Thailand casting ($2-4)
7. metallic maroon body, plastic base, dot

dash wheels, no tempa, Thailand casting
($1-2)

MB44-A REFRIGERATOR TRUCK, issued 1970 (continued)

4. yellow body, turquoise container, black axle covers ($2000+)

MB44-C PASSENGER COACH, issued 1978 (continued)

24. red body, flat cream roof, black base, no windows, train wheels, "431 432" tempa, Macau casting ($2-3)
25. lime & yellow body, flat red roof, blue train wheels, 2 figures with clock tempa, Macau casting ($6-8)(LL)
26. lime & yellow body, flat red roof, blue base, no windows, blue train wheels, 2 figures with clock tempa, China casting ($6-8)(LL)
27. green body, beige flat roof, black base, no windows, train wheels, "British Railways" tempa, China casting ($2-3)(TP)
28. red body, cream flat roof, black base, no windows, train wheels, "431 432" tempa, China casting ($1-2)(MP)
29. red body, cream flat roof, black base with small "Kelloggs" sticker, no windows, train wheels, "431 432" tempa, China casting ($18-25)(DU)(OP)
30. white body, white flat roof, white base, train wheels, no windows, no tempa, China casting ($10-15)(GF)

MB44-D 4 X 4 CHEVY VAN, issued 1982 (USA)
MB68-D 4 X 4 CHEVY VAN, issued 1982 (ROW)
MB26-H 4 X 4 CHEVY VAN, reissued 1991 (USA) (continued)

6. white body, black metal base, "Matchbox Racing" tempa, Macau casting ($2-3)
7. white body, black metal base, "Tokyo Giants/Matsumoto 2" labels, Macau casting ($8-12)(JP)
8. white body, black metal base, "Tokyo Giants/Matsumoto 2" labels, Macau casting ($8-12)(JP)
9. white body, black metal base, "Tokyo Giants/Hara 8" labels, Macau casting ($8-12)(JP)
10. white body, black metal base, "Tokyo Giants/Yamakura 15" labels, Macau casting ($8-12)(JP)
11. white body, black metal base, "Tokyo Giants/Sadaoka 20" labels, Macau casting ($8-12)(JP)
12. white body, black metal base, "Tokyo Giants/Nakahata 24" labels, Macau casting ($8-12)(JP)
13. white body, black metal base, "Tokyo Giants/Nishimoto 26" labels, Macau casting ($8-12)(JP)
14. white body, black metal base, "Tokyo Giants/Egawa 30" labels, Macau casting ($8-12)(JP)
15. white body, black metal base, "Castrol Racing Team" tempa, Macau casting ($7-10)(AU)
16. white body, silver/gray metal base, "Matchbox Motorsports" tempa, Macau casting ($1-2)
17. white body, gray plastic base, "Matchbox Motorsports" tempa, Macau

casting ($1-2)
18. white body, gray plastic base, "Matchbox Motorsports" tempa, Thailand casting ($1-2)
19. white body, black plastic base, no tempa, Thailand casting ($10-15)(GF)

MB44-E CITROEN 15CV, issued 1983 (ROW)

NOTE: all models with clear windows & 5 crown wheels unless noted.

1. black body, chrome base, light gray interior, England casting ($2-3)
2. black body, chrome base, silver/gray interior, England casting ($2-3)
3. black body, chrome base, beige interior, England casting ($2-3)
4. black body, chrome base, white interior, England casting ($2-3)
5. black body, chrome base, silver/gray interior, Macau casting ($2-3)
6. dark blue body, chrome base, silver/gray interior, Macau casting ($2-3)
7. dark blue body, chrome base, red interior, Macau casting ($2-3)
8. dark green body, chrome base, red interior, Macau casting ($8-12)(DY)
9. black body, gray base, cream interior, 5 arch wheels, China casting ($4-6)(GS)

MB44-F DATSUN 280ZX POLICE CAR, issued 1987 (Japan)

1. white/black body, clear windows, opaque red dome lights, tan interior, black base, Japanese lettered tempa, Macau casting ($8-12)(JP)

MB44-G SKODA LR 130, issued 1987 (ROW)

NOTE: all models with powder blue base,

red trunk, black interior, clear windows and 8 dot wheels.

1. white body, "Skoda 44" with "Duckhams" on sides, Macau casting ($2-4)
2. white body, "Skoda 44" without "Duckhams" on sides, Macau casting ($2-4)
3. white body, "Skoda 44" with "Duckhams" on sides, China casting ($2-4)

MB44-H 1921 MODEL T FORD VAN, issued 1990

NOTE: all models with plastic cast base, clear windshield & black plastic spoked wheels.

1. yellow body, red roof, blue chassis, "Bird's Custard Powder" tempa, Macau casting ($1-2)
2. cream body, dark blue roof & chassis, "5th MICA Convention 1990" tempa, Macau casting ($8-10)(UK)
3. cream body, dark blue roof & chassis, "3rd MICA NA Convention 1990" tempa, Macau casting ($8-10)(US)
4. red body with black hood, matt black roof, black chassis, "Royal Mail GR" tempa, Macau casting ($7-10)(UK)
5. white body, powder blue roof & chassis, "Para 90" tempa, Macau casting ($8-10)(UK)
6. white body, powder blue roof & chassis, "Chester Doll Hospital/World's Largest Matchbox Display" tempa, Macau casting ($8-10)(UK)
7. dark green body, black roof, dark charcoal chassis, "Swarfega" tempa, Macau casting ($35-50)(UK)(OP)
8. black body, black roof & chassis, "Mars"

tempa, Macau casting ($10-15)(UK)
9. cream body, red roof, green chassis, "PG Tips" tempa, Macau casting ($15-20)(UK)(OP)
10. cream body, red roof, green chassis, "PG Tips" tempa, China casting ($15-20)(UK)(OP)
11. yellow body, red roof, blue chassis, "Bird's Custard Powder" tempa, China casting ($2-3)
12. black body, black roof & chassis, "4th MICA NA Convention/Detroit Motor City" tempa, China casting ($8-12)(US)
13. red body, red roof, black chassis, "Mars" tempa, China casting ($15-20)(UK)
14. white body, powder blue roof & chassis, "MICA 7-I Could Have Danced All Night" labels, Macau casting ($10-15)(C2)(UK)
15. cream body, dark blue roof & chassis, "Greetings From Philadelphia 1992/ MICA NA" labels ($10-15)(C2)(US)
16. white body, red roof, dark blue chassis, "Lloyds" tempa, China casting ($8-12)(UK)
17. black body, black roof & chassis, "William Lusty" tempa, China casting ($6-8)(UK)

MB45-C KENWORTH AERODYNE, issued 1982 (continued)
NOTE: other versions exist as components to Convoy models. Only models issued as single releases are listed here.
2. white body, amber windows, chrome exhausts, brown & blue stripes tempa, Macau casting ($)
3. pearly silver body, amber windows, chrome exhausts, purple/orange tempa, Macau casting ($2-4)

4. white body, amber windows, chrome exhausts, "Chef Boyardee" labels Macau casting ($35-50)(US)(OP)
5. red body, amber windows, chrome exhausts, yellow/orange/white stripes tempa, Macau casting ($1-2)
6. red body, amber windows, chrome exhausts, orange/yellow/white stripes tempa, Macau casting ($1-2)
7. red body, amber windows, gray exhausts, yellow/orange/white stripes tempa, Macau casting ($1-2)
8. red body, amber windows, gray exhausts, yellow/orange/white stripes tempa, Thailand casting ($1-2)(MP)

MB45-E HIGHWAY MAINTENANCE VEHICLE, issued 1990 (USA)
MB69-G HIGHWAY MAINTENANCE VEHICLE, issued 1990 (ROW)
NOTE: all models with black base, amber windows & 8 spoke wheels.
1. pale lemon body, pale lemon dump & plow, "Int'l Airport Authority 45" tempa, Macau casting ($1-2)
2. red body, gray dump & plow, "Aspen Snow Removal" tempa, Macau casting ($2-4)(AP)
3. pale lemon body, pale lemon dump & plow, "Int'l Airport Authority 45" tempa, China casting ($1-2)
4. dark orange body, red dump & plow, "Int'l Airport Authority 45" tempa, China casting ($8-12)(GS)

MB46-C FORD TRACTOR, issued 1978 (continued)
11. blue body, pearly silver base, white interior, no harrow, gold front & rear hubs, no tempa, Macau casting

($2-3)(TP)
12. yellow body, pearly silver base, black interior, no harrow, orange front & rear hubs, orange stripe tempa, Macau casting ($1-2)(TP)
13. yellow body, pearly silver base, black interior, red harrow, orange front & rear hubs, orange stripe tempa, Macau casting ($1-2)(GS)
14. green body, pearly silver base, white interior, yellow harrow, orange front & rear hubs, no tempa, Macau casting ($1-2)(MC)
15. green body, pearly silver base, black interior, no harrow, orange front & rear hubs, no tempa, Macau casting ($1-2)(TP)
16. green body, pearly silver base, black interior, no harrow, orange front & rear hubs, no tempa, Thailand casting ($1-2)(TP)
17. green body, pearly silver base, black interior, yellow harrow, orange front & rear wheels, no tempa, Thailand casting ($1-2)(MC)

MB46-D HOT CHOCOLATE/BEETLE STREAKER, issued 1982 (USA) (continued)
NOTE: versions 1 & 2 (from book 2) cast with "Hot Chocolate" as base name, versions 3 onward now have base name read "Beetle Streaker". All models listed with "Big Blue" tempa, 5 arch front wheels unless noted.
3. silver blue body, pearly silver base, Macau casting ($3-5)
4. silver blue body, pearly silver base, Hong Kong casting ($3-5)
5. blue body, pearly silver base, Hong Kong

casting ($3-5)
6. blue body, light gray base, Hong Kong casting ($3-5)
7. silver blue body, light gray base, Hong Kong casting ($3-5)
8. dark blue body, pearly silver base, Macau casting ($3-5)
9. blue body, light gray base, 5 spoke front wheels, Hong Kong casting ($3-5)

MB46-F MISSION HELICOPTER, issued 1985 (USA)
MB57-F MISSION HELICOPTER, issued 1985 (ROW)
NOTE: all models issued with blue windows.
1. dark blue body, silver gray base/skis, white tail & blades, orange tempa, Macau casting ($1-2)
2. red body, white base/skis, white tail & blades, "Sheriff Air 1" tempa, Macau casting ($1-2)
3. olive green body, tan base/skis, tan tail & blades, star & emblem tempa, Macau casting ($2-3)(SB)
4. red body, white base/skis, white tail & blades, "Rebels/Rescue/Air 1" tempa, Macau casting ($1-2)(MC)
5. dark blue body, silver/gray base/skis, white tail & blades, bullseye tempa, Macau casting ($2-3)(GS)
6. olive green body, black base/skis, black tail & blades, "AC15" & logo tempa, Macau casting ($4-5)(CM)
7. black body, dark gray base/skis, dark gray tail, yellow blades, "AC99" & logo tempa, Macau casting ($4-5)(CM)
8. red body, white base/skis, white tail & blades, "Sheriff Air 1" tempa, Thailand casting ($1-2)

9. dark blue body, silver/gray base/skis, white tail & blades, bullseye tempa, Thailand casting ($2-3)
10. red body, white base/skis, white tail & blades, "Rebels/Rescue/Air 1" tempa, Thailand casting ($1-2)(MC)
11. olive green body, bright tan base/skis, bright tan tail & tan blades, star & emblem tempa, Thailand casting ($2-3)(SB)
12. olive green body, pink-tan base/skis, pink-tan tail & blades, star & emblem tempa, Thailand casting ($2-3)(SB)
13. white body, dark blue base/skis, white tail & blades, "Police" & crest tempa, Thailand casting ($1-2)
14. green body, white base/skis, white tail & blades, "Polizei" tempa, Thailand casting ($2-3)(CY)
15. black body, white base/skis, white tail & blades, Police" tempa, Thailand casting ($2-3)(CY)

MB47-D JAGUAR SS, issued 1982 (continued)

4. red body & hood, gray base, tan interior, Macau casting ($2-3)
5. red body & hood, black base, tan interior, Macau casting ($2-3)
6. dark blue body, pearly silver hood, black base, tan interior, Macau casting ($3-4)
7. dark green body, red hood, black base, black interior, Thailand casting ($12-15)(UK)(OP) & (US)(GS)

MB47-E SCHOOL BUS, issued 1987

NOTE: all models with black base, clear windows, dot dash wheels unless noted otherwise. Early models cast with small window (two sizes) in lower section of rear door. Later models cast solid with black painted tempa.

1. yellow body, "School District 2" tempa, Macau casting ($1-2)
2. yellow body, "School District 2" tempa, China casting ($1-2)
3. olive body, "Govt Property" tempa, China casting ($8-12)(US)
4. yellow body, "School District 2" tempa with "Chef Boyardee" hood label, China casting ($8-12)(US)(OP)
5. orange-yellow body, "School District 2" tempa with "Chef Boyardee" hood label, China casting ($8-12)(US)(OP)
6. orange-yellow body, "School District 2" tempa, China casting ($1-2)
7. orange-yellow body, red base, "1 + 2 = 3 abc" tempa, green wheels with red hubs, China casting ($6-8)(LL)
8. orange-yellow body, red base, "1 + 2 = 3 abc" tempa, green wheels with blue hubs, China casting ($6-8)(LL)
9. orange-yellow body, "School District 2" tempa, Thailand casting ($4-6)
10. orange-yellow body, "St. Paul Public Schools" tempa, China casting ($25-45)(US)
11. blue body, "Police 88" tempa, China casting ($2-4)(AP)
12. dark blue body, "Hofstra University" tempa, China casting ($7-10)(US)
13. yellow body, "Harvey World Travel" tempa, China casting ($6-8)(AU)
14. dark orange body, "School District 2" tempa, China casting ($4-6)(GS)
15. white/dark blue body, "Penn State-The Loop" tempa, China casting ($6-8)(WR)

MB48-D RED RIDER, issued 1982 (USA) (continued)

3. red body, maltese cross front & rear wheels, Macau casting ($1-2)
4. red body, maltese cross front & 5 crown rear wheels, Macau casting ($1-2)
5. red body, 5 spoke front & 5 crown rear wheels, Macau casting ($1-2)
6. red body, 5 arch front & 5 crown rear wheels, China casting ($1-2)
7. dark red body, 5 arch front & 5 crown rear wheels, China casting ($1-2)

MB48-E UNIMOG WITH PLOW, issued 1982 (ROW) (continued)

4. yellow body, yellow-orange base & plow, black plow stripes, amber windows, no canopy, "Rescue" tempa, England casting ($3-5)
5. yellow body, lemon base & plow, black plow stripes, amber windows, white canopy, "Rescue" tempa, Macau casting ($2-4)
6. red body, red base & plow, white plow stripes, amber windows, white canopy, "UR83" tempa, Macau casting ($2-4)(TP)
7. white body, yellow base & plow, black plow stripes, red windows, no canopy, red/dark blue tempa, Macau casting, includes plastic armament ($5-8)(RB)
8. white body, orange base & plow, black plow stripes, amber windows, orange canopy, "C&S" tempa, Macau casting ($2-4)(TP)
9. white body, orange base & plow, black plow stripes, amber windows, orange canopy, "C&S" tempa, Thailand casting ($2-4)(TP)
10. white body, yellow base & plow, black plow stripes, red windows, no canopy, red/dark blue tempa, Thailand casting, includes plastic armament ($8-12)(JP)(Tomy box)

MB48-F VAUXHALL ASTRA/OPEL KADETTE, issued 1986 (ROW)

NOTE: bases with early models cast "Vauxhall Astra" only and later models cast "Vauxhall Astra/Opel Kadette". All models with clear windows.

1. red body, black base, black interior, 8 dot silver wheels, "GTE" & stripe tempa, Macau casting ($2-4)
2. red body, black base, black interior, 5 arch wheels, "GTE" & stripe tempa, Macau casting ($2-4)
3. white body, black base, black interior, 8 dot silver wheels, "AC Delco 48" tempa, Macau casting ($1-2)
4. white body, black base, black interior, 8 dot white wheels, "AC Delco 48" tempa, Macau casting ($2-4)
5. yellow body, blue base, black interior, 8 dot silver wheels, "Mobile Phone/ Telecom" tempa, Macau casting ($3-5)(GS)
6. black body, black base, black interior, 8 dot silver wheels, "BP 52" & yellow band tempa, Macau base ($8-12)(DU)
7. white body, black base, black interior, 8 dot silver wheels, "AC Delco 48" tempa, China casting ($1-2)
8. white body, black base, blue interior, 8 dot silver wheels, "STP 7/Sphere Drake" tempa, China casting ($1-2)
9. yellow body, yellow base, blue interior, 8 dot silver wheels, no tempa, China casting ($8-12)(GR)(GS)
10. white body, black base, blue interior, 8 dot silver wheels, "STP (no 7)/Sphere Drake" tempa, China casting ($1-2)

MB49-D SAND DIGGER, issued 1983

1. emerald green body, black base, ivory interior, "Sand Digger" tempa, Macau casting ($2-3)
2. very dark green body, black base, ivory interior, "Sand Digger" tempa, Macau casting ($2-3)
3. red body, black base, white interior, "Dune Man" tempa, Macau casting ($2-3)
4. red body, black base, rust red interior, "Dune Man" tempa, Macau casting ($2-3)
5. red body, black base, rust red interior, "Dune Man" tempa, China casting ($2-3)

MB49-E PEUGEOT QUASAR, issued 1986

1. white body & base, chrome interior, smoke gray windows, 8 dot silver wheels, "Quasar" tempa, Macau casting ($2-3)
2. dark blue body, black base, chrome interior, smoke gray windows, starburst wheels, "9" & pink stripes tempa, Macau casting ($3-5)(SF)
3. metallic blue body, black base, chrome interior, smoke gray windows, laser wheels, "9" & pink stripes tempa, Macau casting ($3-5)(LW)
4. white body & base, chrome interior, smoke gray windows, 8 dot white wheels, "Quasar" tempa, Macau casting ($2-3)
5. black body & base, no interior, red modified windows, starburst wheels, lime & orange stripes tempa, Macau casting, includes plastic armament ($5-8)(RB)
6. plum body & base, chrome interior, smoke gray windows with purple tempa, 8 dot silver wheels, "Quasar" tempa, Macau casting ($1-2)

7. plum body & base, chrome interior, smoke gray windows, 8 dot silver wheels, "Quasar" tempa, Macau casting ($1-2)
8. yellow body, lime base, red interior, smoke gray windows, 8 dot blue wheels with yellow hubs, "3" with stripes/flames tempa, Macau casting ($6-8)(LL)
9. yellow body, lime base, red interior, smoke gray windows, 8 dot blue wheels with yellow hubs, "3" with stripes/flames tempa, China casting ($6-8)(LL)
10. orange-yellow body, lime base, red interior, smoke gray windows, 8 dot blue wheels with yellow hubs, "3" with stripes/flames tempa, China casting ($6-8)(LL)
11. plum body & base, chrome interior, smoke gray windows, 8 dot silver wheels, "Quasar" tempa, China casting ($1-2)

MB50-D HARLEY DAVIDSON, issued 1980 (continued)

7. brown body, black handlebars, tan driver, Macau casting ($2-3)
8. dark metallic tan body, black handlebars, brown driver, Macau casting ($2-3)
9. light blue body, chrome handlebars, no driver, Thailand casting ($2-3)(HD)
10. orange body, chrome handlebars, no driver, Thailand casting ($2-3)(HD)

MB50-E CHEVY BLAZER, issued 1984

NOTE: all models issued with white interior & antennae, blue windows & dome lights, maltese cross wheels.

1. white body, chrome base, "Sheriff 7" tempa, Macau casting ($1-2)
2. white body, black base, "Sheriff 7" tempa, Macau casting ($1-2)
3. grape body, black base, orange/red/black

tempa, Macau casting, includes plastic armament ($5-8)(RB)
4. white body, black base, "Sheriff 7" tempa, Thailand casting ($1-2)
5. white body, black base, "Sheriff 7" (without red hood) tempa, Macau casting ($45-60)(BR)

MB51-C COMBINE HARVESTER, issued 1978 (continued)

8. yellow body, no base, solid black wheels, red rotor/chute, "2" with stripes tempa, Macau casting ($1-2)(MC)
9. orange-yellow body, no base, solid black wheels, red rotor/chute, with stripes tempa, Macau casting ($1-2)(MC)
10. lime & blue body, no base, solid red wheels, yellow rotor, lime chute, faces/gears/haystack tempa, Macau casting ($6-8)(LL)
11. lime & blue body, no base, solid red wheels, yellow rotor, lime chute, faces/gears/haystack tempa, Thailand casting ($6-8)(LL)
12. orange-yellow body, no base, solid black rotor/chute, "2" with stripes tempa, Thailand casting ($1-2)(MC)

MB51-D MIDNIGHT MAGIC, issued 1982 (USA) (continued)

4. black body, silver sides, pearly white base, Macau casting ($2-3)
5. black body, silver sides, pearly silver base, Macau casting ($2-3)
6. black body, silver sides, unpainted base, Hong Kong casting ($2-3)

MB51-G CAMARO IROC-Z28, issued 1986 (USA)
MB68-F CAMARO IROC-Z28, issued 1986 (ROW)

NOTE: all models with black base, silver/gray interior & clear windows unless noted otherwise.

1. apple green body, metal base, 8 dot wheels, "IROC Z" tempa, Macau casting ($4-6)
2. apple green body, metal base, 5 arch wheels, IROC Z tempa, Macau casting ($4-6)
3. blue body, metal base, 5 arch wheels, IROC Z on sides only tempa, Macau casting ($2-4)
4. blue body, metal base, 8 dot wheels, "IROC Z" on sides only tempa, Macau casting ($2-4)
5. red body, metal base, starburst wheels, "Carter/Goodyear" tempa, Macau casting ($3-5)(SF)
6. metallic red body, metal base, laser wheels, "Carter/Goodyear" tempa, Macau casting ($3-5)(LW)
7. blue body, metal base, 8 dot wheels, "IROC Z" hood & sides tempa without extra black line, Macau casting ($2-4)
8. blue body, metal base, 8 dot wheels, "IROC Z" hood & sides tempa with extra black line, Macau casting ($2-4)
9. lemon body, metal base, 8 dot wheels, "IROC Z" with stripes tempa, Macau casting ($1-2)
10. metallic copper body, metal base, laser wheels, "Carter/Goodyear" tempa, Macau casting ($4-6)(LW)
11. blue-green body, metal base, 8 dot wheels, "BP Stunt Team" & stripes tempa, Macau casting ($8-12)(DU)

12. lemon body, plastic base, 8 dot wheels, "IROC Z" with stripes tempa, Macau casting ($1-2)
13. lemon body, plastic base, 8 dot wheels, "IROC Z" with stripes tempa, Thailand casting ($1-2)
14. metallic blue body, plastic base, 8 dot wheels, black windows, "350Z" with pink/purple/lime stripes tempa, Thailand casting ($3-5)(TH)
15. black body, plastic base, 8 dot wheels, "Z28" with orange bands & white stripes tempa, Thailand casting ($1-2)

MB52-B POLICE LAUNCH, issued 1976 (continued)
9. black deck, dark gray hull, dark gray figures, light blue windows, tan & gray camouflage tempa, black wheels, Macau casting ($4-5)(CM)
10. white deck, blue hull, red figures, light blue windows, "123" & rope pattern tempa, red wheels, Macau casting ($6-8)(LL)
11. white deck , blue hull, red figures, light blue windows, "123" & rope pattern tempa, red wheels, Thailand casting ($6-8)(LL.)
12. white deck, white hull, white figures, light blue windows, no tempa, China casting ($10-15)(GA)

MB52-D BMW, MI, issued 1983
NOTE: this is a highly modified version to MB52-C in which the hood has been cast closed with spoiler added to the front and rear of the body, so is designated a "new" casting. All models with black plastic base, clear windows, 5 arch wheels unless noted otherwise.

1. white body, black interior, "BMW M1" tempa, Macau casting ($2-3)
2. black body, red interior, "Pirelli 59" tempa, Macau casting ($3-5)
3. yellow body, black interior, "11" with stripes tempa, Macau casting ($1-2)
4. yellow body, black interior, "11" with stripes tempa, China casting ($1-2)
5. yellow body, black interior, "11" with stripes tempa (all black "Bell" logo 7 all red "Champion" logo), China casting ($1-2)
6. bright red body, black interior, "1" with stripes tempa, Macau casting ($1-2)
7. red body, black interior, "1" with stripes tempa, Macau casting ($1-2)
8. dark yellow body, black interior, chrome windows, gray disc w/rubber tires, China casting ($5-8)(WC)
9. chrome plated body, black interior, no tempa, China casting ($12-18)(C2)
10. chrome plated body, black interior, no tempa, China casting ($12-18)(C2)
11. bright red body, black interior, "1" with stripes tempa, Thailand casting ($1-2)

MB52-E ISUZU AMIGO, issued 1991
NOTE: all models with chrome base, gray interior, clear windows & dot dash wheels.
1. metallic blue body, "Isuzu Amigo" tempa, Thailand casting ($5-7)
2. lemon body, pink stripe & design tempa, Thailand casting ($2-3)(DM)
3. red body, "Amigo" with silver/orange stripes tempa, Thailand casting ($2-3)

MB53-D FLARESIDE PICKUP, issued 1982 (continued)
6. orange body, black base, white interior, clear windows, 8 spoke wheels, "326 Baja Bouncer" tempa, Macau casting ($2-4)
7. yellow body, black base, white interior, clear windows, 8 spoke wheels, "Ford 460" tempa, Macau casting ($175-225)
8. yellow body, black base, black interior, clear windows, 8 spoke wheels, "Ford 460" tempa, Macau casting ($35-50)
9. yellow body, black base, black interior, clear windows, racing slicks (chromed letters) wheels, "Ford 460" tempa, Macau casting ($2-3)
10. yellow body, black base, black interior, clear windows, racing slicks (unchromed letters) wheels, "Ford 460" tempa, Macau casting ($1-2)
11. khaki green body, black base, black interior, opaque green windows, dark purple/blue design tempa, racing slicks wheels, Macau casting, includes plastic armament ($5-8)(RB)
12. white body, black base, black interior, "Deb" tempa, racing slicks wheels, Macau casting ($35-50)(UK)(OP)
13. red body, black base, white interior, "326 Baja Bouncer" tempa, 8 spoke wheels, Manaus casting ($75-100)(BR)
14. lime body, yellow base, red interior, jack & wheel design tempa, red racing slicks, Thailand casting ($6-8)(LL)
15. red body, black base, black interior, "Bill Elliot 11" tempa, racing slicks wheels ($5-7)(WR)(TC)

MB53-G FORD LTD TAXI, issued 1992 (USA)
MB56-G FORD LTD TAXI, issued 1992 (ROW)
1. yellow body, chrome base, blue interior, clear windows, 8 dot wheels, "Radio XYZ Cab" & checkers, Thailand casting ($3-5)

MB54-E NASA TRACKING VEHICLE, issued 1982 (continued)
4. white body, black base, red door, clear windows, dot dash wheels, Macau casting ($2-3)
5. white body, black base, maroon door, clear windows, dot dash wheels, Macau casting ($2-3)
6. white body, black base, red door, blue windows, dot dash wheels, Macau casting ($3-5)
7. white body, black base, red door, blue windows, 8 dot wheels, Macau casting ($3-5)
8. white body, black base, maroon door, blue windows, dot dash wheels, Macau casting ($3-5)

MB54-F COMMAND VEHICLE, issued 1984
NOTE: all models with black base & blue windows.
1. red body with white painted roof, metal base, 8 dot silver wheels, silver beacons, red pumper, "Foam Unit" & checker-board tempa, Macau casting ($35-50)
2. yellow body, metal base, 8 dot silver wheels, silver beacons, red pumper, "Foam Unit 3 Metro Airport" tempa, Macau casting ($2-4)
3. red body, metal base, 8 dot silver wheels, silver beacons, red pumper, "Foam Unit 3 Metro Airport" tempa, Macau casting ($2-3)
4. red body, plastic base, dot dash wheels, silver beacons, red pumper, "Foam Unit 3

Metro Airport" tempa, Macau casting ($2-3)

5. red body, plastic base, 8 dot silver wheels, silver beacons, red pumper, "Foam unit 3 Metro Airport" tempa, Macau casting ($2-3)

6. red body, plastic base, 8 dot gold wheels, silver beacons, red pumper, "Foam Unit 3 Metro Airport" tempa, Macau casting ($18-25)

7. white body, plastic base, 8 dot silver wheels, silver beacons, silver radar, "NASA Space Shuttle Command Center" tempa, Macau casting ($3-4)(TC)

8. yellow body, plastic base, 8 dot silver wheels, silver beacons, red pumper, "Foam Unit 3 Metro Airport" tempa, Macau casting ($2-3)(MP)

9. red body, plastic base, 8 dot silver wheels, silver beacons, red pumper, "3" & Japanese lettered tempa, Macau casting ($8-12)(JP)(GS)

10. olive body, plastic base, 8 dot black wheels, black beacons & black radar, "9" & red/white stripes tempa, Macau casting ($4-5)(CM)

11. olive body, plastic base, 8 dot silver wheels, black beacons & black radar, "9" & red/white stripes tempa, Macau casting ($18-25)(CM)

12. black body, plastic base, 8 dot black wheels, dark gray beacons, dark gray radar, "LS150" & yellow stripes tempa, Macau casting ($4-5)(CM)

13. yellow body, plastic base, 8 dot silver wheels, black beacons, red pumper, "Foam Unit 3 Metro Airport" tempa, Macau casting ($2-4)

14. red body, plastic base, 8 dot silver wheels, black beacons, red pumper,

"Foam Unit 3 Metro Airport" tempa, Macau casting ($2-4)

15. white body, plastic base, 8 dot silver wheels, black beacons, silver radar, "NASA Space Shuttle Command Center" tempa, Macau casting ($2-4)

16. red body, plastic base, 8 dot silver wheels, black beacons, red pumper, "Foam Unit 3 Metro Airport" tempa, Thailand casting ($1-2)

17. yellow body, plastic base, 8 dot silver wheels, black beacons, red pumper, "Foam Unit 3 Metro Airport" tempa, Thailand casting ($1-2)

18. white body, plastic base, 8 dot silver wheels, black beacons, silver radar, "NASA Space Shuttle Command Center" tempa, Thailand casting ($1-2)

19. fluorescent orange body, plastic base, 8 dot silver wheels, silver pumper, "Foam Unit/City Airport/ Emergency Rescue" tempa, Thailand casting ($1-2)(EM)

MB54-G CHEVY LUMINA, issued 1989
NOTE: the body casting for MB54-G is based on a redesigned body casting of the MB10-D Buick LeSabre with different size side window and redesigned front grille. All models with clear windows, Goodyear slicks wheels.

1. matt pink & white body, white base, black interior, "Superflo 46" tempa, Macau casting ($6-8)(DT)

2. dark pink & white body, white base, black interior, "Superflo 46" tempa, Macau casting ($6-8)(DT)

3. matt green & lime body, green base, black interior, "City Chevrolet 46" tempa, Macau casting ($6-8)(DT)

4. bright green & lime body, green base, black interior, "City Chevrolet 46" tempa, Macau casting ($6-8)(DT)

5. bright orange & blue body, orange base, black interior, "Hardees 18" tempa, Macau casting ($6-8)(DT)

6. black body, plastic base, red interior "Exxon 51" without roof signature tempa, Macau casting ($6-8)(DT)

7. black body, black base, red interior, "Exxon 51" with roof signature tempa, Macau casting ($6-8)(DT)

8. black body, green base, gray interior, "Mello Yello 51" tempa, Macau casting ($6-8)(DT)

9. white body, white base, white interior, no tempa, China casting ($6-8)(GF)

MB54-H CHEVY LUMINA, issued 1990
NOTE: all models with black base.

1. dark pink & white body, clear windows, black interior, Goodyear slicks wheels, "Superflo 46" tempa, Macau casting ($6-8)(DT)

2. green & lime body, clear windows, black interior, Goodyear slicks wheels, "City Chevrolet" tempa, Macau casting ($6-8)(DT)

3. bright orange & blue body, clear windows, black interior, Goodyear slicks wheels, "Hardees 18" tempa, Macau casting ($6-8)(DT)

4. black body, clear windows, red interior, Goodyear slicks wheels, large "Exxon 51" (with signature) tempa, Macau casting ($6-8)(DT)

5. black body, clear windows, red interior, Goodyear slick wheels, small "Exxon 51" (without signature) tempa, Macau casting ($25-40)(ST)

6. black body, clear windows, gray interior, Goodyear slicks wheels, "Mello Yello 51" tempa, Macau casting ($6-8)(DT)

7. black body, clear windows, red interior, Goodyear slicks wheels, "Goodwrench 3/ GM" without "Western Steer" & plain trunk tempa, Macau casting ($12-15)(WR)

8. dark blue body, clear windows, gray interior, Goodyear slicks wheels, "Matchbox Motorsports 35" tempa, Macau casting ($2-3)

9. dark blue body, clear windows, gray interior, Goodyear slicks wheels "Matchbox Motorsports 35" tempa, China casting ($2-3)

10. black body, clear windows, red interior, Goodyear slicks wheels, "Goodwrench 3/ GM" without "Western Steer", with trunk tempa, China casting ($7-10)(WR)

11. white body, clear windows, black interior, Goodyear slicks wheels, "PG Tags" tempa, Macau casting ($85-110)(UK)(OP)

12. chrome plated body, clear windows, gray interior, Goodyear slicks wheels, no tempa, Macau casting ($12-18)(C2)

13. white & fluorescent yellow body, blue-chrome & black windows, no interior, lightning wheels, blue spatter & lightning bolts tempa, China casting ($7-10)(LT)(PS)

14. white & black body, chrome & black windows, no interior, lightning wheels, pink spatter & lightning bolts tempa, China casting ($7-10)(LT)(PS)

15. black body, clear windows, red interior, black disc w/rubber tires, "Goodwrench 3/GM" with "Western Steer" tempa, China casting ($5-7)(WR)(TC)

16. orange-yellow body, clear windows, black interior, black disc w/rubber tires, "Kodak Film 4 Racing" tempa, China casting ($5-7)(WR)(TC)
17. yellow body, clear windows, black interior, Goodyear slicks wheels, "MAC Tool Distributors 10" tempa, China casting ($15-20)(WR)
18. red & yellow body, clear windows, black interior, Goodyear slicks wheels, "Matchbox Racing 7" tempa, China casting ($2-3)(HS)
19. white & metallic blue body, clear windows, red interior, Goodyear slicks wheels, "Matchbox Racing 1" tempa, China casting ($2-3)(HS)
20. white body, clear windows, gray interior, Goodyear slicks wheels, "Team Goodyear 11" tempa, China casting ($2-3)(HS)
21. orange body, clear windows, black interior, Goodyear slicks wheels, "Team Goodyear 22" tempa, China casting ($2-3)(HS)
22. black body, clear windows, red interior, Goodyear slicks wheels, "Champion 4" tempa, China casting ($2-3)(HS)
23. yellow body, clear windows, gray interior, Goodyear slicks wheels, "Champion 3" tempa, China casting ($2-3)(HS)
24. orange-yellow body, clear windows, black interior, Goodyear slicks wheels, "Kodak Film 4 Racing" tempa, China casting ($3-5)(WR)
25. fluorescent orange & white body, clear windows, black interior, gray disc w/ rubber tires, "Purolator 10" tempa, China casting ($4-6)(WR)(TC)
26. dark pink & white body, clear windows, black interior, Goodyear slicks wheels, "Superflo 46" tempa, China casting ($3-5)(DT)
27. green & lime body, clear windows, black interior, Goodyear slicks wheels, "City Chevrolet 46" tempa, China casting ($3-5)(DT)
28. bright orange & blue body, clear windows, black interior, Goodyear slicks wheels, "Hardees 18" tempa, China casting ($3-5)(DT)
29. black body, clear windows, red interior, Goodyear slicks tempa, "Exxon 51" tempa, China casting ($3-5)(DT)
30. black body, clear windows, gray interior, Goodyear slicks tempa, "Mello Yello 51" tempa, China casting ($3-5)(DT)
31. black body, clear windows, red interior, gray disc w/rubber tires, "Goodwrench 3/ GM" with "Western Steer" tempa, China casting ($6-8)(WR)(TC)
32. black body, clear windows, red interior, gray disc w/rubber tires, "Goodwrench 3/ GM" without "Western Steer" tempa, China casting ($15-25)(WR)(Winross 3 piece)
33. matt black body, clear windows, red interior, gray disc w/rubber tires, "3" & "GM Parts" tempa, China casting ($15-25)(WR)(Winross 3 piece)
34. white & green body, clear windows, silver/gray interior, gold disc w/rubber tires, "Hendricks 25" tempa, China casting ($6-8)(WR)(TC)
35. fluorescent green body, clear windows, silver/gray interior, Goodyear slicks wheels, "Matchbox Motorsport 35" tempa, China casting ($1-2)
36. black body, clear windows, red interior, Goodyear slicks wheels, "Goodwrench 3/

GM/Mom N Pops" with "Western Steer" tempa, China casting ($6-8)(WR)(OP)
37. white & orange body, clear windows, orange interior, Goodyear slicks wheels, "Ferree Chevrolet 49" tempa, China casting ($3-5)(WR)
38. yellow & orange body, black /chrome windows, no interior, lightning wheels, red spatter & lightning bolts tempa, China casting ($3-4)(LT)(PS)
39. green & white body, black/chrome windows, no interior, lightning wheels, purple spatter & lightning bolts tempa, China casting ($3-4)(LT)(PS)
40. dark purple & white body, clear windows, silver/gray interior, Goodyear slicks wheels, "Phil Parsons Racing 29/ Matchbox" tempa, China casting ($12-15)(WR)
41. dark purple & white body, clear windows, silver/gray interior, Goodyear slicks wheels, "White Rose Collectibles 29/Matchbox" tempa, China casting ($3-5)(WR)
42. maroon body, clear windows, silver/gray interior, Goodyear slicks wheels, "Penrose 44/Fire Cracker Sausage/Big Mama" tempa, China casting ($6-8)(WR)
43. lime & black body, clear windows, silver/gray interior, Goodyear slicks wheels, "Interstate Batteries 18" tempa, China casting ($3-5)(WR)
44. metallic blue & white body, clear windows, red interior, Goodyear slicks wheels, "Raybestos 12" tempa China casting ($3-5)(WR)
45. metallic blue & white body, clear windows, red interior, Goodyear slicks wheels, "Raybestos 12" without "Tic Tac" logo, China casting ($75-100)(WR)

46. fluorescent orange & white body, clear windows, black interior, 9 spoke Goodyear slicks wheels, "Purolator 10" tempa, China casting ($3-5)(WR)
47. fluorescent orange & white body, clear windows, black interior, Goodyear slicks wheels, "Purolator 10" tempa, China casting ($3-5)(WR)
48. black body, clear windows, gray interior, Goodyear slicks wheels, "Stanley Tools 92" tempa, China casting ($3-5)(WR)
49. black body, clear windows, gray interior, 9 spoke Goodyear slicks wheels, "Stanley Tools 92" tempa, China casting ($3-5)(WR)
50. maroon body, clear windows, silver/gray interior, Goodyear slicks wheels, "Slim Jim 44" tempa ($6-8)(WR)
51. fluorescent green body, clear windows, silver/gray interior, Goodyear slicks wheels, "Matchbox Motorsports 35" without black grille tempa, China casting ($1-2)
52. yellow body, clear windows, silver/gray interior, Goodyear slicks wheels, "MAC Tools 7" tempa, China casting ($7-10)(WR)
53. lemon & white body, clear windows, gray interior, yellow disc w/rubber tires, "Texas Pete/Lozito's 87" tempa, Thailand casting ($4-6)(WR)
54. black body, clear windows, red interior, 9 spoke Goodyear slick wheels, "Goodwrench 3/GM" with "Western Steer" tempa, China casting ($3-5)(WR)

MB55-D FORD CORTINA, issued 1979 (continued)
18. red body, silver/gray base, white interior, clear windows, white & orange

flames tempa, dot dash wheels, doors cast, China casting ($150-200)(CH)

MB55-F RACING PORSCHE, issued 1983 (USA)
MB41-E RACING PORSCHE, issued 9183 (ROW)

NOTE: all models with black base & clear windows unless noted otherwise.

1. powder blue body, black interior, 8 dot silver wheels, "Elf 71 Sachs" tempa, Macau casting ($2-4)
2. baby blue body, black interior, 8 dot silver wheels, "Elf 71 Sachs" tempa, Macau casting ($2-4)
3. white body, black interior, 8 dot silver wheels, "Cadbury's Buttons" tempa, Macau casting ($8-12)(UK)
4. red body, black interior, starburst wheels, "Autotech 35" tempa, Macau casting ($3-5)(SF)
5. metallic red body, black interior, laser wheels, "Autotech 35" tempa, Macau casting ($3-5)(LW)
6. white body, black interior, 8 dot gold wheels, "Porsche 10" tempa, Macau casting ($3-5)
7. white body, black interior, 8 dot silver wheels, "Porsche 10" tempa, Macau casting ($2-4)
8. baby blue body, black interior, 8 dot gold wheels, "Elf 71 Sachs" tempa, Macau casting ($40-65)
9. black body, black interior, 8 dot silver wheels, "11 Ox Racing Team" tempa, Macau casting ($8-12)(HK)
10. red body, tan interior, 8 dot silver wheels, "41 Porsche" tempa, Macau casting ($1-2)
11. red body, tan interior, 8 dot silver

wheels, "41 Porsche" tempa, Thailand casting ($1-2)
12. red body, tan interior, 8 dot silver wheels, "Porsche" logo tempa, Thailand casting ($2-4)(MC)
13. lemon body, no interior, gray disc/w rubber tires, chrome windows, detailed trim tempa, China casting ($5-8)(WC)
14. white body, no interior, lightning wheels, blue-chrome & black windows, orange stripes with "935" tempa, China casting ($2-4)(LT)
15. black body, no interior, lightning wheels, black & chrome windows, lime stripes with "935" tempa, China casting ($2-4)(LT)
16. iridescent cream body, gray interior, gray disc w/rubber tires, chrome windows, detailed trip tempa, China casting ($5-8)(WC)
17. yellow body, black interior, 8 dot silver wheels, "Porsche 10" tempa, Thailand casting ($2-4)(MP)
18. powder blue body, black interior, dot dash wheels, "FAR Porsche 71 Sachs" tempa, silver/gray base, Manaus casting ($35-50)(BR)
19. powder blue body, black interior, 8 dot wheels, "FAR Porsche 71 Sachs" tempa, silver/gray base, Manaus casting ($35-50)(BR)

MB55-G MERCURY SABLE WAGON, issued 1987 (USA)
MB33-F MERCURY SABLE WAGON, issued 1989 (ROW)

1. white body, gray base, silver/gray interior, clear windows, 8 dot wheels, gray side stripe tempa, Macau casting ($2-4)

2. white body, gray base, silver/gray interior, clear windows, 8 dot wheels, gray side stripe tempa, China casting ($2-4)

MB56-C MERCEDES 450SL, issued 1979 (continued)
13. white & light green body, pearly silver base, tan interior, blue windows, black beacon, Macau casting ($2-4)

MB56-D PETERBILT TANKER, issued 1982 (USA)
MB 5-E PETERBILT TANKER, issued 1984 (ROW) (continued)

6. black body, yellow tank, amber windows, chrome exhausts, "Supergas" tempa, Macau casting ($2-4)
7. black body, orange-yellow tank, amber windows, chrome exhausts, "Supergas" tempa, Macau casting ($2-4)
8. white body, gray tank, amber windows, chrome exhausts, "Shell" tempa, Macau casting ($2-4)
9. red body, chrome tank, clear windows, chrome exhausts, "Getty" tempa, Macau casting ($2-4)
10. white body, yellow tank, amber windows, chrome exhausts, "Supergas" tempa, Macau casting ($25-40)
11. white body, gray tank, amber windows, chrome exhausts, "Ampol" tempa, Macau casting ($7-10)(AU)
12. white body, gray tank, clear windows, chrome exhausts, "Shell" tempa, Macau casting ($2-4)
13. black body, black tank, clear windows, chrome exhausts, "Amoco" tempa, Macau casting ($3-5)
14. black body, chrome tank, clear

windows, chrome exhausts, "Amoco" tempa, Macau casting ($60-75)
15. black body, white tank, clear windows, chrome exhausts, "Amoco" tempa, Macau casting ($35-50)
16. white body, white tank, clear windows, chrome exhausts, "Amoco" tempa, Macau casting ($2-4)
17. red body, chrome tank, clear windows, chrome exhausts, "Amoco" on tank & "Getty" on door tempa, Macau casting ($40-65)
18. white body, white tank, clear windows, chrome exhausts, no tempa, Macau casting ($40-55)
19. blue body, white tank, clear windows, chrome exhausts, no tempa, England casting ($3-5)
20. olive body, olive tank, clear windows, black exhausts & base, "Gas" tempa, Macau casting ($4-5)(CM)
21. white body, chrome tank, clear windows, gray exhausts, "Shell" tempa, Macau casting ($1-2)
22. lime body, red tank, clear windows, red exhausts, blue base, gas pump & stripes tempa, red wheels with yellow hubs, Macau casting ($6-8)(LL)
23. lime body, red tank, clear windows, red exhausts, blue base, gas pump & stripes tempa, red wheels with yellow hubs, Thailand casting ($6-8)(LL)
24. lime body, red tank, clear windows, red exhausts, blue base, gas pump & stripes tempa, red wheels (plain hubs), Thailand casting ($6-8)(LL)
25. white body, chrome tank, clear windows, gray exhausts, "Shell" tempa, Thailand casting ($1-2)
26. white body, white tank, blue windows,

gray exhausts, no tempa, Thailand casting ($10-15)(GF)

27. black body, black tank, clear windows, chrome exhausts, "Indy Racing Fuel" labels ($10-15)(IN)

28. black body, black tank, clear windows, gray exhausts & base, "Matchbox" & "Getty" tempa, Manaus casting, gold hubs ($35-50)(BR)

29. black body, black tank, clear windows, gray exhausts & base, "Matchbox" & "Getty" tempa, Manaus casting, silver hubs ($35-50)(BR)

30. white body, chrome tank, clear windows, gray exhausts, "Shell" without yellow on cab tempa, Thailand casting ($1-2)

31. white body, chrome tank, clear windows, gray exhausts, black base, "Shell" with "IC" logo on door tempa, China casting ($10-15)(IC)

MB57-D 4X4 MINI PICK-UP, issued 1982 (USA) (continued)

3. dark powder blue body, black metal base, blue windows, "Mountain Man" tempa, Hong Kong casting ($3-5)

4. light powder blue body, black metal base, blue windows, "Mountain Man" tempa, Macau casting ($2-3)

5. baby blue body, black metal base, blue windows, "Mountain Man" tempa, Macau casting ($2-3)

6. baby blue body, black plastic base, blue windows, "Mountain Man" tempa, Macau casting ($1-2)

7. baby blue body, black plastic base, blue windows, "Mountain Man" tempa, Thailand casting ($1-2)

MB57-E CARMICHAEL COMMANDO, issued 1982 (ROW) (continued)

3. red body, black base, gray interior, "Fire" tempa, England casting ($15-18)

4. red body, charcoal base, gray interior, "Fire" tempa, England casting ($15-18)

5. red body, black base, gray interior, "Fire" tempa, Macau casting ($15-18)

MB57-H MACK AUXILIARY POWDER TRUCK, issued 1991 (USA)
MB50-G MACK AUXILIARY POWDER TRUCK, issued 1991 (ROW)

NOTE: all models with chrome base, red windows, white roof with "Floodlight Heavy Rescue" tempa, 8 spoke wheels.

1. yellow body, chrome & yellow roof fixtures, small door shield body tempa, China casting ($2-3)

2. fluorescent orange body, chrome & yellow roof fixtures, "Fire Rescue Unit 2" & checkers body tempa, China casting ($1-2)

3. white body, chrome & white roof fixtures, no roof or body tempa, China casting ($10-15)(GF)

4. red body, chrome & yellow roof fixtures, "Fire Rescue Unit 2" & checkers body tempa, China casting ($8-12)(GS)

MB58-D RUFF TREK, issued 1983

NOTE: all models with black base, black rear load & 8 spoke wheels unless noted otherwise.

1. tar body, red interior, amber windows, "Ruff Trek" tempa, Macau casting ($2-4)

2. white body, red interior, amber windows, "Ruff Trek" tempa, Macau casting ($2-4)

3. white body, red interior, amber windows, "217" tempa, Macau casting ($3-4)(TM)

4. white body, red interior, clear windows, "217" tempa, Macau casting ($3-4)(TM)

5. white body, black interior, clear windows, "Brut/Faberge" tempa, Macau casting ($3-4)(TM)

6. white body, red interior, clear windows, "Brut/Faberge" tempa, Macau casting ($40-60)(TM)

7. dark blue body, red interior, clear windows, "STP/Goodyear" tempa, Macau casting ($45-60)(TM)

8. white body, red interior, clear windows, "7 Up" tempa, Macau casting ($3-4)(TM)

9. brown body, red interior, olive green windows, red/yellow/blue tempa, green rear load, Macau casting, includes plastic armament ($5-8)(RB)

10. white body, black interior, clear windows, flames tempa, Macau casting ($8-12)(GS)(JB)

11. yellow body, black interior, clear windows, "Matchbox Rescue Team Support" tempa, Macau casting ($3-5)(MC)

12. yellow body, black interior, clear windows, "Matchbox Rescue Team Support" tempa, Thailand casting ($3-5)(MC)

13. brown body, red interior, dark olive windows, red/yellow/blue tempa, dark green load, Thailand casting, includes plastic armament ($8-12)(Tomy Box)(JP)

MB58-E MERCEDES BENZ 300SE, issued 1986

NOTE: all models with black base, clear windows & 8 dot wheels.

1. silver blue body, dark blue interior, no tempa, Macau casting ($1-2)

2. silver blue body, dark blue interior, no tempa, Thailand casting ($1-2)

3. white body, tank interior, green stripe & silver star with "Polizei 5075" tempa, Thailand casting ($1-2)

4. white body, tan interior, green stripe (no star) with "Polizei 5075" tempa, Thailand casting ($1-2)

MB59-B MERCURY FIRE CHIEF, issued 1971 (continued)

39. white body, pearly silver base, blue windows, bar lights, gray interior, starburst wheels, "State Police" tempa, Macau casting ($3-5)(SF)

40. white body, black base, blue windows, bar lights, gray interior, starburst wheels, "State Police" tempa, Macau casting ($7-10)(SF)

41. white body, pearly silver base, blue windows, bar lights, gray interior, laser wheels, "State Police" tempa, Macau casting ($3-5)(LW)

42. black body, black base, blue windows, bar lights, gray interior, starburst wheels, "Halley's Comet" tempa, Macau casting ($7-10)(MP)

43. metallic blue body, pearly silver base, blue windows, bar lights, gray interior, starburst wheels, yellow/blue/red tempa, Macau casting, includes plastic armament ($5-8)(RB)

MB59-D PORSCHE 928, issued 1980 (continued)

27. black body, pearly silver base, red interior, clear windows, "Porsche" & stripes tempa, 5 arch wheels, Macau casting ($2-4)

28. black body, pearly silver base, red interior, light amber windows, "Porsche"

& stripes tempa, 5 arch wheels, Macau casting ($2-4)

29. light metallic gray body & base, red interior, clear windows, purple/blue tempa, 5 arch wheels, Macau casting ($3-5)

30. pearly silver body & base, red interior, clear windows, "Martini Racing Porsche" tempa, 5 arch wheels, Macau casting ($7-10)(JP)

31. white & metallic blue body, white base, red interior, clear windows, "28" with "Cale Jenkins" tempa, starburst wheels, Macau casting ($3-5)(SF)

32. white & metallic blue body, white base, red interior, clear windows, "28" without "Cale Jenkins" tempa, starburst wheels, Macau casting ($3-5)(SF)

33. white & metallic blue body, white base, red interior, clear windows, "28" with "Cale Jenkins" tempa, laser wheels, Macau casting ($3-5)(LW)

34. white & metallic blue body, white base, red interior, clear windows, "28" without "Cale Jenkins" tempa, laser wheels, Macau casting ($3-5)(LW)

35. orange body, pearly silver base, red interior, clear windows, "Lufthansa" tempa, Macau casting ($4-6)(GS)

36. charcoal gray body & base, no interior, chrome windows, detailed trim tempa, gray disc w/rubber tires, doors cast shut, Macau casting ($5-8)(WC)

37. charcoal gray body & base, no interior, chrome windows, detailed trim tempa, gray disc w/rubber tires, doors cast shut, Thailand casting ($5-8)(WC)

MB59-F T-BIRD TURBO COUPE, issued 1987 (USA)
MB61-E T-BIRD TURBO COUPE, issued 1987 (ROW)
MB28-J T-BIRD TURBO COUPE, reissued 1992 (ROW)

1. plum body & base, clear windows, silver/gray interior, 8 dot wheels, "Turbo Coupe" tempa, Macau casting ($1-2)

2. metallic gold body & base, clear windows, silver/gray interior, laser wheels, "56 Motorcraft" tempa, Macau casting ($3-5)(LW)

3. light pear green body & base, clear windows, silver/gray interior, 8 dot wheels, "Turbo Coupe" tempa, Macau casting ($3-4)(SC)

4. dark brown body & base, clear windows, silver/gray interior, 8 dot wheels, "Turbo Coupe" tempa, Macau casting ($3-4)(SC)

5. pink body & base, clear windows, silver/gray interior, 8 dot wheels, "Turbo Coupe" tempa, Macau casting ($3-4)(SC)

6. dark purple body & base, clear windows, silver/gray interior, 8 dot wheels, "Turbo Coupe" tempa, Macau casting ($3-4)(SC)

7. silver/gray body & base, chrome windows, red interior, gray disc w/rubber tires, no tempa, Macau casting ($5-8)(WC)

8. silver/gray body & base, chrome windows, black interior, gray disc w/ rubber tires, no tempa, Macau casting ($55-70)(WC)

9. plum body & base, clear windows, silver/gray interior, 8 dot wheels, "Turbo Coupe" tempa, Thailand casting ($1-2)

10. bright plum body & base, clear windows, silver/gray interior, 8 dot

wheels, "Turbo Coupe" tempa, Thailand casting ($1-2)

11. metallic blue body, fluorescent pink base, clear windows, silver/gray interior, 8 dot wheels, purple/pink stripes tempa, Thailand casting ($1-2)

MB60-D PISTON POPPER, issued 1982 (continued)

7. orange body, red interior, unpainted base, "Sunkist" tempa, Hong Kong casting ($4-6)

8. orange body, red interior, black base, "Sunkist" tempa, Macau casting ($)

MB60-F PONTIAC FIREBIRD RACER, issued 1985 (USA)
MB12-F PONTIAC FIREBIRD RACER, issued 1985 (ROW)

NOTE: all models with blue base & clear windows unless noted.

1. yellow body, metal base, red interior, 8 dot silver wheels, "Son of A Gun 55" tempa, Macau casting ($4-6)

2. yellow body, metal base, red interior, 8 dot silver wheels, "Pirelli 56" tempa, Macau casting ($4-6)

3. yellow body, metal base, red interior, 8 dot gold wheels, "Pirelli 56" tempa, Macau casting ($4-6)

4. yellow body, metal base, red interior, 8 dot gold wheels, "Son of A Gun 55" tempa, Macau casting ($4-6)

5. yellow body, metal base, red interior, dot dash wheels, "Son of A Gun 55" tempa, Macau casting ($4-6)

6. yellow body, metal base, red interior, dot dash wheels, "Pirelli 56" tempa, Macau casting ($4-6)

7. yellow body, metal base, red interior, 8 dot wheels, "Pirelli 56" (plain sides) tempa, Macau casting ($4-6)

8. powder blue body, metal base, gray interior, starburst wheels, "10" with blue/yellow tempa, Macau casting ($3-5)(SF)

9. metallic blue body, metal base, gray interior, laser wheels, "10" with blue/yellow tempa, Macau casting ($3-5)(LW)

10. white body, metal base, red interior, 5 arch wheels, "Fast Eddies 15" tempa, Macau casting ($2-4)

11. white body, metal base, red interior, 8 dot silver wheels, "Fast Eddies 15" tempa, Macau casting ($2-4)

12. white body, metal base, red interior, 8 dot silver wheels, "6 Horse Racing Team" tempa, Macau casting ($8-12)(HK)

13. dark brown body, metal base, red interior, 8 dot silver wheels, "Fast Eddies 15" tempa, Macau casting ($7-10)(SC)

14. light pea green body, metal base, red interior, 8 dot silver wheels, "Fast Eddies 15" tempa, Macau casting ($7-10)(SC)

15. white body, plastic base, red interior, 8 dot silver wheels, "Fast Eddies 15" tempa, Macau casting ($3-5)

16. powder blue body, plastic base, gray interior, starburst wheels, "10" with blue/yellow tempa, Macau casting ($7-10)(SF)

17. yellow body, red plastic base, red interior, starburst wheels, "10" with red/white stripes, Macau casting ($15-20)(SF)

18. metallic blue body, plastic base, gray interior, laser wheels, "10" with yellow/blue stripes, Macau casting ($7-10)(LW)

19. white body, plastic base, red interior, starburst wheels, "Fast Eddies 15" tempa, China casting ($6-8)(GS)

MB60-G FORD TRANSIT, issued 1986 (ROW)
MB57-G FORD TRANSIT, issued 1990 (USA)
NOTE: all models with dark gray base, gray interior, 8 dot wheels unless noted otherwise.
1. red body, left hand drive, "Motorsport" tempa, Macau casting ($2-4)
2. red body, right hand drive, no tempa, Macau casting ($30-45)(UK)
3. red body, left hand drive, no tempa, Macau casting ($2-4)
4. red body, left hand drive, "Motorsport" tempa, dot dash wheels, Macau casting ($3-5)
5. white body, left hand drive, red cross & stripe with gray rectangle tempa, Macau casting ($2-3)
6. white body, left hand drive, "Federal Express" tempa, Macau casting ($1-2)
7. yellow body, left hand drive, "British Telecom" tempa, Macau casting ($3-5)(GS)
8. red body, left hand drive, "Australia Post" tempa, China casting ($6-8)(AU)
9. white body, left hand drive, red cross & stripe with gray rectangle tempa, China casting ($2-4)
10. white body, left hand drive "Unichem" tempa, Macau casting ($7-10)(UK)
11. white body, right hand drive, "JCB Job Site" tempa, Macau casting ($6-8)(GS)
12. white body, left hand drive, "Federal Express" tempa, China casting ($1-2)
13. white body, right hand drive, "XP Express Parcels" tempa, China casting ($6-8)(UK)
14. red body, right hand drive-black interior, red base, "Royal Mail" tempa, Macau

casting ($5-8)(UK)(MP)
15. orange-red body, right hand drive, "Australia Post-We Deliver" tempa, China casting ($6-8)(AU)
16. white body, left hand drive, "Wigwam" tempa, China casting ($7-10)(GR)
17. lime green body, left hand drive, no tempa, China casting ($12-15)(DU)
18. white body, right hand drive, "Ormond St. Appeal" tempa, China casting ($6-8)(UK)
19. white body, left hand drive, "Wella" tempa, China casting ($6-8)(GR)
20. white body, right hand drive, "Australia Telecom" tempa, China casting ($6-8)(AU)
21. white body, left hand drive, "XP Express Parcels" tempa, China casting ($4-6)(UK)
22. white body, right hand drive, "Federal Express" tempa, China casting ($3-5)
23. white body, left hand drive, "Peter Cox Preservation" tempa, China casting ($7-10)(UK)
24. white body, left hand drive, "Kiosk" tempa, China casting ($8-12)(SW)
25. red body, left hand drive, "Blick" tempa, China casting ($8-12)(SW)
26. orange body, left hand drive, "Ovomaltine" tempa, China casting ($8-12)(SW)
27. silver/gray body, left hand drive, "Isotar/ Perform/Powerplay" tempa, China casting ($8-12)(SW)
28. white body, left hand drive, "Kelloggs" tempa, China casting ($60-75)(GR)(AS)(OP)
29. white body, right hand drive, "Kelloggs" tempa, China casting ($60-75)(GR)(AS)(OP)

30. yellow body, left hand drive, "Ryder" tempa, China casting ($1-2)
31. white body, left hand drive, "DCS" tempa, China casting ($85-110)(SC)
32. white body, left hand drive, white interior & base, amber windows, no tempa, China casting ($10-15)(GF)
33. white body, left hand drive, "Supertoys" tempa, China casting ($20-25)(IR)
34. metallic green body, left hand drive, "Taronga Zoomobile" tempa, China casting ($6-8)(AU)
35. white body, left hand drive, "McKesson" tempa, China casting ($50-75)(US)
36. yellow body, left hand drive, "Cadbury's Flakes" tempa, China casting ($1-2)
37. white body, left hand drive, "Garden Festival Wales" tempa, China casting ($5-8)(WL)

MB61-C PETERBILT WRECKER, issued 1982 (continued)
5. white body, black booms, amber windows, chrome exhausts & base, black "9" tempa, Macau casting ($2-3)
6. white body, black booms, clear windows, chrome exhausts & base, black "9" tempa, Macau casting ($2-3)
7. white body, black booms, amber windows, chrome exhausts & base, blue "9" tempa, Macau casting ($1-2)
8. white body, blue booms, amber windows, chrome exhausts & base, blue "9" tempa, Macau casting ($1-2)
9. white body, blue booms, clear windows, chrome exhausts & base, blue "911" tempa, Macau casting ($1-2)
10. white body, orange booms, clear windows, chrome exhausts & base,

"SFPD" & star tempa, Macau casting ($1-2)
11. white body, dull orange booms, clear windows, chrome exhausts & base, "SFPD" & star tempa, Macau casting ($1-2)
12. orange body, dark green booms, dark green windows, gray exhausts & base, black stripes tempa, Macau casting, includes plastic armament ($5-8)(RB)
13. olive body, black booms, clear windows, black exhausts & base, "8" with red/white stripes tempa, Macau casting ($4-5)(CM)
14. white body, orange booms, clear windows, gray exhausts, chrome base, "SFPD" & star tempa, Macau casting ($1-2)
15. white body, orange booms, clear windows, gray exhausts, chrome base, "SFPD" & star tempa, Thailand casting ($1-2)
16. orange body, dark green booms, dark green windows, gray exhausts & base, black stripes tempa, Thailand casting ($8-12)(Tomy box)(JP)
17. red body, black booms, clear windows, gray exhausts & base, "Police" tempa, Manaus casting ($35-50)(BR)
18. white body, black booms, clear windows, gray exhausts, chrome base, black "9" tempa, Manaus casting ($35-50)(BR)

MB61-E NISSAN 300ZX, issued 1990 (USA)
MB37-H NISSAN 300ZX, issued 1990 (ROW)
1. yellow body, smoke gray windows, white interior, 8 dot wheels, outlined "300ZX" tempa, Macau casting ($1-2)

2. chrome plated body, smoke gray windows, white interior, 8 dot wheels, no tempa, Macau casting ($12-18)(C2)

3. chrome plated body, smoke gray windows, white interior, 8 dot wheels, no tempa, Thailand casting ($12-18)(C2)

4. yellow body, smoke gray windows, white interior, 8 dot wheels, outlined "300ZX" tempa, Thailand casting ($1-2)

5. iridescent cream body, chrome windows, pink interior, gray disc w/rubber tires, detailed trim tempa, China casting ($5-8)(WC)

6. iridescent cream body, chrome windows, dark gray interior, gray disc w/rubber tires, detailed trim tempa, China casting ($5-8)(WC)

7. iridescent cream body, chrome windows, white interior, gray disc w/rubber tires, detailed trim tempa, China casting ($5-8)(WC)

8. iridescent pink-cream body, blue-chrome & black windows, no interior, peach/silver lightning wheels, "Turbo Z" tempa, China casting ($2-3)(LT)

9. iridescent pink-cream body, blue-chrome & black windows, no interior, pink/silver lightning wheels, "Turbo Z" tempa, China casting ($2-3)(LT)

10. metallic blue body, chrome & black windows, no interior, lightning wheels, "300ZX" tempa, China casting ($2-3)(LT)

11. bright orange body, blue-chrome & black windows, no interior, lightning wheels, "300ZX" tempa, China casting ($2-3)(LT)

12. lemon body, black & chrome windows, no interior, lightning wheels, "Turbo Z" tempa, China casting ($2-3)(LT)

13. yellow body, clear windows, white interior, 8 dot wheels, outlined "300ZX" tempa, China casting ($1-2)

14. yellow body, clear windows, white interior, 8 dot wheels, solid "300ZX" tempa, China casting ($1-2)

15. metallic red body, chrome windows, dark gray interior, gray disc w/rubber tires, detailed trim tempa, China casting ($5-8)(WC)

16. metallic red body, black & chrome windows, no interior, silver/yellow lightning wheels, "Turbo Z" tempa, China casting ($5-8)(LT)(OP)(US)

17. metallic red body, black & chrome windows, no interior, pink/red lightning wheels, "Turbo Z" tempa, China casting ($5-8)(LT)(OP)(US)

18. white/lime/pink body, black & chrome windows, no interior, lightning wheels, "Z" & "Turbo" tempa, China casting ($2-3)(LT)

19. white/orange/black body, black & chrome windows, no interior, lightning wheels, "Z" & "Turbo" tempa, China casting ($2-3)(LT)

MB62-D CHEVROLET CORVETTE, issued 1979 (continued)

10. black body, red base, gray interior, clear windows, yellow/orange stripes tempa, 5 arch front & 5 crown rear wheels, Macau casting ($2-4)

11. black body, silver/gray base, gray interior, clear windows, "The Force" tempa, 5 arch front & 5 crown rear wheels, Macau casting ($2-4)

12. black body, pearly silver base, gray interior, clear windows, "The Force" tempa, 5 arch front & 5 crown rear wheels, Macau casting ($2-4)

13. green body, pearly silver base, gray interior, clear windows, "Brut/Faberge" tempa, 5 arch front & 5 crown rear wheels, Macau casting ($6-8)(TM)

14. black body, pearly silver base, red interior, clear windows, starburst wheels, "Turbo Vette" tempa, Macau casting ($3-5)(SF)

15. black body, pearly silver base, red interior, clear windows, laser wheels, "Turbo Vette" tempa, Macau casting ($3-5)(LW)

16. metallic red body, pearly silver base, red interior, clear windows, laser wheels, "Turbo Vette" tempa, Macau casting ($7-10)(LW)

17. black body, red plastic base, red interior, clear windows, 8 spoke wheels, "The Force" tempa, Manaus casting ($35-50)(BR)

NOTE: the following models with plastic base, clear windows, gray disc with rubber tires and China casting.

18. blue body, white base, dark blue interior, "Dodgers 1992" tempa ($3-5)(WR)

19. blue body, pumpkin orange base, dark blue interior, "Royals 1992" tempa ($3-5)(WR)

20. blue body, dark blue base, black interior, "Mariners 1992" tempa ($3-5)(WR)

21. blue body, dark silver/gray base, gray interior, "Rockies 1992" tempa ($3-5)(WR)

22. blue body, bright orange base, black interior, "Mets 1992" tempa ($3-5)(WR)

23. dark blue body, white base, red interior, "Red Sox 1992" tempa ($3-5)(WR)

24. dark blue body, gray base, black interior, "Twins 1992" tempa ($3-5)(WR)

25. metallic blue body, red base, black interior, "Braves 1992" tempa ($3-5)(WR)

26. bright blue body, dark blue base, black interior, "Blue Jays 1992" tempa ($3-5)(WR)

27. powder blue body, white base, dark blue interior, "Expos 1992" tempa ($3-5)(WR)

28. green body, dark yellow base, dark yellow interior, "Athletics 1992" tempa ($3-5)(WR)

29. yellow body, red base, red interior, "Angels 1992" tempa ($3-5)(WR)

30. orange body, silver/gray base, dark blue interior, "Tigers 1992" tempa ($3-5)(WR)

31. gray body, black base, red interior, "Giants 1992" tempa ($3-5)(WR)

32. white body, yellow base, dark blue interior, "Brewers 1992" tempa ($3-5)(WR)

33. white body, dark blue base & interior, "Indians 1992" tempa ($3-5)(WR)

34. white body, dark blue base & interior, "Padres 1992" tempa ($3-5)(WR)

35. white body, blue base, orange interior, "Astros 1992" tempa ($3-5)(WR)

36. white body, turquoise base, orange interior, "Marlins 1992" tempa ($3-5)(WR)

37. white body, red base, red interior, "Yankees 1992" tempa ($3-5)(WR)

38. red body, white base, black interior, "Reds 1992" tempa ($3-5)(WR)

39. red body, white base, dark blue interior, "Phillies 1992" tempa ($3-5)(WR)

40. red body, white base, dark blue interior, "Cubs 1992" tempa ($3-5)(WR)

41. red body, white base, red interior, "Cardinals 1992" tempa ($3-5)(WR)

42. red body, blue base, dark blue interior,

"Rangers 1992" tempa ($3-5)(WR)
43. black body, white base, black interior, "Pirates 1992" tempa ($3-5)(WR)
44. black body, white base, orange interior, "Orioles 1992" tempa ($3-5)(WR)
45. black body, silver/gray base, red interior, "White Sox 1992" tempa ($3-5)(WR)
46. white body, red base, red interior, "Cooperstown 1993" tempa ($3-5)(WR)

MB62-F ROLLS ROYCE SILVER CLOUD, issued 1986 (USA)
MB31-F ROLLS ROYCE SILVER CLOUD, issued 1986 (ROW)
1. silver/gray body, chrome base, dark gray interior, clear windows, dot dash wheels, England casting ($8-10)(JB)
2. cream body, chrome base, dark gray interior, clear windows, dot dash wheels, England casting ($1-2)
3. cream body, chrome base, dark gray interior, clear windows, dot dash wheels, Macau casting ($1-2)
4. green-gold body, chrome base, black interior, chrome windows, gray disc w/ rubber tires, Macau casting ($5-8)(WC)
5. green-gold body, chrome base, black interior, chrome windows, gray disc w/ rubber tires, Thailand casting ($5-8)(WC)
6. cream body, chrome base, dark gray interior, clear windows, dot dash wheels, Thailand casting ($1-2)

MB62-G VOLVO 760, issued 1986 (ROW)
NOTE: all models with black base and 8 dot wheels unless noted.
1. pearly silver body, black interior, clear windows, Macau casting ($2-3)
2. dark silver/gray body, black interior, clear

windows, China casting ($2-3)
3. plum body, black interior, clear windows, China casting ($2-3)
4. dark plum body, black interior, clear windows, China casting ($2-3)
5. white body & base, white interior, amber windows, China casting ($10-15)(GF)

MB62-H OLDSMOBILE AEROTECH, issued 1989 (USA)
MB64-F OLDSMOBILE AEROTECH, issued 1989 (ROW)
NOTE: all models with gray plastic base, smoke windows & 8 dot wheels.
1. silver/gray body, "Quad 4/Aerotech/ Oldsmobile" tempa, Macau casting ($1-2)
2. silver/gray body, "Quad 4/Aerotech/ Oldsmobile" tempa, Thailand casting ($1-2)
3. fluorescent orange body, "Aerotech" tempa, Thailand casting ($1-2)

MB63-E SNORKEL, issued 1982 (USA)
MB13-E SNORKEL, issued 1982 (ROW) (continued)
3. red body, black base, black metal base, insert, "Metro Fire" tempa, Macau casting ($1-2)
4. red body, black base, black plastic base insert, "Metro Fire" tempa, Macau casting ($1-2)
5. red body, black base, black plastic base insert, "Metro Fire" tempa, China casting ($1-2)
6. dull red body, black base, black plastic base insert, "Fire Dept." & shield tempa, China casting ($1-2)
7. red body, black base, black plastic base insert, Japanese lettered tempa, China casting ($8-12)(JP)

8. greenish-yellow body, black base, black plastic base insert, "Fire Dept." & shield tempa, China casting ($3-4)(AP)
9. fluorescent orange body, black base, black plastic base insert, "Rescue Unit 1 Fire" & checkers tempa, China casting ($1-2)
10. white body, white base, white plastic base insert, no tempa, China casting ($10-15)(GF)
11. red body, black base, black plastic insert, "Rescue Unit 1 Fire" & checkers tempa, China casting ($8-12)(GS)
12. fluorescent orange body, black base, black plastic base insert, " Rescue Unit 1 Fire", checkers & "IC" logo tempa, China casting ($10-15)(IC)

MB64-D CATERPILLAR BULLDOZER, issued 1979
MB9-G CATERPILLAR BULLDOZER, reissued 1986 (ROW) (continued)
15. yellow body, yellow blade, pearly silver metal base, lemon rollers, black canopy, "C CAT" tempa, Macau casting ($1-2)
16. yellow body, yellow blade, pearly silver metal base, lemon rollers, black canopy, "C CAT" tempa, China casting ($1-2)
17. yellow body, yellow blade, pearly silver metal base, lemon rollers, black canopy, no tempa, Macau casting ($1-2)
18. yellow body, gray blade, pearly silver metal base, lemon rollers, black canopy, no tempa, Macau casting ($2-3)(MP)
19. red body, yellow blade, blue metal base, blue rollers, yellow tread, blue canopy, yellow stripes tempa, Macau casting ($6-8)(LL)
20. orange body, orange blade, pearly silver metal base, orange rollers, black canopy,

"Losinger" tempa, Macau casting ($8-12)(SW)
21. yellow body, yellow blade, light gray plastic base, yellow rollers, black canopy, no tempa, Thailand casting ($1-2)
22. yellow body, yellow blade, dark gray plastic base, yellow rollers, black canopy, no tempa, Thailand casting ($1-2)
23. yellow body, dark gray blade, light gray plastic base, yellow rollers, black canopy, no tempa, Thailand casting ($1-2)
24. red body, lime blade, yellow plastic base, lemon rollers, yellow treads, blue canopy, gravel design tempa, Thailand casting ($6-8)(LL)
25. orange-yellow body, yellow blade, light gray plastic base, yellow rollers, black canopy, no tempa, Thailand casting ($1-2)
26. orange-yellow body, yellow blade, light gray plastic base, yellow rollers, red canopy, red stripes (blade) tempa, Thailand casting ($1-2)
27. orange-yellow body, yellow blade, silver/gray plastic base, yellow rollers, red canopy, red stripes (blade) tempa, Thailand casting ($1-2)

MB65-B AIRPORT COACH, issued 1977 (continued)
27. white body & roof, "Alitalia" tempa, amber windows, light yellow interior, pearly silver base, Macau casting ($3-5)
28. orange body, white roof, "Lufthansa" tempa, amber windows, light yellow interior, pearly silver base, Macau casting ($3-5)(GS)
29. orange body, white roof, "Pan Am" tempa, amber windows, light yellow interior, pearly silver base, Macau casting

($3-5)(GS)

30. white body & roof, "Stork SB" tempa, amber windows, light yellow interior, pearly silver base, Macau casting ($8-12)(AU)(OP)

31. red body, white roof, "Virgin Atlantic" tempa, amber windows, light yellow interior, pearly silver base, Macau casting ($3-5)(GS)

32. blue body, white roof, "Australian" tempa, amber windows, light yellow interior, pearly silver base, Macau casting ($7-10)(AU)

33. blue body, white roof, "Girobank" labels, blue windows, light yellow interior, pearly silver base, Macau casting ($7-10)(UK)

34. white body, green roof, "Alitalia" tempa, amber windows, white interior, white base, Thailand casting ($5-7)(MP)

35. white body, blue roof, "KLM" tempa, amber windows, white interior, white base, Thailand casting ($5-7)(MP)

36. white body, blue roof, "SAS" tempa, amber windows, white interior, white base, Thailand casting ($5-7)(MP)

37. white body & roof, "Lufthansa" tempa, amber windows, white interior, pearly silver base, Thailand casting ($5-7)(MP)

38. red body, white roof, "Virgin Atlantic" tempa, amber windows, white interior, pearly silver base, Thailand casting ($3-5)

39. red body, white roof, "TWA" tempa, amber windows, white interior, white plastic base, Manaus casting ($35-50)(BR)

MB65-C BANDAG BANDIT, issued 1982 (USA) (continued)

7. black body, yellow & white stripes tempa, "Tyrone" & no stripes on airfoil, 5 crown front & dot dash rear wheels, Macau casting ($1-2)

8. black body, yellow & white stripes tempa, "Tyrone" & 4 stripes on airfoil, 5 crown front & dot dash rear wheels, Macau casting ($1-2)

9. black body, yellow & white stripes tempa, plain airfoil, maltese cross front & dot dash rear wheels, China casting ($4-6)

MB65-D F.1 RACER, issued 1985 (USA)

NOTE: all models with 8 dot front wheels, racing special rear wheels unless noted otherwise.

1. dark blue body, black metal base, red driver, red "Goodyear" airfoil, 5 arch front wheels, chromed lettered rear wheels, chrome exhausts, "20 Bosch STP" tempa, Macau casting ($4-6)

2. dark blue body, black metal base, red driver, red "Goodyear" airfoil, 5 arch front wheels, unchromed rear wheels, chrome exhausts, "20 Bosch STP" tempa, Macau casting ($4-6)

3. yellow body, black metal base, red driver, red "Goodyear" airfoil, chrome exhausts, "Matchbox Racing Team" tempa, Macau casting ($2-3)

4. yellow body, black metal base, dark red driver, dark red "Goodyear" airfoil, chrome exhausts, "Matchbox Racing Team" tempa, Macau casting ($2-3)

5. pink body, black metal base, red driver red "Goodyear" airfoil, chrome exhausts, "Matchbox Racing Team" tempa, Macau casting ($3-4)(SC)

6. light pear green body, black metal base, red driver, red "Goodyear" airfoil, chrome exhausts, "Matchbox Racing Team" tempa, Macau casting ($3-4)(SC)

7. light peach body, black metal base, red driver, red "Goodyear" airfoil, chrome exhausts, "Matchbox Racing Team" tempa, Macau casting ($3-4)(SC)

8. red body, black plastic base, red driver, red "Goodyear" airfoil, black exhausts, "Matchbox Racing Team" tempa, Macau casting ($3-4)(SC)

9. orange body, black plastic base, red driver, red "Goodyear" airfoil, black exhausts, "Matchbox Racing Team" tempa, Macau casting ($3-4)(SC)

10. yellow body, black plastic base, red driver, red "Goodyear" airfoil black exhausts, "Matchbox Racing Team" tempa, Macau casting ($1-2)

11. white body, red plastic base, red driver, red plain airfoil, blue wheels with yellow front hubs, lime exhausts, "123456" & flames tempa, Macau casting ($6-8)(LL)

12. white body, red plastic base, red driver, red plain airfoil, blue wheels with yellow front hubs, lime exhausts, "123456" & flames tempa, Thailand casting ($6-8)(LL)

13. yellow body, black plastic base, red driver, red "Goodyear" airfoil, black exhausts, "Matchbox Racing Team" tempa, Thailand casting ($1-2)

14. white/hot pink/blue body, black plastic base, pink driver, pink "Rain-X" airfoil, black exhausts, "Amway/Speedway 22" tempa, China casting ($3-5)(IN)

15. dark blue/white body, black plastic base, dark blue driver, dark blue "Valvoline" airfoil, black exhausts, "Valvoline 5"

tempa, China casting ($3-5)(IN)

16. orange-yellow/blue body, black plastic base, yellow driver, yellow "Kraco" airfoil, black exhausts, "Kraco/Otter Pops 18" tempa, China casting ($3-5)(IN)

17. lemon/black body, black plastic base, lemon driver, lemon "Goodyear" airfoil, black exhausts, "Indy 11" tempa, China casting ($3-5)(IN)

18. lemon/black body, black plastic base, lemon driver, lemon "Indy" airfoil, black exhausts, Indy 11" tempa, China casting ($4-6)(IN)

19. black body, black plastic base, black driver, black "Havoline" airfoil, black exhausts, "Havoline 86" tempa, China casting ($3-5)(IN)

20. chrome plated body, black plastic base, red driver, red "Goodyear" airfoil, black exhausts, no tempa, Thailand casting ($12-18)(C2)

21. yellow body, black plastic base, red driver, red "Goodyear" airfoil, black exhausts, "Matchbox Racing Team" tempa, China casting ($1-2)

22. orange-yellow/blue body, black plastic base, orange-yellow driver, orange-yellow "Kraco" airfoil, black exhausts, "Kraco 18" tempa, Thailand casting ($3-5)(IN)

23. yellow body, black plastic base, maroon driver, maroon "Goodyear" airfoil, black exhausts, "Matchbox Racing Team" tempa, Thailand casting ($1-2)

24. white/hot pink/blue body, black plastic base, pink driver, pink "Rain-X" airfoil, black exhausts, "Amway/Speedway 22" tempa, Thailand casting ($3-5)(IN)

25. blue/white body, black plastic base, blue driver, blue "Mitre 10" airfoil, black

exhausts, "Mitre 10/Larkham/Taubmans 3" tempa, Thailand casting ($6-8)(AU)(TC)

MB66-D SUPER BOSS, issued 1982 (continued)

7. white body, no "Detroit Diesel" on roof, dot dash rear wheels, tempa & 4 stripes on spoiler, Macau casting ($1-2)
8. white body, with "Detroit Diesel" on roof, dot dash rear wheels, decal & 4 stripes on spoiler, Macau casting ($1-2)
9. white body, with "Detroit Diesel" on roof, dot dash rear wheels, decal & no stripes on spoiler, Macau casting ($1-2)
10. white body, no "Detroit Diesel" on roof, dot dash rear wheels, decal & 4 stripes on spoiler, Macau casting ($1-2)
11. white body, no "Detroit Diesel" on roof, dot dash rear wheels, tempa & 4 stripes on spoiler, China casting ($1-2)
12. white body, with "Detroit Diesel" on roof, dot dash rear wheels/maltese cross front wheels, plain spoiler, China casting ($1-2)
13. tan body, orange/yellow/black stripes tempa, dot dash rear wheels, silver-gray armament replaces spoiler, Macau casting, includes plastic armament ($5-8)(RB)
14. white body, no "Detroit Diesel" on roof, dot dash rear wheels, plain spoiler, China casting ($2-3)
15. white body, no tempa, dot dash rear wheels, China casting ($10-15)(GF)

MB66-E SAUBER GROUP C RACER, issued 1984
MB46-E SAUBER GROUP C RACER, issued 1984

1. red body, clear windows, ivory interior, black airfoil, 8 dot silver wheels, "BASF Cassettes" tempa, Macau casting ($3-5)
2. red body, clear windows, ivory interior, black airfoil, 8 dot gold wheels, "BASF" Cassettes" tempa, Macau casting ($3-5)
3. white body, clear windows, ivory interior, black airfoil, 8 dot silver wheels, "Jr. Collectors Club" tempa, Macau casting ($8-12)(AU)
4. white body, clear windows, ivory interior, black airfoil, 8 dot gold wheels, "61 Sauber Castrol" tempa, Macau casting ($1-2)
5. white body, clear windows, ivory interior, black airfoil, 8 dot silver wheels, "61 Sauber Castrol" tempa, Macau casting ($1-2)
6. yellow body, clear windows, ivory interior, blue airfoil, starburst wheels, orange & blue tempa, Macau casting ($3-5)(SF)
7. yellow body, clear windows, ivory interior, blue airfoil, laser wheels, orange & blue tempa, Macau casting ($3-5)(LW)
8. black body, black windows, ivory interior, black airfoil, starburst wheels, "Carguantua" tempa, Macau casting ($15-18)(PS)
9. pink-red body, dark blue windows, no interior or airfoil, starburst wheels, black/white/yellow tempa, Macau casting, includes plastic armament ($5-8)(RB)
10. white/orange body, clear windows, ivory interior, orange airfoil, 8 dot silver wheels, "Bisotherm/Baustein" tempa, Macau casting ($8-12)(SW)
11. red body, clear windows, black interior, red airfoil, 8 dot silver wheels, "Royal Mail Swiftair" tempa, Macau casting ($5-7)(UK)(MP)
12. robin egg blue, clear windows, dark blue interior, black airfoil, 8 dot silver wheels, "Grand Prix 46' tempa, Macau casting ($4-6)(MP)
13. white body, clear windows, ivory interior, black airfoil, 8 dot silver wheels, "Grand Prix 46" tempa, Thailand casting ($1-2)
14. chrome plated body, clear windows, ivory interior, black airfoil, 8 dot silver wheels, no tempa, Macau casting ($12-18)(C2)
15. fluorescent pink/blue body, chrome & black windows, black interior, blue airfoil, lightning wheels, "Matchbox" & flames tempa, China casting ($3-4)(LT)
16. fluorescent orange/yellow body, blue-chrome & black windows, black interior, fluorescent yellow airfoil, lightning wheels, lightning bolts tempa, China casting ($3-4)(LT)
17. fluorescent yellow/orange body, blue-chrome & black windows, black interior, fluorescent yellow airfoil, lightning wheels, lightning & flames tempa, China casting ($3-4)(LT)
18. blue/florescent pink body, chrome & black windows, black interior, blue airfoil, lightning & flames tempa, China casting ($3-4)(LT)
19. white body, clear windows, ivory interior, 8 dot silver wheels, "Grand Prix 46" tempa, China casting ($1-2)
20. yellow/red body, clear windows, red interior, 8 dot silver wheels, "Matchbox U.S.A. 11th Annual Convention & Toy Show 1992" tempa, China casting ($10-12)(US)

MB66-F ROLLS ROYCE SILVER SHADOW, issued 1987 (ROW)
MB55-H ROLLS ROYCE SILVER SHADOW, issued 1990 (USA)
NOTE: all models with black base, clear windows & dot wheels.

1. tan body, tan interior, no tempa, Macau casting ($2-4)
2. green-gold body, tan interior, no tempa, Macau casting ($3-5)
3. metallic red body, tan interior, no tempa, Macau casting ($1-2)
4. metallic red body, tan interior, no tempa, Thailand casting ($1-2)
5. metallic red body, cream interior, crest tempa, Thailand casting ($1-2)

MB67-C DATSUN 260Z, issued 1978 (continued)
NOTE: previous English based models with "Lesney" cast, subsequent issues with English bases have "Matchbox International" cast. Doors now all cast shut.

21. black body, black base, white interior, clear windows, no tempa, England casting ($3-5)
22. black body, black base, black interior, opaque windows, no tempa, England casting ($8-12)
23. silver/gray body, black base, black interior, clear windows, 2 tone blue tempa, England casting ($2-4)
24. black body, black base, black interior, milky opaque windows, no tempa, England casting ($8-12)

25. silver/gray body, black base, black interior, clear windows, no tempa, England casting ($5-8)
26. silver/gray body, black base, black interior, opaque windows, no tempa, England casting ($8-12)
27. silver/gray body, black base, black interior, clear windows, green & blue tempa, China casting ($125-175)(CH)
28. silver/gray body, silver/gray base, black interior, clear windows, two-tone blue tempa, England casting ($4-6)
29. silver/gray body, silver/gray base, black interior, clear windows, no tempa England casting ($3-5)

MD67-D FLAME OUT, issued 1983 (USA)

1. white body, red windows, maltese cross rear wheels, Macau casting ($4-6)
2. white body, red windows, 5 crown rear wheels, Macau casting ($4-6)

MB67-F LAMBORGHINI COUNTACH, issued 1985 (USA)
MB11-F LAMBORGHINI COUNTACH, issued 1985 (ROW)

1. red body & base, tan interior, clear windows, 5 arch silver wheels, "Lamborghini" logo tempa, Macau casting ($2-3)
2. red body & base, tan interior, clear windows, 8 dot silver wheels, "Lamborghini" logo tempa, Macau casting ($2-3)
3. red body & base, tan interior, clear windows, 8 dot gold wheels, "Lamborghini" logo tempa, Macau casting ($8-12)
4. black body & base, tan interior, clear

windows, 5 arch silver wheels, "5" with stripes tempa, Macau casting ($2-3)
5. black body & base, tan interior, clear windows, 8 dot gold wheels, "5" with stripes tempa, Macau casting ($3-5)
6. white body & base, dark blue interior, clear windows, starburst wheels, "LP500S" & stripes tempa, Macau casting ($3-5)(SF)
7. metallic pearly silver body & base, dark blue interior, clear windows ($3-5)(LW)
8. red body & base, tan interior, clear windows, 5 arch silver wheels, green "15" & "BP" tempa, Macau casting ($8-12)(DU)
9. yellow body & base, tan interior, clear windows, 5 arch silver wheels, "10 Tiger Racing Team" tempa, Macau casting ($8-12)(HK)
10. yellow body & base, white interior, clear windows, 5 arch silver wheels, "Countach" & "Lamborghini" logo tempa, Macau casting ($2-3)
11. canary yellow body & base, white interior, chrome windows, gray disc w/ rubber tires, "LP500" tempa, Macau casting ($5-8)(WC)
12. white body & base, dark blue interior, clear windows, starburst wheels, "LP500" with stripes tempa, Macau casting ($3-4)(SF)
13. black body base, white interior, clear windows, starburst wheels, "LP500" with stripes tempa, Macau casting ($12-15)(SF)
14. chrome plated body & base, white interior, clear windows, 5 arch wheels, no tempa, Macau casting ($12-18)(C2)
15. red body & base, white interior, clear windows, 5 arch wheels, "Countach"

tempa, Thailand casting ($1-2)
16. red body & base, white interior, chrome windows, gray disc w/rubber tires, "Countach" & detailed trim, China casting ($5-8)(WC)
17. red body & base, white interior, clear windows, 5 arch wheels, "Countach" tempa, China casting ($1-2)
18. white body & base, white interior, clear windows, 5 arch wheels, no tempa, China casting ($10-15)(GF)
19. iridescent cream body & base, white interior, black windows, 5 arch wheels, "Countach" & logo tempa, China casting ($3-5)(TH)

MB67-G IKARUS COACH, issued 1986 (ROW)

NOTE: all models with black base & dot dash wheels. Interior is color of the body as this is one piece.

1. white body, orange roof, smoke windows, "Voyager" tempa, Macau casting ($2-4)
2. white body, orange roof, clear windows, "Voyager" tempa, Macau casting ($2-4)
3. white body, orange roof, amber windows, "Voyager" tempa, Macau casting ($2-4)
4. cream body, cream roof, smoke windows, "Ikarus" tempa, Macau casting ($2-4)
5. white body, red roof, clear windows, "Gibraltor" tempa, Macau casting ($5-7)(SP)
6. white body, green roof, smoke windows, "City Line Tourist" tempa, China casting ($2-4)
7. white & orange body, white roof, smoke windows, "Airport Limousine 237" tempa, China casting ($8-12)(JP)(GS)
8. white body, orange roof, smoke windows, "Voyager" tempa, China casting ($2-3)

9. white body, white roof, smoke windows, "Canary Island" tempa, China casting ($7-10)(UK)
10. beige body, brown roof, smoke windows, "Marti" tempa, China casting ($8-12)(SW)
11. white body, red roof, smoke windows, "Gibraltor" tempa, China casting ($5-7)(SP)
12. white body, green roof, smoke windows, 上傷友誼汽車服務公司 "/2384584" & "FT" tempa, China casting ($30-50)(CH)
13. white body, white roof, smoke windows, "Espana" tempa, China casting ($1-2)

MB68-C CHEVY VAN, issued 1979 (continued)

NOTE: models with 5 crown front & dot dash rear wheels, although intermixed front to rear and vice versa or all same pattern exist.

18. pearly silver body, black base, blue windows, "Vampire" tempa, Macau casting ($3-5)
19. yellow body, pearly silver base, blue windows, "Pepsi Challenge" tempa, Macau casting ($4-6)(TM)
20. white/maroon body, pearly silver base, blue windows, "Dr. Pepper" tempa, Macau casting ($7-10)(TM)
21. bright yellow body, pearly silver base, blue windows, "STP Son of a Gun" tempa, Macau casting ($50-65)(TM)
22. black body, pearly silver base, blue windows, "Goodwrench Racing Team Pit Crew" tempa, Macau casting ($4-6)(WR)(TC)
23. black body, silver/gray base, blue windows, "Goodwrench Racing Team Pit Crew" tempa, Thailand casting

($4-6)(WR)(TC)

24. orange-yellow body, silver/gray base, blue windows, "Kodak Film 4 Racing" tempa, Thailand casting ($4-6)(WR)(TC)
25. white body, silver/gray base, blue windows, gold hubs, "25" and green design tempa, Thailand casting ($4-6)(WR)(TC)
26. black body, silver/gray base, blue windows, "Goodwrench 5 Time National Champion Dale Earnhardt" tempa, Thailand casting ($4-6)(WR)(TC)
27. bright yellow body, silver/gray base, blue windows, "Pennzoil 30" tempa, Thailand casting ($4-6)(WR)(TC)
28. blue/fluorescent orange body, silver/gray base, blue windows, "43 STP Oil Treatment" tempa, Thailand casting ($4-6)(WR)(TC)
29. black body, silver/gray base, blue windows, "Pontiac Excitement 2" tempa, Thailand casting ($4-6)(WR)(TC)
30. black/green body, silver/gray base, blue windows, "Mello Yello 42" tempa, Thailand casting ($4-6)(WR)(TC)
31. fluorescent orange/white body, silver/gray base, blue windows, "Purolator 10" tempa, Thailand casting ($4-6)(WR)(TC)

MB68-E DODGE CARAVAN, issued 1984 (USA)
MB64-E DODGE CARAVAN, issued 1984 (ROW)
NOTE: model with cast door and "Las Vegas" base listed at end of variations. Early models cast "1983", later ones cast "1984". All models with clear windows and chrome base unless noted otherwise.
1. burgundy body, maroon door & interior, black stripe tempa, England casting

($15-20)(US)

2. silver/gray body, gray door, maroon interior, black stripe tempa, England casting ($6-8)
3. silver/gray body, gray door, brown-red interior, black stripe tempa, England casting ($6-8)
4. black body, black door, maroon interior, no tempa, England casting ($4-6)
5. black body, black door, maroon interior, silver stripe tempa, England casting ($3-5)
6. black body, black door, maroon interior, silver stripe tempa, Macau casting ($2-4)
7. black body, black door, maroon interior, silver & gold stripes tempa, China casting ($2-4)
8. white body, white door, maroon interior, "Pan Am" tempa, Macau casting ($3-5)(GS)
9. white body, white door, maroon interior, "Caravan" with stripes tempa, Macau casting ($2-4)
10. white body, white door, maroon interior, "Fly Virgin Atlantic" tempa, Macau casting ($3-5)(GS)
11. black body, black door, maroon interior, silver stripe with "Adidas" hood label, England casting ($250-300)(US)
12. black body, black door, maroon interior, green & yellow stripes tempa, Macau casting ($8-12)(DU)
13. white body, white door, maroon interior, "NASA Shuttle Personnel" tempa, Macau casting ($3-5)(MP)
14. light gray/dark navy body, light gray door, red interior, "British Airways" tempa, Thailand casting ($3-5)(MC)
15. red body, red door, black interior & base, "Red Arrows/Royal Air Force"

tempa, Thailand casting ($3-5)(MC)
16. black body, black door, red interior, gray base, silver & gold stripe tempa, Manaus casting ($35-50)

MB68-G TV NEWS TRUCK, issued 1989 (USA)
MB73-D TV NEWS TRUCK, issued 1989 (ROW)
NOTE; all models with frosted amber windows, dot dash wheels.
1. dark blue body, silver/gray roof, orange roof, orange roof units, black base, "75 News/MB TV Mobile One' tempa, Macau casting ($1-2)
2. dark blue body, silver/gray roof, orange roof units, black base, "75 News/MB TV Mobile One" tempa, Thailand casting ($1-2)
3. dark blue body, silver/gray roof, pale orange units, black base, "75 News/MB TV Mobile One" tempa, Thailand casting ($1-2)
4. white body, white roof, orange roof units, white base, no tempa, Thailand casting ($10-15)(GF)
5. white body, white roof, brown roof units, white base, no tempa, Thailand casting ($10-15)(GF)
6. white body, dark blue roof, orange roof units, black base, "Sky Satellite Television" tempa, Thailand casting ($1-2)

MB69-C ARMOURED TRUCK, issued 1978 (continued)
8. dark olive body, pearly silver base, blue windows, 5 crown wheels, "Dresdner Bank" tempa, Macau casting ($60-75)(GR)

MB69-D '33 WILLYS STREET ROD, issued 1982 (USA) (continued)
5. dark blue body, pearly silver metal base, orange flames tempa, dot dash rear wheels, Macau casting ($1-2)
6. dark blue body, pearly silver metal base, red flames tempa, dot dash rear wheels, Macau casting ($1-2)
7. dark blue body, light gray metal base, red flames tempa, dot dash rear wheels, Hong Kong casting ($2-3)
8. dark blue body, light gray metal base, red flames tempa, 5 spoke rear wheels, Hong Kong casting ($2-3)
9. dark blue body, black plastic base, red flames tempa, dot dash rear wheels, Macau casting ($1-2)
10. dark blue body, black plastic base, red flames tempa, dot dash rear wheels, China casting ($1-2)
11. black body, black plastic base, red flames tempa, dot dash rear wheels, China casting ($7-10)(US)(OP)

MB69-F VOLVO 480ES, issued 1988 (ROW)
NOTE: all models with gray interior & smoke windows.
1. pearly silver body, gray metal base, laser wheels, "Volvo 480" with dual green stripes tempa, Macau casting ($3-5)(LW)
2. white body, gray metal base, 8 dot wheels, "Volvo" & "480ES" tempa, Macau casting ($2-3)
3. white body, gray metal base, 8 dot wheels, "480ES" tempa, Macau casting ($2-3)
4. white body, gray plastic base, 8 dot wheels, "480ES" tempa, Macau casting ($3-4)

5. white body, gray plastic base, 8 dot wheels, "Volvo" & "480ES" tempa, China casting ($5-7)

MB70-C SELF PROPELLED GUN, issued 1976 (continued)

4. black body & base, tan treads, gray & yellow camouflage tempa, Macau casting ($4-5)(CM)
5. olive green body & base, tan treads, tan & black camouflage tempa, Macau casting ($4-5)(CM)
6. olive green body & base, tan treads, tan & black camouflage tempa, China casting ($6-8)(CM)

MB70-D FERRARI 308GTB, issued 1981 (continued)

7. orange-red body, pearly silver base, 5 arch wheels, black interior, clear windows, "Ferrari" & logo tempa, Macau casting ($2-3)
8. orange-red body, pearly silver base, 5 spoke wheels, black interior, clear windows, "Ferrari" & logo tempa, Macau casting ($2-3)
9. dark red body, pearly silver base, 5 spoke wheels, black interior, clear windows, "Ferrari" & logo tempa, Macau casting ($2-3)
10. red body, blue base, 5 arch wheels, white interior, clear windows, "Pioneer 39" tempa, Macau casting ($2-3)
11. yellow body, red base, starburst wheels, black interior, clear windows, "Ferrari 308GTB" tempa, Macau casting ($3-5)(SF)
12. yellow body, red base, laser wheels, black interior, clear windows, "Ferrari 308GTB" tempa, Macau casting

($3-5)(LW)
13. orange body, blue base, 5 arch wheels, white interior, clear windows, "12 Rat Racing Team" tempa, Macau casting ($8-12)(HK)
14. red body, red base, 5 arch wheels, black interior, clear windows, "Data East/Secret Service" tempa, Macau casting ($65-80)(CN)
15. red body, red base, gray disc w/rubber tires, black interior, chrome windows, detailed trim tempa, Macau casting ($5-8)(WC)
16. red body, red base, 5 arch wheels, black interior, clear windows, rectangular "Ferrari" logo on hood, Macau casting ($3-4)(MP)
17. red body, red base, 5 arch wheels, black interior, clear windows, rectangular "Ferrari" logo on hood, Thailand casting ($2-3)(MC)
18. white body, white base, 5 arch wheels, black interior, green windows, no tempa, Thailand casting ($10-15)(GF)
19. lemon body, lemon base, 5 arch wheels, black interior, clear windows, geometric designs tempa, Thailand casting ($2-3)(DM)

MB70-E FORD CARGO SKIP TRUCK, issued 1989 (USA)
MB45-D FORD CARGO SKIP TRUCK, issued 1987 (ROW)

NOTE: all models with blue windows, 8 spoke wheels.
1. yellow body, pearly gray metal dump, black base, black arms, orange stripe tempa, Macau casting ($1-2)
2. blue body, red metal dump, yellow base, yellow arms, white stripe tempa, Macau

casting ($6-8)(LL)
3. yellow body, pearly gray metal dump, black base, black arms, orange stripe tempa, Thailand casting ($1-2)
4. yellow body, gray plastic dump, black base, black arms, orange stripe tempa, Macau casting ($1-2)
5. yellow body, gray plastic dump, black base, black arms, orange stripe tempa, Thailand casting ($1-2)
6. yellow body, red plastic dump, black base, black arms, no tempa, Thailand casting ($1-2)

MB71-C CATTLE TRUCK, issued 1976
MB12-I CATTLE TRUCK, reissued 1992 (ROW) (continued)

66. powder blue body, chocolate stakes, red windows, black base, black cows, black tow hook, 5 crown wheels, Macau casting ($2-4)(TP)
67. red body, yellow stakes, red windows, yellow base, no cows, red tow hook, 5 crown blue wheels with yellow hubs, cow head tempa, Macau casting ($6-8)(LL)
68. green body, yellow stakes, red windows, black base, black cows, black tow hook, 5 crown wheels, Macau casting ($1-2)(TP)
69. green body, yellow stakes, red windows, black base, black cows, black tow hook, 5 crown wheels, Thailand casting ($1-2)(TP)
70. green body, yellow stakes, red windows, black base, black cows, no tow hook, 5 crown wheels, Thailand casting ($1-2)

MB71-D 1962 CORVETTE, issued 1982 (USA) (continued)

NOTE: all models with chrome interior &

clear windows unless noted.
8. white body & base, red flames tempa, 5 arch silver wheels, Macau casting ($2-4)
9. white body & base, orange flames tempa, 5 arch silver wheels, Macau casting ($2-4)
10. white body & base, red flames tempa, 5 arch silver wheels, Hong Kong casting ($2-4)
11. white body & base, orange flames tempa, 5 arch silver wheels, Hong Kong casting ($2-4)
12. bright blue body & base, "Firestone" tempa, 5 arch silver wheels, Macau casting ($7-10)(JP)
13. orange body & base, "11" & white stripe tempa, small starburst wheels, Macau casting ($3-5)(SF)
14. orange body & base, "11" & white stripe tempa, large starburst wheels, Macau casting ($3-5)(SF)
15. red body & base, "454 Rat" tempa, 5 arch gold wheels, Macau casting ($2-3)
16. red body & base, "454 Rat" tempa, 5 arch silver wheels, Macau casting ($2-3)
17. metallic copper body & base, "11" & white stripe tempa, laser wheels, Macau casting ($3-5)(LW)
18. metallic green body & base, "11" & white stripe tempa, laser wheels, Macau casting ($3-5)(LW)
19. rose red body & base, "Heinz 57" tempa, 5 arch silver wheels, Macau casting ($25-35)(US)
20. turquoise body & base, white roof tempa, gray disc w/rubber tires, China casting ($5-8)(WC)
21. white body & base, no tempa, 5 arch silver wheels, China casting ($10-15)(GF)

22. red body & base, "454 Rat" tempa, 5 arch silver wheels, China casting ($1-2)
23. metallic copper body & base, "4" & black roof tempa, black disc w/rubber tires, China casting ($5-8)(WP)

MB71-G PORSCHE 944 TURBO, issued 1989 (USA)
MB59-E PORSCHE 944 TURBO, issued 1989 (ROW)

1. red body, black base, tan interior, clear windows, 8 dot wheels, doors open, "944 Turbo" & logo tempa, Macau casting ($1-2)
2. black body, black base, no interior, chrome windows, gray disc w/rubber tires, "944 Turbo" & detailed trim tempa, Macau casting ($5-8)(WC)
3. red body, black base, tan interior, clear windows, 8 dot wheels, doors open, "944 Turbo/Credit Charge" tempa, Macau casting ($7-10)(UK)
4. black body, black base, tan interior, clear windows, 8 dot wheels, doors open, "944 Turbo" & logo tempa, Macau casting ($2-3)(KS)
5. white body, black base, tan interior, clear windows, 8 dot wheels, doors open, "Duckhams" tempa, Macau casting ($15-20)(UK)(OP)
6. white body, black base, tan interior, clear windows, 8 dot wheels, doors open, "Duckhams" tempa, Thailand casting ($15-20)(UK)(OP)
7. black body, black base, no interior, chrome windows, gray disc w/rubber tires, doors cast, "944 Turbo" & detailed trim tempa, Thailand casting ($5-8)(WC)
8. metallic green body, black base, tan interior, clear windows, 8 dot wheels,

doors open, "944 Turbo" & logo tempa, Thailand casting ($1-2)

MB72-E DODGE DELIVERY TRUCK, issued 1982 (ROW) (continued)

7. red body, white container, blue windows, "Kellogg's" tempa, Macau casting ($4-6)
8. white body, white container, blue windows, "Street's Ice Cream" tempa, Macau casting ($6-8)(AU)
9. yellow body, yellow container, blue windows, "Hertz" tempa, Macau casting ($2-4)
10. orange-yellow body & container, blue windows, "Hertz" tempa, Macau casting ($2-4)
11. red body, red container, blue windows, "Royal Mail Parcels" tempa, Macau casting ($5-7)(UK)
12. red body, white container, blue windows, "Kelloggs/Milch-Lait-Latte" tempa, Macau casting ($15-20)(GR)(SW)
13. white body, white container, blue windows, "Jetspress Road Express" tempa, Macau casting ($6-8)(AU)
14. green body, white container, blue windows, "Minties" tempa, Macau casting ($6-8)(AU)
15. orange-yellow body, yellow container, blue windows, "Risi" tempa, Macau casting ($15-25)(SP)
16. blue body, blue container, blue windows, "Mitre 10" tempa, Macau casting ($6-8)(AU)
17. red body, dark yellow container, blue windows, "Nestles Chokito" tempa, Macau casting ($6-8)(AU)
18. red body & base, red container, red windows, "Kit Kat" tempa, Macau casting ($7-10)(UK)(OP)

19. blue body & base, blue container, blue windows, "Yorkie" tempa, Macau casting ($7-10)(UK)(OP)
20. white body, white container, blue windows, "Pirelli Gripping Stuff" tempa, Macau casting ($3-5)(TC)
21. red body, white container, blue windows, "Matchbox U.S.A. Sheraton Inc 1989" tempa, Macau casting ($8-12)(US)
22. white body, white container, blue windows, "XP Express Parcels Systems" tempa, Macau casting ($3-5)(TC)
23. dark green body, orange container, blue windows, "C Plus Orange" tempa, Macau casting ($10-15)(CN)
24. light gray/dark navy body, light gray container, blue windows, "British Airways Cargo" tempa, Thailand casting ($3-5)(MC)
25. white body, white container, blue windows, "Wigwam" tempa, Thailand casting ($7-10)(CY)(DU)
26. white body, white container, blue windows, "XP Express parcels Systems" tempa, Thailand casting ($3-5)(TC)
27. red body, white container, blue windows, "Big Top Circus" tempa, Thailand casting ($3-5)(TP)

MB72-F SAND RACER, issued 1984 (USA)

1. white body, pearly silver base, black interior & rollbar, "Goodyear 211" tempa, Macau casting ($18-25)

MB72-G AIRPLANE TRANSPORTER, issued 1985 (USA)
MB65-E AIRPLANE TRANSPORTER, issued 1985 (ROW)

1. yellow body, blue windows, "Rescue" & checkers tempa, plastic plane with red wings & white undercarriage, Macau casting ($6-8)
2. yellow body, blue windows, "Rescue" & checkers tempa, plastic plane with white wings & red undercarriage, Macau casting ($6-8)
3. yellow body, blue windows, "Rescue" & checkers tempa, plastic plane with white wings & red undercarriage, China casting ($6-8)
4. white body, blue windows, "NASA" tempa, plastic plane with white wings & undercarriage, Macau casting ($3-5)(MP)
5. olive body, blue windows, black & tan camouflage tempa, black hubs, plastic plane with olive wings & undercarriage, China casting ($35-50)(CM)

MB72-I CADILLAC ALLANTE, issued 1987 (USA)
MB65-F CADILLAC ALLANTE, issued 1987 (ROW)

1. silver/gray body, red interior, clear windows, 8 dot wheels, no tempa, Macau casting ($1-2)
2. black body, red interior, clear windows, laser wheels, red & silver stripes tempa, Macau casting ($3-5)(LW)
3. pink body, gray interior, clear windows, 8 dot wheels, "Cadillac" tempa, Macau casting ($1-2)
4. light pink body, gray interior, clear windows, 8 dot wheels, "Cadillac" tempa, China casting ($1-2)

5. silver/gray body, red interior, clear windows, 8 dot wheels, no tempa, China casting ($1-2)
6. charcoal body, red interior, chrome windows, gray disc w/rubber tires, no tempa, Macau casting ($5-8)(WC)
7. white body, red interior, clear windows, gray disc w/rubber tires, no tempa, China casting ($100-150)(US)
8. bright pink body, gray interior, clear windows, 8 dot wheels, "Cadillac" tempa, Thailand casting ($1-2)
9. pink body, white interior, clear windows, 8 dot wheels, green zigzags & blue stripes tempa, Thailand casting ($1-2)
10. metallic red body, gray interior, clear windows, 8 dot wheels, "Official Pace Car 76th Indy" tempa, Thailand casting ($7-10)(IN)

MB72-K DODGE ZOO TRUCK, issued 1992 (ROW)
1. white body, silver/gray cage with brown lions, red/orange/yellow stripes tempa, Thailand casting ($3-5)(MC)

MB73-B WEASEL, issued 1974 (continued)
8. olive green body & base, black base inset, black hubs, tan & black camouflage tempa, Macau casting ($4-5)(CM)
9. black body & base, black base insert, black hubs, yellow & gray camouflage tempa, Macau casting ($4-5)(CM)

MB73-C MODEL A FORD, issued 1979
MB55-I MODEL A FORD, reissued 1991 (USA)
NOTE: some variations may be found with front wheels (5 crown) used on the rear

axles and vice versa.
7. beige body, brown fenders, clear windows, no tempa, 5 crown front & dot dash rear wheels, Macau casting ($2-4)
8. red body, black fenders, clear windows, no tempa, 5 crown front & dot dash rear wheels, Macau casting ($2-4)
9. black body & fenders, clear windows, flames tempa, 5 arch front & racing special rear wheels, Macau casting ($2-4)
10. purple body, yellow fenders, clear windows, stripes tempa, 5 arch front & racing special rear wheels, Macau casting ($3-5)
11. yellow body, red fenders, clear windows, "Pava" tempa, 5 arch front & racing special rear wheels, Macau casting ($8-12)(DU)
12. red body, dark green fenders, clear windows, no tempa, 5 crown front & dot dash rear wheels, Macau casting ($15-20)(UK)(OP)
13. red body, dark green fenders, clear windows, no tempa, 5 crown front & dot dash rear wheels, Thailand casting ($15-20)(UK)(OP)
14. red body, dark green fenders, clear windows, no tempa, 5 crown front & dot dash rear wheels, China casting ($15-20)(UK)(OP)
15. purple body, yellow chassis, clear windows, stripes tempa, 5 arch front racing special rear wheels, Thailand casting ($2-4)
16. orange-yellow body, white fenders, clear windows, "GT" & yellow jacket tempa, 5 crown front & dot dash rear wheels, Thailand casting ($6-8)(WR)
17. powder blue body, red chassis, clear windows, clown tempa, 5 arch front &

racing special yellow wheels, Thailand casting ($7-10)(MC)

MB73-E MERCEDES TRACTOR, issued 1990 (USA)
MB27-E MERCEDES TRACTOR, issued 1990 (ROW)
1. pea green upper body, olive plastic chassis, olive interior, pear green wheels, no tempa, Macau casting ($1-2)
2. pea green upper body, olive plastic chassis, olive interior, pea green wheels, no tempa, Thailand casting ($1-2)
3. green upper body, pale yellow plastic chassis, pale yellow interior, green wheels, no tempa, Thailand casting ($2-4)(TP)(TC)
4. pea green upper body, olive plastic chassis, olive interior, pea green wheels, "MB Trac" tempa, Thailand casting ($1-2)

MB74-D ORANGE PEEL, issued 1982 (USA) (continued)
2. white body, black base, light orange "Orange Peel" tempa, 5 arch front wheels, Macau casting ($3-5)
3. white body, black base, light orange "Orange Peel" tempa, 5 arch front wheels, Hong Kong casting ($3-5)
4. white body, black base, light orange, "Orange Peel" tempa, maltese cross front wheels, Macau casting ($3-5)

MB74-F MUSTANG GT, issued 1984 (USA)
NOTE: all models with black base & interior, clear windows, 5 arch front wheels.
1. light orange body, racing special rear

wheels, yellow/blue stripes tempa, Macau casting ($2-4)
2. dark orange body, racing special rear wheels, yellow/blue stripes tempa, Macau casting ($2-4)
3. dark orange body, racing special rear wheels, orange/blue stripes tempa, Macau casting ($2-4)
4. dark orange body, dot dash rear wheels, yellow/blue stripes tempa, China casting ($2-4)
5. dull orange body, dot dash rear wheels, yellow/blue stripes tempa, China casting ($2-4)
6. pearly silver body, racing special rear wheels, purple/yellow stripes tempa, China casting ($2-4)

MB74-H GRAND PRIX RACING CAR, issued 1987 (USA)
MB14-H GRAND PRIX RACING CAR, issued 1987 (ROW)
1. white/powder blue body, dark blue metal base, red driver, black airfoil, "15/Goodyear/Shell" tempa, Macau casting ($3-4)
2. red body, black metal base, white driver, black airfoil, "27 Fiat" tempa, Macau casting ($1-2)
3. red body, black metal base, white driver, black airfoil, "27 Fiat" tempa, Thailand casting ($1-2)
4. yellow body, yellow metal base, red driver, dark yellow airfoil, "Pennzoil 2" tempa, China casting ($3-4)(IN)
5. dark orange/white body, white metal base, red driver, orange airfoils, "Indy 4" tempa, China casting ($3-4)(IN)
6. dark yellow body & metal base, red driver, orange-yellow airfoil, "Squirt"

tempa, Thailand casting ($18-25)(US)(OP)

7. chrome plated body, black metal base, white driver, black airfoil, no tempa, Thailand casting ($12-18)(C2)

8. red body, black metal base, white driver, black airfoil, "27 Fiat" tempa, China casting ($1-2)

9. red body, black plastic base, white driver, black airfoil, "27 Fiat" tempa, China casting ($7-10)

10. red body, red metal base, white driver, black airfoil, "Scotch/Target" tempa, Thailand casting ($3-4)(IN)

11. blue body, blue metal base, yellow driver, blue airfoil, "Panasonic 7" tempa, Thailand casting ($3-4)(IN)

12. white/blue body, white metal base, red driver, blue airfoil, "Indy 76" tempa, Thailand casting ($3-4)(IN)

13. white/black body, black metal base, red driver, black airfoil, "Havoline/K-Mart 6" tempa, Thailand casting ($3-4)(IN)

14. yellow body, yellow metal base, red driver, yellow airfoil, "Pennzoil 4" tempa, Thailand casting ($3-4)(IN)

15. orange/lavender/white body, lavender base, white driver, white airfoil, "Indy" tempa, Thailand casting ($4-6)(IN)

MB75-D HELICOPTER, issued 1982 (continued)

17. pearly gray body, orange base, gray interior/skis, amber windows, "-600-" tempa, Macau casting ($2-4)(CY)

18. white body, orange base, black interior/skis, amber windows, blue "Rescue" tempa, Macau casting ($2-4)

19. white body, black base, gray interior/skis, amber windows, blue "Rescue"

tempa, Macau casting ($2-4)

20. black body, black base, gray interior/skis, amber windows, "Air Car" tempa, Macau casting ($2-4)(CY)

21. white body, red base, gray interior/skis, amber windows, "Fire Dept." tempa, Macau casting ($2-4)(MP)

22. white body, red base, gray interior/skis, amber windows, "NASA" tempa, Macau casting ($2-4)(MP)

23. white body, red base, black interior/skis, amber windows, "Virgin Atlantic" tempa, Macau casting ($2-4)(GS)

24. white body, yellow base, red interior/skis, amber windows, "JCB" tempa, Macau casting ($4-6)(GS)

25. white body, lemon base, gray interior/skis, amber windows, "Fire Dept." tempa, Macau casting ($2-4)(MP)

26. white body, dull orange base, gray interior/skis, amber windows, black "Rescue" tempa, Macau casting ($2-4)

27. red body, white base, gray interior/skis, amber windows, "Red Rebels" tempa, Macau casting ($2-4)(MC)

28. white body, black base, gray interior/skis, amber windows, Japanese lettered tempa, Macau casting ($8-12)(JP)(GS)

29. red body, white base, gray interior/skis, amber windows, "Fire Dept." tempa, Macau casting ($2-4)(TC)

30. white body, lemon base, gray interior/skis, amber windows, "Fire Dept." tempa, Thailand casting ($2-3)(MP)

31. black body, black base, gray interior/skis, amber windows, "Air Car" tempa, Thailand casting ($2-3)(CY)

32. white body, black base, gray interior/skis, amber windows, "Rescue" tempa, Thailand casting ($2-3)

33. white body, yellow base, lime interior/skis, amber windows, "123456" tempa, blue & yellow blades, Macau casting ($6-8)(LL)

34. white body, yellow base, lime interior/skis, amber windows, "123456" tempa, blue & yellow blades, Thailand casting ($6-8)(LL)

35. red body, white base, gray interior/skis, amber windows, "Red Rebels" tempa, Thailand casting ($1-2)(MC)

36. red body, white base, gray interior/skis, clear windows, "Royal Air Force" tempa, Thailand casting ($2-3)(MC)

37. red body, white base, gray interior/skis, amber windows, "Fire Dept." tempa, Thailand casting ($1-2)(MP)

38. white body, fluorescent orange base, gray interior/skis, amber windows, "Air Rescue 10" & "IC" logo tempa, Thailand casting ($10-15)(IC)

MB75-E FERRARI TESTAROSSA, issued 1986

1. red body & base, black interior, clear windows, 8 dot wheels, "Ferrari" logos tempa, Macau casting ($1-2)

2. black body & base, black interior, clear windows, star burst wheels, silver with logos tempa, Macau casting ($3-5)(SF)

3. metallic pearly silver body & base, black interior, clear windows, laser wheels, gold with logos tempa, Macau casting ($3-5)(LW)

4. yellow body, black base, no interior, dark blue modified windows, starburst wheels, red/blue/yellow tempa, Macau casting, includes plastic armament ($5-8)(RB)

5. yellow body & base, black interior, clear windows, 8 dot wheels, "9 Rabbit Racing

Team" tempa, Macau casting ($8-12)(HK)

6. metallic red body & base, black interior, clear windows, laser wheels, silver with logos tempa, Macau casting ($3-5)(LW)

7. orange-red body, black base, black interior, clear windows, 8 dot wheels, "Ferrari" logo tempa, Macau casting ($1-2)

8. dark red body, black base, black interior, clear windows, 8 dot wheels, "Ferrari" logo tempa, Macau casting ($3-4)(SC)

9. orange body, black base, black interior, clear windows, 8 dot wheels, "Ferrari" logo tempa, Macau casting ($3-4)(SC)

10. rose red body, black base, black interior, clear windows, 8 dot wheels, "Ferrari" logo tempa, Macau casting ($3-4)(SC)

11. red body & base, black interior, clear windows, 8 dot wheels, "Redoxon" tempa, Macau casting ($20-35)(HK)

12. red body & base, black interior, chrome windows, gray disc w/rubber tires, detailed trim tempa, Macau casting ($5-8)(WC)

13. red body, lime base, lime interior, clear windows, 8 dot yellow wheels with blue hubs, "1" with face & checkered flag tempa, Macau casting ($6-8)(LL)

14. red body & base, black interior, clear windows, 8 dot wheels, "Ferrari" logo tempa, Thailand casting ($1-2)

15. red body & base, black interior, chrome windows, gray disc w/rubber tires, detailed trim tempa, Thailand casting ($5-8)(WC)

16. white body & base, black interior, chrome windows, gray disc w/rubber tires, detailed trim tempa, Thailand casting ($5-8)(WC)

17. red body & base, black interior, clear windows, starburst wheels, silver with logos tempa, Macau casting ($8-12)(SF)
18. red body & base, black interior, clear windows, 8 dot wheels, "Ferrari" logo tempa, China casting ($1-2)
19. white body & base, black interior, clear windows, 8 dot wheels, no tempa, China casting ($10-15)(GF)
20. red body & base, black interior, clear windows, 8 dot wheels, "Lloyds/Ferrari" tempa, China casting ($7-10)(UK)
21. red body & base, black interior, clear windows, 8 dot wheels, "Ferrari" logo (all yellow background) tempa, China casting ($1-2)

NOTE: as of the end of 1992, the following models were not included in the normal 1-75 range. These were promotional castings offered by White Rose Collectibles. The models are identified by a three digit "frame number" which is cast inside the body. The MB212 and MB216 are scheduled for 1993 inclusion into the 1-75 line.

MB212 FORD THUNDERBIRD, issued 1992 (White Rose)

NOTE: all models with clear windows, black base & Goodyear slicks unless noted otherwise.

1. red body, red interior, "Motorcraft Quality Parts 15" tempa, China casting ($3-5)(WR)
2. black body, silver/gray interior, "Havoline"/Texaco star & "28" without "MAC" logo tempa, China casting ($3-5)(WR)
3. yellow body, red interior, "92 Matchbox Lightning" & assorted logos tempa,

China casting ($20-25)(US)
4. white/orange-red body, black interior, "Hooters 7" without "Naturally Fresh" tempa, China casting ($5-8)(WR)
5. dark red body, black interior, "Phillips 66/Trop Artic" tempa, China casting ($3-5)(WR)
6. bright red body, black interior, "Phillips 66/Trop Artic" tempa, China casting ($3-5)(WR)
7. blue body, black interior, "Maxwell House 22" tempa, China casting ($3-5)(WR)
8. bright red body, black interior, "White Rose Collectibles 92" tempa, China casting ($15-25)(WR)
9. red body, black interior, "Motorcraft Quality Parts 15" tempa, China casting ($3-5)(WR)
10. black body, silver/gray interior, "Havoline"/Texaco star & "28" with "MAC" logo tempa, China casting ($3-5)(WR)
11. black body, silver/gray interior, "Phillips 66/Trop Artic" tempa, China casting ($3-5)(WR)
12. red body, red interior, "Bill Elliot 11" tempa, China casting ($3-5)(WR)
13. white body, silver/gray interior, "Baby Ruth 1" tempa, China casting ($)(WR)
14. white/orange-red body, black interior, "Hooters 7" with "Naturally Fresh" tempa, China casting ($4-6)(WR)
15. brown/red body, black interior, "Snickers 8" tempa, China casting ($3-5)(WR)
16. brown/red body, black interior, "Snickers 8" tempa, 9 spoke Goodyear slicks, China casting ($3-5)(WR)
17. red body, red interior, "Bill Elliot 11"

tempa, black disc w/rubber tires tempa, China casting ($4-6)(WR)(TC)
18. red/white body, black interior, "Melling 9" tempa, China casting ($3-5)(WR)
19. green body, black interior, "Quaker State 26" tempa, Thailand casting ($3-5)(WR)

MB216 PONTIAC GRAND PRIX, issued 1992 (WHITE ROSE)

NOTE: all models with clear windows & black base.

1. bright yellow body, black interior, Goodyear slicks wheels, "Pennzoil 30" tempa, China casting ($3-5)(WR)
2. bright yellow body, black interior, gray disc w/rubber tires, "Pennzoil 30" tempa, China casting ($4-6)(WR)(TC)
3. blue body, black interior, Goodyear slicks wheels, "STP 43" tempa, China casting ($3-5)(WR)
4. blue body, black interior, blue disc w/rubber tires, "STP 43" tempa, China casting ($4-6)(WR)(TC)
5. blue body, black interior, black disc w/rubber tires, "Rumple 70/Son's" tempa, China casting ($4-6)(WR)(TC)
6. yellow body, gray interior, Goodyear slicks wheels, "White House Apple Juice 4" tempa, China casting ($4-6)(WR)(OP)
7. yellow body, gray interior, 9 spoke Goodyear slicks wheels, "White House Apple Juice 41" tempa, China casting ($4-6)(WR)(OP)
8. brown/red body, black interior, Goodyear slicks tempa, "Pontiac Excitement 2" tempa, China casting ($3-5)(WR)
9. black body, black interior, black disc w/rubber tires, "Pontiac Excitement 2" tempa, China casting ($4-6)(WR)(TC)
10. black/green body, black interior,

Goodyear slicks wheels, "Mello Yello 42" tempa, China casting ($3-5)(WR)
11. black/green body, black interior, black disc w/rubber tires, "Mello Yello 42" tempa, China casting ($4-6)(WR)(TC)
12. blue/white body, gray interior, Goodyear slicks wheels, "Evinrude 89" tempa, China casting ($4-6)(WR)
13. matt yellow body, black interior, Goodyear slicks wheels, "Country Time 68" tempa, China casting ($3-5)(WR)
14. white/yellow body, red interior, Goodyear slicks wheels, "Medford Speed Shop/Valtrol 48" tempa, China casting ($4-6)(WR)

MB217 D.I.R.T. MODIFIED, issued 1992 (WHITE ROSE)

1. powder blue body, red interior, chrome base, "White Rose Collectibles 1" tempa, Thailand casting ($60-75)(WR)
2. orange-yellow body, red interior, chrome base, "7X Turbo Blue" tempa, Thailand casting ($5-7)(WR)
3. orange-yellow body, blue interior, blue base, "91 Wheels" tempa, Thailand casting ($5-7)(WR)
4. orange & white body, red interior, chrome base, "9 Kinney" tempa, Thailand casting ($5-7)(WR)
5. red body, red interior, chrome base, "72 Auto Palace" tempa, Thailand casting ($5-7)(WR)
6. brown body, black interior, chrome base, "1 Phil's Chevrolet" tempa, Thailand casting ($5-7)(WR)
7. white body, red interior, chrome base, "6 Freightliner" tempa, Thailand casting ($5-7)(WR)

MB245 CHEVY PANEL VAN, issued 1992 (WHITE ROSE)

NOTE: all models with clear windows, 5 crown wheels, silver/gray base and China casting.

1. white body, brown chassis, "Browns 1992" tempa ($3-5)(WR)
2. white body, red-brown chassis, "Redskins 1992" tempa ($3-5)(WR)
3. white body, orange chassis, "Buccaneers 1992" tempa ($3-5)(WR)
4. white body, red chassis, "Falcons 1992" tempa ($3-5)(WR)
5. white body, turquoise chassis, "Dolphins 1992" tempa ($3-5)(WR)
6. white body, dark blue chassis, "Patriots 1992" tempa ($3-5)(WR)
7. white body, dark blue chassis, "Bills 1992" tempa ($3-5)(WR)
8. white body, dark blue chassis, "Colts 1992" tempa ($3-5)(WR)
9. white body, bright blue chassis, "Oilers 1992" tempa ($3-5)(WR)
10. white body, green chassis, "Jets 1992" tempa ($3-5)(WR)
11. white body, green chassis, "Eagles 1992" tempa ($3-5)(WR)
12. yellow body, purple chassis, "Vikings 1992" tempa ($3-5)(WR)
13. yellow body, black chassis, "Steelers 1992" tempa ($3-5)(WR)
14. yellow body, dark blue chassis, "Chargers 1992" tempa ($3-5)(WR)
15. yellow body, mid blue chassis, "LA Rams 1992" tempa ($3-5)(WR)
16. bright yellow body, green chassis, "Packers 1992" tempa ($3-5)(WR)
17. bright yellow body, red chassis, "Chiefs 1992" tempa ($3-5)(WR)
18. bright yellow body, red-brown chassis,

"Cardinals 1992" tempa ($3-5)(WR)
19. orange body, black chassis, "Bengals 1992" tempa ($3-5)(WR)
20. orange body, dark blue chassis, "Bears 1992" tempa ($3-5)(WR)
21. orange body, mid blue chassis, "Broncos 1992" tempa ($3-5)(WR)
22. gold body, red chassis, "49ers 1992" tempa ($3-5)(WR)
23. green-gold body, black chassis, "Saints 1992" tempa ($3-5)(WR)
24. silver/gray body, black chassis, "Raiders 1992" tempa ($3-5)(WR)
25. silver/gray body, dark blue chassis, "Cowboys 1992" tempa ($3-5)(WR)
26. silver/gray body, mid blue chassis, "Seahawks 1992" tempa ($3-5)(WR)
27. silver/gray body, bright blue body, "Lions 1992" tempa ($3-5)(WR)
28. red body, dark blue chassis, "Giants 1992" tempa ($3-5)(WR)
29. fluorescent orange body, lavender chassis, "York Fair 1992" tempa ($6-8)(WR)

MBXX DODGE DAYTONA, issued 1983

1. burgundy body, black interior, clear windows, silver-gray England base, casting closed, base cast "Expressly for Dodge Las Vegas" ($75-100)(VG)

MBXX DODGE CARAVAN, issued 1983

1. burgundy body, maroon interior, clear windows, chrome England base, casting closed, base cast "Expressly for Dodge Las Vegas" ($75-100)(VG)

DODGE DAYTONA & DODGE CARAVAN PLAQUE

1. above two models mounted by velcro to

wooden plaque with gold plate denoting models made for Las Vegas and includes Lee Iacocca signature ($500-750)(VG)

SUPER GT'S

Super GT'S are a form of miniatures. These models are based on older 1970s castings in which interiors are no longer fitted, the windows are solid in color (opaque) and opening features are cast closed. The original name for the line was to be the "Budget Range" and models are still designated with "BR" numbers. All models are in paired numbers ie. BR1/2, BR3/4, etc. The line was introduced in 1985 and ran through 1988. This line consists of varied tempa colors and wheels varieties. Model names do not appear on the castings. Model names are taken from the original model's name in the 1970s. In 1990, some models were reissued in neon colors as "Neon Racers". These are denoted with "NR".

BR 1/2 Iso Grifo
BR 3/4 Gruesome Twosome
BR 5/6 Datsun 126X
BR 7/8 Siva Spyder
BR 9/10 Lotus Europa
BR11/12 Saab Sonett
BR13/14 Hairy Hustler
BR15/16 Monteverdi Hai
BR17/18 Fire Chief
BR19/20 Ford Group 6
BR21/22 Alfa Carabo
BR23/24 Vantastic
BR25/26 Ford Escort
BR27/28 Lamborghini Marzal
BR29/30 Maserati Bora
BR31/32 Fandango
BR33/34 Hi-Tailer

BR35/36 Porsche 910
BR37/38 Ford Capri
BR39/40 DeTomasso Pantera

BR 1/2 ISO GRIFO, issued 1985

1. cream body, 8 dot wheels, black windows, red & black tempa, England base ($3-5)
2. cream body, 8 dot wheels, glow windows, red & black tempa, England base ($12-15)
3. pale yellow body, 8 dot wheels, black windows, red & black tempa, England base ($7-10)
4. pale yellow body, dot dash wheels, black windows, red & black tempa, England base ($7-10)
5. yellow body, 8 dot wheels, black windows, red & black tempa, England base ($7-10)
6. yellow body, dot dash wheels, black windows, red & black tempa, England base ($7-10)
7. light blue body, 8 dot wheels, black windows, black & purple tempa, England base ($4-6)
8. met. blue body, 8 dot wheels, black windows, lime & maroon with "57" tempa, England base ($3-5)
9. met. blue body, 8 dot wheels, glow windows, lime & red with "57" tempa, England base ($12-15)
10. met. blue body, 8 dot wheels, black windows, lemon & brown with "57" tempa, England base ($4-6)
11. met. blue body, 8 dot wheels, black windows, yellow & red with "57" tempa, England base ($4-6)
12. met. blue body, 8 dot wheels, black windows, red & black tempa, England

base ($7-10)

13. met. blue body, 8 dot wheels, black windows, purple & black tempa, England base ($7-10)

14. met. blue body, dot dash wheels, black windows, purple & black tempa, England base ($7-10)

15. met. blue body, 8 dot wheels, black windows, maroon & black tempa, England base ($7-10)

16. met. blue body, 8 dot wheels, black windows, lime & red with "57" tempa, China base ($7-10)

17. bright green body, 8 dot wheels, black windows, white & black with "57" tempa, China base ($3-5)

18. cream body, 8 dot wheels, black windows, red & black tempa, China base ($6-8)

19. yellow body, 8 dot wheels, black windows, red & black tempa, China base ($3-5)

BR 3/4 GRUESOME TWOSOME, issued 1985

NOTE: all models below with black windows, 8 dot wheels unless noted.

1. yellow body, blue & orange with "Gruesome" tempa, England base ($4-6)

2. yellow body, green & orange with "Gruesome" tempa, England base ($4-6)

3. powder blue body, white arrow design tempa, England base ($4-6)

4. powder blue body, white arrow design tempa, no tempa on tail, England base ($4-6)

5. dark blue body, white arrow design tempa, England base ($5-8)

6. yellow body, blue & orange with "Gruesome" tempa, China base ($7-10)

7. white body, red & black with "Gruesome" tempa, China tempa ($3-5)

8. white body, red & black with "Gruesome" tempa, glow windows, China base ($20-25)

9. gray body, dark blue arrow design tempa, China base ($3-5)

10. fluorescent orange body, yellow & black tempa, China base ($4-6)(NR)

BR 5/6 DATSUN 126X, issued 1985

1. dark blue body, 5 arch wheels, black windows, gold & white tempa, England base ($4-6)

2. dark blue body, 8 dot wheels, black windows, gold & white tempa, England base ($4-6)

3. silver body, 5 arch wheels, black windows, green & red stripes tempa, England base ($4-6)

4. silver body, 8 dot wheels, black windows, green & red stripes tempa, England base ($4-6)

5. silver-gray body, 5 arch wheels, black windows, green & red stripes tempa, England base ($4-6)

6. metallic silver body, 5 arch wheels, black windows, green & red stripes tempa, China base ($7-10)

7. powder blue body, 5 arch wheels, black windows, white & blue stripes tempa, China base ($3-5)

8. beige body, 5 arch wheels, black windows, 2 tone blue tempa, China base ($3-5)

9. beige body, 5 arch wheels, glow windows, 2 tone blue tempa, China base ($20-25)

BR 7/8 SIVA SPYDER, issued 1985

NOTE: all models fitted with black windows

1. orange-yellow body, 8 dot wheels, black & white "Turbo" tempa, England base ($4-6)

2. orange-yellow body, dot dash wheels, black & white "Turbo" tempa, England base ($7-10)

3. yellow body, 8 dot wheels, black & red "Turbo" tempa, England base ($7-10)

4. yellow body, 8 dot wheels, black & white "Turbo" tempa, England base ($7-10)

5. white body, 8 dot wheels, blue & lemon tempa, England base ($4-6)

6. white body, 8 dot wheels, blue & yellow tempa, England base ($4-6)

7. white body, dot dash wheels, blue & yellow tempa, England base ($4-6)

8. blue body, 8 dot wheels, white & black tempa, England base ($7-10)

9. gray body, 8 dot wheels, white & black tempa, England base ($7-10)

10. gray body, 8 dot wheels, gray & black tempa, England base ($7-10)

11. white body, 8 dot wheels, blue & yellow tempa, China base ($7-10)

12. dark red body, 8 dot wheels, white & gold tempa, China base ($3-5)

13. orange-yellow body, 8 dot wheels, black & white "Turbo" tempa, China base ($7-10)

14. yellow body, 8 dot wheels, black & red "Turbo" tempa, China base ($3-5)

BR 9/10 LOTUS EUROPA, issued 1985

NOTE: all models with black windows & 8 dot wheels.

1. white body, red & black with "1" tempa, England base ($4-6)

2. met. blue body, lemon leaf pattern tempa, England base ($4-6)

3. met. blue body, cream leaf pattern tempa, England base ($4-6)

4. met. blue body, yellow leaf pattern tempa, England base ($4-6)

5. purple body, gold leaf pattern tempa, China base ($3-5)

6. white body, red & black with "1" tempa, China base ($7-10)

7. blue body, yellow & black with "1" tempa, China base ($3-5)

BR 11/12 SAAB SONETT, issued 1985

NOTE: all models with black windows.

1. orange body, 8 dot wheels, white & red with "5" tempa, England base ($4-6)

2. light tan body, 8 dot wheels, white & red triangle tempa, England base ($4-6)

3. dark tan body, 8 dot wheels, white & red triangle tempa, England base ($4-6)

4. light tan body, dot dash wheels, white & red triangle tempa, England base ($4-6)

5. light tan body, 8 dot wheels, white & red with "5" tempa, England base ($7-10)

6. blue body, 8 dot wheels, white & red triangle tempa, England base ($7-10)

7. light tan body, 8 dot wheels, white & red triangle tempa, China base ($7-10)

8. powder blue body, 8 dot wheels, yellow & black triangle tempa, China base ($3-5)

9. orange body, 8 dot wheels, white & red with "5" tempa, China base ($7-10)

10. green body, 8 dot wheels, white & gold with "5" tempa, China base ($3-5)

BR 13/14 HAIRY HUSTLER, issued 1985

NOTE: all models with black windows.

1. lemon body, 5 crown wheels, black & red "L" & reverse "F" with "7" on roof and

tail stripes tempa, England base ($4-6)

2. yellow body, 5 crown wheels, black & red "L" & reverse "F" with "7" on roof and tail stripes tempa, England base ($4-6)

3. yellow body, 5 crown wheels, black & red "L" & reverse "F" tempa only, England base ($7-10)

4. yellow body, 5 arch front & 5 crown rear wheels, black & red "L" & reverse "F" with "7" on roof and tail stripes tempa, England base ($4-6)

5. yellow body, 5 arch wheels, black & red "L" & reverse "F" with "7" on roof and tail stripes tempa, England base ($4-6)

6. yellow body, 5 crown front & 5 arch rear wheels, black & red "L" & reverse "F" with "7" on roof and tail stripes tempa, England base ($4-6)

7. yellow body, 8 dot wheels, black & red "L" & reverse "F" with "7" on roof and tail stripes tempa, England base ($4-6)

8. green body, 5 crown wheels, yellow & black with "2" tempa, England base ($4-6)

9. green body, 5 arch front & 5 crown rear wheels, yellow & black with "2" tempa, England base ($4-6)

10. dark green body, 5 crown wheels, yellow & black with "2" tempa, England base ($7-10)

11. gray body, 5 crown wheels, red & black with "2" tempa, China base ($3-5)

12. yellow body, 5 crown wheels, black & red "L" & reverse "F" with "7" on roof and tail stripes tempa, China base ($7-10)

13. red body, 5 crown wheels, white & yellow "L" & reverse "F" with "7" on roof and tail stripes tempa, China base ($3-5)

BR15/16 MONTEVERDI HAI, issued 1985

1. green body, 8 dot wheels, black windows, no tempa, England base ($12-15)

2. green body, 8 dot wheels, black windows, white & gold tempa, England base ($4-6)

3. green body, dot dash wheels, black windows, white & gold tempa, England base ($6-8)

4. green body, 8 dot wheels, glow windows, white & gold tempa, England base ($12-15)

5. black body, 8 dot wheels, glow windows, white & gold tempa, England base ($15-20)

6. yellow body, 8 dot wheels, black windows, white & gold tempa, England base ($50-65)

7. light tan body, 8 dot wheels, black windows, blue & white with "28" tempa, England base ($4-6)

8. light tan body, 8 dot wheels, black windows, green & white with "28" tempa, England base ($6-8)

9. mid tan body, 8 dot wheels, black windows, black & white with "28" tempa, England base ($6-8)

10. dark tan body, 8 dot wheels, black windows, green & white with "28" tempa, England base ($6-8)

11. gold body, 8 dot wheels, black windows, blue & white with "28" tempa, England base ($4-6)

12. white body, 8 dot wheels, black windows, orange & black with "28" tempa, China base ($30-45)

13. orange body, 8 dot wheels, black windows, red & blue with "28" tempa, China base ($3-5)

14. green body, 8 dot wheels, black windows, white & gold tempa, China base ($7-10)

15. blue body, 8 dot wheels, black windows, red & lime tempa, China base ($3-5)

16. fluorescent pink body, 8 dot wheels, black windows, black & yellow tempa, China base ($4-6)(NR)

BR17/18 FIRE CHIEF, issued 1985
NOTE: all models fitted with black windows.

1. white body, 5 crown wheels, "Police" tempa, England base ($4-6)

2. white body, 5 arch front & 5 crown rear wheels, "Police" tempa, England base ($4-6)

3. white body, 5 arch front & rear wheels, "Police" tempa, England base ($4-6)

4. red body, 5 crown wheels, white "Rescue" tempa, England base ($4-6)

5. red body, 5 arch front & 5 crown rear wheels, white "Rescue" tempa, England base ($4-6)

6. red body, 5 crown wheels, white "Rescue" tempa, China base ($7-10)

7. orange body, 5 crown wheels, black "Rescue" tempa, China base ($3-5)

8. white body, 5 crown wheels, "Police" tempa, China base ($7-10)

9. blue body, 5 crown wheels, "Police" tempa, China base ($3-5)

BR19/20 FORD GROUP 6, issued 1985
NOTE: all models with black windows.

1. white body, 8 dot wheels, red & black "18" tempa, England base ($4-6)

2. white body, dot dash wheels, pink & black "18" tempa, England base ($4-6)

3. yellow body, 8 dot wheels, red & black "18" tempa, England base ($7-10)

4. red body, 8 dot wheels, white & yellow "2" tempa, England base ($4-6)

5. red body, dot dash wheels, white & yellow "2" tempa, England base ($4-6)

6. red body, 8 dot wheels, white & yellow "2" tempa, China base ($7-10)

7. gray body, 8 dot wheels, purple & white "2" tempa, China base ($3-5)

8. white body, 8 dot wheels, red & black "18" tempa, China base ($7-10)

9. dark red body, 8 dot wheels, blue & black "18" tempa, China base ($3-5)

10. fluorescent lime body, 8 dot wheels, yellow & black with "18" tempa, China base ($4-6)(NR)

BR21/22 ALFA CARABO, issued 1985
All models fitted with black windows and 8 dot wheels unless noted. Various shades of the panel tempa on the silver model exist.

1. silver body, lt. & dk. blue panels tempa, England base ($4-6)

2. silver body, lt. blue & dk. green panels tempa, England base ($4-6)

3. silver body, no tempa, England base ($12-15)

4. green body, dk. blue & white panels tempa, England base ($12-15)

5. beige body, lt. & dk. blue panels tempa, England base ($25-40)

6. yellow body, blue & green panels tempa, England base ($25-40)

7. yellow body, lt. & dk. blue panels tempa, England base ($25-40)

8. silver body, white & blue panels tempa, England base ($7-10)

9. orange body, blue & white with "6" tempa, England base ($4-6)

10. orange body, blue & white with "6"

tempa, 5 crown wheels, England base ($7-10)
11. orange body, no tempa, England base ($12-15)
12. orange body, dk. & lt. blue panels tempa, England base ($12-15)
13. orange body, green & lt. blue panels tempa, England base ($12-15)
14. pearly silver with black & dk. blue panels tempa, China base ($7-10)
15. powder blue body, yellow & white panels tempa, China base ($3-5)
16. green body, gold & white with "6" tempa, China base ($3-5)

BR23/24 VANTASTIC, issued 1986

1. beige body, blue & red "Starfire" tempa, black windows, China base ($3-5)
2. blue body, red & yellow with "51" tempa, black windows, China base ($3-5)

BR25/26 FORD ESCORT, issued 1985

NOTE: all models with black windows and 8 dot wheels.
1. red body, gold & white with "10" tempa, England base ($4-6)
2. red body, blue & white with "10" tempa, England base ($4-6)
3. blue body, blue & white with "10" tempa, England base ($5-8)
4. beige body, lt. blue & mid blue with "9" tempa, England base ($4-6)
5. beige body, lt. blue & green with "9" tempa, England base ($4-6)
6. yellow body, lt. & dk. blue with "9" tempa, England base ($12-15)
7. blue body, yellow & dk. blue with "9" tempa, China base ($3-5)
8. purple body, white & black with "10" tempa, China base ($3-5)

BR27/28 LAMBORGHINI MARZAL, issued 1985

NOTE: all models with black windows and 8 dot wheels unless noted.
1. green body, white & yellow with "8" tempa, dot dash wheels, England base ($7-10)
2. green body, white & yellow with "8" tempa, England base ($4-6)
3. green body, black & gray with "16" tempa, England base ($12-15)
4. blue body, white & yellow with "8" tempa, England base ($7-10)
5. cream body, black & gray with "16" tempa, England base ($4-6)
6. white body, black & gray with "16" tempa, England base ($12-15)
7. cream body, black & gray with "16" tempa, China base ($7-10)
8. yellow body, blue & gold with "16" tempa, China base ($3-5)
9. green body, white & yellow with "8" tempa, China base ($7-10) 10. red body, black & red with "8" tempa, China base ($3-5)

BR29/30 MASERATI BORA, issued 1985

NOTE: all models with black windows & 8 dot wheels unless noted. The tempa on the yellow model exists in varying shades.
1. yellow body, blue & orange tempa, England base ($4-6)
2. yellow body, green & orange tempa, England base ($4-6)
3. yellow body, green & orange tempa, dot dash wheels, England base ($7-10)
4. yellow body, purple & orange tempa, England base ($4-6)
5. yellow body, blue & brown tempa, England base ($4-6)
6. light blue body, dk. blue & white with "70" tempa, England base ($4-6)
7. light blue body, black & white with "70" tempa, England base ($4-6)
8. light blue body, black & white with "70" tempa, dot dash wheels, England base ($7-10)
9. dark blue body, black & white with "70" tempa, England base ($6-8)
10. dark blue body, black & white with "70" tempa, dot dash wheels, England base ($7-10)
11. light blue body, black & white with "70" tempa, China base ($7-10)
12. beige body, 2 tone blue with "70" tempa, China base ($3-5)
13. yellow body, blue & orange tempa, China base ($7-10)
14. powder blue body, purple & black tempa, China base ($3-5)
15. fluorescent lime body & white with "70" tempa, China base ($4-6)(NR)

BR31/32 FANDANGO, issued 1985

NOTE: all models with black windows.
1. yellow body, 5 crown wheels, blue & red with "19" tempa, England base ($4-6)
2. yellow body, 5 arch front & 5 crown rear wheels, blue & red with "19" tempa, England base ($4-6)
3. maroon body, 5 crown wheels, white & yellow with "217" & white prop tempa, England base ($4-6)
4. maroon body, 5 crown wheels, white & yellow with "217" & unpainted prop tempa, England base ($4-6)
5. maroon body, 5 arch front & 5 crown rear wheels, white & yellow with "217" & white prop tempa, England base ($4-6)
6. maroon body, 5 crown wheels, white & yellow with "217" & white prop tempa, China base ($7-10)
7. yellow body, 5 crown wheels, red & black with "217" & red prop tempa, China base ($3-5)
8. lemon body, 5 crown wheels, blue & red with "19" tempa, China base ($7-10)
9. gray body, 5 crown wheels, gray & red with "19" tempa, China base ($3-5)

BR33/34 HI-TAILER, issued 1986

1. lemon body, black & red with "Super 61" tempa, China base ($3-5)
2. yellow body, black & red with "Super 61" tempa, China base ($3-5)
3. white body, black & red with "45" tempa, China base ($3-5)

BR35/36 PORSCHE 910, issued 1985

NOTE: all models with black windows.
1. dark blue body, 8 dot wheels, blue & white with "49" tempa, England base ($4-6)
2. light blue body, 8 dot wheels, blue & white with "49" tempa, England base ($4-6)
3. light blue body, dot dash wheels, blue & white with "49" tempa, England base ($4-6)
4. light blue body, dot dash wheels, black "3" & "Drive" tempa, England base ($7-10)
5. light blue body, 8 dot wheels, black "3" & "Drive" tempa, England base ($7-10)
6. silver body, 8 dot wheels, black "3" & "Drive" tempa, England base ($4-6)
7. silver-gray body, 8 dot wheels, black "3" & "Drive" tempa, England base ($4-6)
8. silver body, 8 dot wheels, black 7 white

with "49" tempa, England base ($7-10)
9. silver body, dot dash wheels, black & white with "49" tempa, England base ($7-10)
10. pearly silver body, dot dash wheels, black "3" & "Drive" tempa, China base ($7-10)
11. white body, 5 arch wheels, red "3" & "Drive" tempa, China base ($3-5)
12. light blue body, 5 arch wheels, black with "49" tempa, China base ($7-10)
13. lime body, 5 arch wheels, gold & white with "49" tempa, China base ($3-5)
14. fluorescent orange body, 5 arch wheels, dk. blue & yellow with "49" tempa, China base ($4-6)(NR)

BR37/38 FORD CAPRI, issued 1986
1. cream body, red-brown & green with "48" tempa, China base ($3-5)
2. blue body, red & white with "8" tempa, China base ($3-5)

BR39/40 DETMASSO PANTERA, issued 1986
1. maroon body, orange & white with "4X4" tempa, China base ($3-5)
2. orange body, blue & brown with "Ace" tempa, China base ($3-5)
3. fluorescent yellow body, blue-green & orange with "4X4" tempa, China base ($4-6)(NR)

ORIGINALS
   In 1988, Matchbox Toys decided to celebrate the 40th anniversary of Matchbox toys by recreating five miniatures that were made back in the early 1950s. The recreated toys were based on the 1A Road Roller, 4A Tractor, 5A London Bus, 7A Milk Float and

9A Dennis Fire Engine. All were produced as closely as possible to original colors. These were then packaged in a limited edition gift box either in English or German text. A French version was also supposed to have been issued in gold plate but this was canceled. In 1991, these same five models were reissued as singles under the name "Originals" and packaged with recreated boxes but in new colors. These were numbered as MX101 through MX105 on the boxes. In late 1992, a second series of originals were also released based on the 6A Quarry Truck, 13A Wreck Truck, 19B MGA, 26A Concrete Truck and 32A Jaguar XK120.

MX101-A ROAD ROLLER, issued 1988
1. green body, red rollers, double post roof pillars ($3-5)(GS)
2. dark blue body, red rollers, double post roof pillars ($2-4)
3. dark blue body, red rollers, single post roof pillars ($2-4)

MX102-A TRACTOR, issued 1988
1. red body, silver metal wheels ($3-5)(GS)
2. green body, silver metal wheels ($2-4)

MX103-A LONDON BUS, issued 1988
1. red body, "Buy Matchbox Series" tempa, silver metal wheels ($3-5)(GS)
2. red body, "Matchbox Originals" tempa, silver metal wheels ($2-4)
NOTE: Version 2 may be found with two box styles depicting "Matchbox Originals" or "players Please" on the box art. The "Players Please" is much rarer. An intact piece is valued ($12-15)

MX104-A MILK FLOAT, issued 1988
1. orange body, brown horse, silver metal wheels ($3-5)(GS)
2. powder blue body, brown horse, silver metal wheels ($2-4)

MX105-A DENNIS FIRE ENGINE, issued 1988
1. red body, red flywheel, silver metal wheels ($3-5)(GS)
2. red body, yellow flywheel, silver metal wheels ($2-4)

MX106-A QUARRY TRUCK, issued 1992
1. blue body, gray dump, silver metal wheels ($2-4)

MX107-A WRECK TRUCK, issued 1992
1. red body, yellow boom and hook, silver metal wheels ($2-4)

MX108-A MGA, issued 1992
1. dark green body and base, cream driver, silver metal wheels ($2-4)

MX109-A CONCRETE TRUCK, issued 1992
1. orange body, gray barrel, silver metal wheels ($2-4)

MX110-A JAGUAR XK120, issued 1992
1. black body and base, silver metal wheels ($2-4)

TWIN PACKS
   Originally introduced as Two Packs in 1976, this diecast line features paired models. Over the years the line's name has

been changed to the "900 Series" and "Hitch N Haul" and in 1984 became known as "Twin Packs". Along with the standard Twinpacks range there were several other two model packs including the 1984 Matchmates and several "one-shot" sets.

TP101-A Matra Rancho & Pony Trailer
TP102-A Ford Escort & Glider Transporter
TP103-A Cattle Truck & Trailer
TP104-A Steam Locomotive & Passenger Coach
TP105-A Datsun & Boat
TP106-A Renault & Motorcycle Trailer
TP107-A Datsun & Caravan
TP108-A Tractor & Trailer
TP109-A Citroen & Boat
TP110-A Matra Rancho & Inflatable
TP111-A Ford Cortina & Pony Trailer
TP112-A Unimog & Trailer
TP113-A Porsche & Caravan
TP114-A Volkswagen & Pony Trailer
TP115-A Ford Escort & Boat
TP116-A Jeep Cherokee & Caravan
TP117-A Mercedes G Wagon & Pony Trailer
TP118-A BMW Cabriolet & Gilder Transporter
TP119-A Flareside Pickup & Seafire
TP120-A Volkswagen & Inflatable
TP121-A Land Rover & Seafire
TP122-A Porsche & Glider Transporter
TP123-A BMW Cabriolet & Caravan
TP124-A Locomotive & Passenger Coach
TP125-A Shunter & Side Tipper
TP126-A Mercedes Tractor & Trailer
TP127-A BMW Cabriolet & Inflatable
TP128-A Circus Set
TP129-A Isuzu Amigo & Seafire
TP130-A Land Rover & Pony Trailer

TP131-A Mercedes G Wagon & Inflatable

M-01-A Citroen Matchmates
M-02-A Ford Matchmates
M-03-A Jaguar Matchmates
M-04-A Jeep Matchmates
M-05-A Corvette Matchmates
M-06-A Kenworth Matchmates

CS-81 Bulldozer & Tractor Shovel
CS-82 Peterbilt Quarry Truck & Excavator
CS-83 Road Roller & Faun Crane
EM-81 Snorkel & Foam Pumper
EM-83 Auxiliary Power Truck &
    Ambulance

010301 Dodge Instant Winner Game
010120 Limited Edition Christmas Offer
32520 Days of Thunder Race Car Challenge
**NOTE: Prices quoted on Twinpacks are
for SEALED BLISTERPACKS. Loose
models may not command the same
price if removed from the package.**

TP101-A MATRA RANCHO & PONY TRAILER, issued 1984
1. MB37-E Matra Rancho in blue with blue base & MB43-A beige Pony Trailer ($8-12)
2. MB37-E Matra Rancho in blue with black base & MB43-A beige Pony Trailer ($8-12)
3. MB37-E Matra Rancho in blue with yellow base & MB43-A beige Pony Trailer ($15-25)
4. MB37-E Matra Rancho in yellow with yellow base & MB43-A beige Pony Trailer ($8-12)
5. MB37-E Matra Rancho in blue with white base & MB43-A beige Pony Trailer ($8-12)

TP102-A FORD ESCORT & GLIDER TRANSPORTER, issued 1984
1. MB9-C Ford Escort in light green, "Seagull" labels with Glider Transporter in dark green with "Seagull" labels ($15-18)
2. MB9-C Ford Escort in dark green, "Seagull" labels with Glider Transporter in dark green with "Seagull" labels ($7-10)

TP103-A CATTLE TRUCK & TRAILER, issued 1984
1. MB71-C Cattle Truck & trailer in red with beige stakes, brown cows ($8-10)
2. MB71-C Cattle Truck & trailer in red with beige stakes, coffee cows ($8-10)
3. MB71-C Cattle Truck & trailer in yellow with chocolate stakes red-brown cows ($8-10)
4. MB71-C Cattle Truck & trailer in powder blue with chocolate stakes, black cows, Macau base ($8-12)
5. MB71-C Cattle Truck & trailer in green with yellow stakes, black cows, Thailand base ($7-10)

TP104-A STEAM LOCOMOTIVE & PASSENGER COACH, issued 1984
1. MB43-C Loco in red with "4345" labels: MB44-C Coach with flat tan roof with "431 432" labels ($10-15)
2. MB43-C Loco in red with "4345" labels; MB44-C Coach with flat beige roof with "431 432" labels ($10-15)
3. MB43-C Loco in red with "4345" labels; MB44-C Coach with flat beige roof with "5810 6102" labels ($15-18)
4. MB43-C Loco in red with "4345" labels; MB44-C Coach with flat tan roof with "5810 6102" labels ($15-18)
5. MB43-C Loco in red with "4345" labels; MB44-C Coach with raised beige roof with "5810 6102" labels ($15-18)
6. MB43-C Loco in red with "4345" labels; MB44-C Coach with raised beige roof with "431 432" labels ($15-18)

TP105-A DATSUN & BOAT, issued 1984
1. MB67-C Datsun in black with no tempa & MB9-A Boat with blue deck, white hull & blue trailer ($10-15)
NOTE: Datsun with clear or opaque windows.

TP106-A RENAULT & MOTORCYCLE TRAILER, issued 1984
1. MB21-C Renault in white with green tempa; yellow trailer with red cycles ($10-15)
2. MB21-C Renault in white with pink & yellow tempa; yellow trailer with red cycles ($10-15)
3. MB21-C Renault in white with pink & yellow tempa; yellow trailer with green cycles ($15-18)
4. MB21-C Renault in white with pink & yellow tempa; yellow trailer with black cycles ($15-18)
5. MB21-C Renault in pearly silver with "Scrambler" tempa; pearly silver trailer with black cycles, Macau ($15-18)

TP107-A DATSUN & CARAVAN, issued 1984
1. MB67-C Datsun in silver-gray with 2 tone blue tempa; MB31-C Caravan in white with "Mobile 500" tempa ($10-15)

TP108-A TRACTOR & TRAILER, issued 1984
1. MB46-C Ford Tractor in blue with 40-C Hay Trailer in red ($8-12)
2. MB46-C Ford Tractor in green with 40-C Hay Trailer in yellow ($8-12)
3. MB46-C Ford Tractor in yellow with 40-C Hay Trailer in yellow ($8-12)

TP109-A CITROEN & BOAT, issued 1984
1. MB12-D Citroen in white with "Marine" tempa; MB9-A Boat with blue deck, white hull, blue trailer & no labels ($15-18)
2. MB12-D Citroen in white with "Marine" tempa; MB9-A Boat with blue deck, white hull, blue trailer & "8" label ($15-20)
3. MB12-D Citroen in white with "Marine" tempa; MB9-A Boat with white deck, blue hull, blue trailer & "8" label ($15-20)
4. MB12-D Citroen in white with "Marine" tempa; MB9-A Boat with white deck, blue hull, orange trailer & "8" label ($15-20)
5. MB12-D Citroen in white with "Ambulance" tempa; MB9-A Boat with blue deck, white hull, orange trailer & no label ($25-40)
6. MB12-D Citroen in white with "Ambulance" tempa; MB9-A Boat with white deck, blue hull, orange trailer & "8" label ($15-20)
7. MB12-D Citroen in white with "Ambulance" tempa; MB9-A Boat with white deck, white hull, orange trailer & "8" label ($15-20)
8. MB9-C Ford Escort in dark green with

"Seagull" labels; MB9-A Boat with white deck, blue hull, orange trailer & "8" label ($50-75)

TP110-A MATRA RANCHO & INFLATABLE, issued 1984
1. MB37-E Matra Rancho in navy blue with white base & tempa; Inflatable with orange deck, white hull and no tempa, black trailer ($10-15)
2. MB37-E Matra Rancho in navy blue with white base & tempa; Inflatable with orange deck, white hull and "SR" tempa, black trailer ($10-15)
3. MB37-E Matra Rancho in black with white base & tempa; Inflatable with orange deck, white hull & "SR" tempa, black trailer ($10-15)
4. MB37-E Matra Rancho in navy blue with white base & tempa; Inflatable with yellow deck, white hull and "SR" tempa, black trailer ($12-18)
5. MB37-E Matra Rancho in orange with black base & tempa; Inflatable with black deck, orange hull & "2" tempa on white trailer ($10-15)

TP111-A FORD CORTINA & PONY TRAILER, issued 1984
1. MB55-D Ford Cortina in tan with black stripe; MB43-A Pony Trailer in beige with horse head label ($8-12)
2. MB55-D Ford Cortina in red with black stripe; MB43-A Pony Trailer in beige with "Silver Shoes" tempa ($8-12)

TP112-A UNIMOG & TRAILER, issued 1984
1. MB48-E Unimog & MB2-A Trailer in yellow with white canopies and "Alpine

Rescue" tempa ($10-15)
2. MB48-E Unimog & MB2-A Trailer in red with white canopies and "Unfall Rettung" tempa ($10-15)
3. MB48-E Unimog & MB2-A Trailer in white with orange canopies with "GES" tempa ($10-15)

TP113-A PORSCHE & CARAVAN, issued 1985
1. MB3-C Porsche in black with gold tempa; MB31-C Caravan in white with "Mobile 500" tempa ($15-18)

TP114-A VOLKSWAGEN & PONY TRAILER, issued 1985
1. MB7-C VW Golf in black with red tempa; MB43-A Pony Trailer in beige with "Silver Shoes" tempa ($15-18)

TP115-A FORD ESCORT AND BOAT, issued 1987
1. MB117-D/MB37-F Ford Escort in white with "XR3i" tempa; MB9-A Boat with white deck, white hull, black trailer & "Seaspray" tempa ($8-12)
2. MB17-D/MB37-F Ford Escort in met. blue with spatter tempa; MB9-A Boat with blue deck, blue hull, black trailer & spatter tempa ($8-12)

TP116-A JEEP CHEROKEE & CARAVAN, issued 1987
1. MB27-D Jeep Cherokee in beige with "Holiday Club" tempa; MB31-C Caravan in beige with "500" tempa ($8-12)

TP117-A MERCEDES G WAGON & PONY TRAILER, issued 1987
1. MB30-E Mercedes in white with

"Polizei" checkered tempa; MB43-A Pony Trailer in white with "Polizei" checkered tempa ($8-12)
2. MB30-E Mercedes in white with green "Polizei" tempa; MB43-A Pony Trailer in green with "Polizei" tempa ($8-12)

TP118-A BMW CABRIOLET & GLIDER TRANSPORTER, issued 1987
1. MB39-D BMW in red with "Gliding Club" tempa; Glider transporter in red with "Auto Glide" tempa ($12-15)

TP119-A FLARESIDE PICKUP & SEAFIRE, issued 1987
1. MB53-D Flareside in yellow with "Ford" tempa; MB5-B Seafire in yellow with blue hull & "460" tempa ($8-12)

TP120-A VOLKSWAGEN & INFLATABLE, issued 1989
1. MB33-D Volkswagen in dark gray; Inflatable with orange deck, black hull, white trailer with red & blue tempa ($8-12)

TP121-A LAND ROVER & SEAFIRE, issued 1989
1. MB35-F Land Rover in white with "Country" tempa; MB5-A Seafire with white deck, red hull, red tempa with white plastic trailer with blue & red tempa ($8-12)
2. MB35-F Land Rover in white with "Bacardi" tempa; MB5-A Seafire with white deck, white hull, "Bacardi" tempa with white metal trailer (from MB9-A)($50-75)(OP)(UK)

TP122-A PORSCHE & GLIDER TRANSPORTER, issued 1989
1. MB3-C Porsche in dark blue with yellow tempa; Transporter in dark blue with white glider and yellow tempa ($7-10)
2. MB3-C Porsche in yellow with spatter tempa; Transporter in yellow with hot pink glider and spatter tempa ($8-12)

TP123-A BMW CABRIOLET & CARAVAN, issued 1989
1. MB39-D BMW in silver blue with dark blue stripe; MB31-C Caravan in gray with orange tempa ($8-12)
2. MB39-D BMW in white with geometric tempa; MB31-C Caravan in white with geometric design tempa ($8-12)

TP124-A LOCOMOTIVE & PASSEN-GER COACH, issued 1991
1. MB43-C Loco in green with "British Railways" tempa; MB44-C Passenger Coach in green with "British Railways" tempa ($8-12)

TP125-A SHUNTER & TIPPER, issued 1991
1. MB24-C Shunter in yellow; ex-TP20 Side Tipper in yellow with red tipper ($8-10)

TP126-A MERCEDES TRACTOR & TRAILER, issued 1991
1. MB73-E Mercedes Tractor in yellow & green; 40-C Hay Trailer in yellow ($8-10)

TP127-A BMW CABRIOLET & INFLATABLE, issued 1991
1. MB39-D BMW in white with red & blue tempa; Inflatable in dark blue with gray

hull, white trailer with red & blue tempa ($8-10)

TP128-A CIRCUS SET, issued 1992
1. MB72-E dodge Truck in red with white container with "Big Top Circus" tempa; MB2-A Mercedes Trailer in red with white canopy with "Big Top Circus" tempa ($8-12)

TP129-A ISUZU AMIGO & SEAFIRE, issued 1992
1. MB52-E Isuzu Amigo in red; MB5-A Seafire in red with white hull with "Surf Rider" tempa, white plastic trailer ($7-10)

TP130-A LAND ROVER & PONY TRAILER, issued 1992
1. MB35-F Land Rover in white; MB43-A Pony Trailer in white with red roof and red/black stripes tempa ($7-10)
2. MB35-F Land Rover in green; MB43-A Pony Trailer in white with horse silhouette tempa ($7-10)

TP131-A MERCEDES G WAGON & INFLATABLE, issued 1992
1. MB30-E Mercedes in white with orange roof & "Marine Rescue" tempa; Inflatable in fluorescent orange with gray hull with "Rescue" tempa & white trailer ($7-10)

MATCHMATES

M-01-A CITROEN MATCHMATES, issued 1984
1. MB12-D Citroen in white with "Marine" tempa & MB44-E Citroen in black ($10-15)

M-02-A FORD MATCHMATES, issued 1984
1. MB38-E Model A Van in white/red/blue with "Pepsi" tempa & MB73-C Model A Ford in tan/brown ($10-15)

M-03-A JAGUAR MATCHMATES, issued 1984
1. MB22-E Jaguar XK120 in green & MB47-D Jaguar SS100 in red ($10-15)

M-04-A JEEP MATCHMATES, issued 1984
1. MB5-D Jeep in brown; MB20-C Jeep in black ($10-15)

M-05-A CORVETTE MATCHMATES, issued 1984
1. MB14-D 1984 Corvette in red & MB21-D Corvette in pearly silver ($10-15)

M-06-A KENWORTH MATCHMATES, issued 1984
1. MB41-D Kenworth in black & MB45-C Kenworth in pearly silver ($10-15)

CS-81 BULLDOZER & TRACTOR SHOVEL, issued 1992
1. MB64-D Bulldozer in yellow with red roof; MB29-C Tractor Shovel in yellow with red shovel, with accessories ($6-8)

CS-82 PETERBILT QUARRY TRUCK & EXCAVATOR, issued 1992
1. MB30-E Peterbilt in yellow with red dump; MB32-D Excavator in yellow with red scoop, with accessories ($6-8)

CS-83 ROAD ROLLER & FAUN CRANE, issued 1992
1. MB40-E Road Roller in yellow; MB42-E Faun Crane in yellow with red cab, with accessories ($6-8)

EM-81 SNORKEL & FOAM PUMPER, issued 1992
1. MB63-E Snorkel in neon orange & MB54-E Command Vehicle in neon orange, with accessories ($6-8)

EM-83 AUXILIARY POWER TRUCK & AMBULANCE, issued 1992
1. MB57-H Power Truck in neon orange; MB25-E Ambulance in white, with accessories ($6-8)

010301 DODGE INSTANT WINNER GAME, issued 1984
1. MB28-E Dodge Daytona in maroon; MB68-E Dodge Caravan in black ($20-25)

010120 LIMITED EDITION CHRIST-MAS OFFER, issued 1984
1. assorted vehicles esp. English models using a Christmas stocking sleeve ($12-15 each)

32520 DAYS OF THUNDER RACE CAR CHALLENGE, issued 1990
1. 2X MB54-G Chevy Lumina in "Superflo" & "City Chevrolet" liveries ($10-15)
2. 2X MB54-G Chevy Lumina in "Hardees" & "Mello Yello" liveries ($10-15)

THEME & GIFT PACKS

This section covers three pack and multiple pack sets issued in themes or for other purposes, issuing three vehicles together. Like the Gift sets, different prefixes are used as well as serial numbers with a few unnumbered sets. Prefixes used include MP (Multipack), MB (Matchbox), EM (Emergency), CS (Construction), SB (Skybuster) & GF (Graffic Traffic).

MP-101 EMERGENCY SET, issued 1988
1. contains MB8-G, MB21-E, MB60-F

MP-102 CONSTRUCTION SET, issued 1988
1. contains MB19-D, MB64-D, MB70-E

MP-103 AIRPORT FIRE SET, issued 1988
1. contains MB18-C, MB54-F, MB75-D

MP-104-A 4X4 SET, issued 1988
1. contains MB5-D, MB20-C, MB37-F

MP-105-A DRAGSTER SET, issued 1988
1. contains MB26-D, MB48-D, MB69-D

MP-106-A PORSCHE SET, issued 1988
1. contains MB3-C, MB7-F, MB55-F

MP-107-A FARM SET, issued 1989
1. contains MB46-C, MB51-C, MB71-C (1989)
2. contains MB46-C, MB51-C, MB71-C (Boot's packaging) (1991)

MP-108-A NASA SET, issued 1989
1. contains MB54-F, MB68-E, MB72-F in NASA liveries MP-109-A

MP-109-A FERRARI SET, issued 1990
1. MB24-F, MB70-D, MB75-E

MP-803-A FERRARI SET, issued 1989
1. MB24-F, MB70-D, MB75-E in special red packaging
2. MB24-F, MB70-D, MB75-E in special gray packaging (Italy)

MP-804-A PROSCHE SET, issued 1989
1. MB3-C, MB7-F, MB71-G in black with special red packaging

MP-805-A HOBBY SET, issued 1992 (Holland/Austria)
1. contains MB27-D, MB31-C & TP-21 trailer
2. contains MB31-C, MB39-D & TP-110 raft
3. contains MB35-F, MB5-B, MB43-A
4. contains MB3-C, MB9-A, & TP7 Glider

MB-170-A BUY 2 GET 1 FREE AT WOOLWORTH'S, issued 1987
1. contains three assorted miniatures (contents varies)

MB-803-A MOTORPLAY, issued 1989 (Early Learning Centre pack)
1. contains MB19-D, MB64-D, MB70-E
2. contains MB8-G, MB21-F, MB60-G
3. contains MB4-E, MB17-C, MB66-G

831-A SUPER VALUE PACK FREE SKYBUSTER, issued 1991
1. contains four miniatures & 1 Skybuster (contents vary)

MB-838-A TRAINPLAY, issued 1991 (Early Learning Centre pack)
1. contains MB43-C, 2X MB44-C

MB860-A BUY 2 GET 1 FREE, issued 1990
1. contains three assorted miniatures (contents vary)

MB861-A SONDERPREIS, issued ? (German)
1. contains three assorted miniatures (contents vary)

MB862-A THREE PIECE SET, issued 1991
1. contains three assorted miniatures (contents vary)

1101 METRO POLICE DEPT., issued 1987 (U.S.)
1. contains MB10-C, MB46-F, MB50-E
2. contains MB16-E, MB46-F, MB50-E

1102 METRO FIRE DEPT., issued 1987 (U.S.)
1. contains MB18-C, MB25-F, MB63-E

1103 SPORTS RACERS, issued 1987 (U.S.)
1. contains MB2-E, MB55-F, MB67-F
2. contains MB10-D, MB32-E, MB51-G

1104 4X4 MOUNTAIN ADVENTURE, issued 1989 (U.S.)
1. contains MB5-D, MB13-D, MB39-E
2. contains MB5-D, MB20-C, MB37-F

1105 HOT ROD DRIVE-IN, issued 1989 (U.S.)
1. contains MB4-D, MB40-C, MB73-C
2. contains MB2-G, MB4-D, MB59-D
3. contains MB2-G, MB66-F, MB73-C

1106 TRUCK STOP, issued 1989 (U.S.)
1. contains MB8-F, MB30-E, MB45-C
2. contains MB8-H, MB30-E, MB45-C

0312 PUFFY STICKERS, issued 1983
1. contains MB11-E, MB34-C, MB55-F (Track Burners)
2. contains MB7-D, MB13-D, MB22-D (Off Road 4X4s)
3. contains MB4-D, MB40-C, MB71-D (Street Racers)

2501 ROADBLASTERS, issued 1987
1. contains an assortment of 3 miniatures with armament (contents vary)

010779 FREE RIDE BUY 2 GET 1 FREE, issued 1985
1. contains three assorted miniatures (contents vary)

010787 HALLEY'S COMET BUY 2 GET 1 FREE, issued 1985
1. contains two assorted miniatures with either MB12-E, MB34-C, 59-B Halley's

32900 DREAM MACHINES, issued 1992
1. contains MB14-G, MB43-F, MB53-F in special colors
2. contains MB5-D, MB22-F, MB70-D in special colors

25823 SUPER VALUE PACK WITH FREE POSTER, issued 1990
1. contains five assorted miniatures with poster (contents vary)

SB-807 BUY 2 SKYBUSTERS, GET 2 CARS FREE, issued 1990
1. contains 2 assorted miniatures & 2 assorted miniatures (contents vary)

GF-180 GRAFFIC TRAFFIC THREE PACK, issued 1991
1. contains 3 assorted miniatures (contents vary)

LONDON LIFE, issued 1987 (U.K.)
1. contains MB4-E, MB17-C, MB66-G

LE MANS, issued 1987 (U.K.)
1. contains MB55-F, MB66-E, MB67-G

CONSTRUCTION, issued 1987 (U.K.)
1. contains MB30-E, MB32-D, MB64-D

ROYAL MAIL BY PROMOD, issued 1987 (U.K.)
1. contains MB38-E, MB60-G, MB66-E in Royal Mail delivery

GERMAN FIVE PACK CHRISTMAS STOCKING, issued 1989
1. contains five assorted miniatures (contents vary)

SUPER GT CHRISTMAS STOCKING, issued 1991
1. contains five assorted Super GT's (contents vary)

SUPER GT THREE & FOUR PACKS, issued 1987

1. contains three assorted Super GT's (contents vary)
2. contains four assorted Super GT's (contents vary)

CONVOY

With the inception of Convoys in 1982, the last ten years have made Convoys one of the most popular Matchbox series ever. The series branched off in 1989 to include special promotionals by White Rose Collectibles (see elsewhere).

this list is continued from book #2

CY-3-B DAF Box Truck (AU)
CY-4-B Scania Box Truck (AU)
CY-10-A Racing Transporter
CY-11-A Kenworth Helicopter Transporter
CY-12-A Kenworth Aircraft Transporter
CY-13-A Peterbilt Fire Engine
CY-14-A Kenworth Boat Transporter
CY-15-A Peterbilt Tracking Vehicle
CY-16-A Scania Box Truck
CY-17-A Scania Petrol Tanker
CY-18-A Scania Double Container
CY-19-A Peterbilt Box Truck
CY-20-A Kenworth Articulated Dumper Truck
CY-21-A DAF Aircraft Transporter
CY-22-A DAF Boat Transporter
CY-23-A Scania Covered Truck
CY-24-A DAF Box Truck
CY-25-A DAF Container Truck
CY-26-A DAF Double Container Truck
CY-27-A Mack Container Truck
CY-28-A Mack Double Container

CY-29-A Mack Aircraft Transporter
CY-30-A Grove Crane
CY-31-A Mack Pipe Truck
CY-32-A Mack Tractor Shovel Transporter
CY-33-A Mack Helicopter Transporter
CY-34-A Peterbilt Emergency Center
CY-35-A Mack Tanker
CY-36-A Kenworth Transporter
CY-104-A Kenworth Superstar Transporter
CY-105-A Kenworth Tanker
CY-106-A Peterbilt Dumper Truck
CY-107-A Mack Superstar Transporter
CY-108-A DAF Airplane Transporter
CY-109-A Ford Aeromax Superstar Transporter
CY-110-A Kenworth Superstar Transporter
CY-111-A Racing Transporter
CY-203-A Peterbilt Lowloader with Excavator
CY-803-A Scania Lowbed with Dodge Truck

CY-1-A KENWORTH CAR TRANSPORTER, issued 1982 (continued)

5. red cab with white/yellow/black stripes, red trailer with beige ramp & no stripe, Macau casting ($6-8)
6. red cab with white/yellow/black stripes, red trailer with beige ramp with stripe, Macau casting ($6-8)
7. yellow cab with chrome exhausts & blue & purple tempa, dark blue trailer with yellow ramp, Macau casting ($6-8)
8. yellow cab with gray exhausts & blue & purple tempa, dark blue trailer with yellow ramp, Macau casting ($6-8)
9. yellow cab with gray exhausts & blue & purple tempa, dark blue trailer with yellow ramp, Thailand casting ($6-8)
10. blue cab with chrome exhausts & yellow & orange tempa, blue trailer with yellow ramp, Thailand casting ($4-6)

CY-2-A KENWORTH ROCKET TRANSPORTER, issued 1982 (continued)

4. pearly silver cab with chrome exhausts, white plastic rocket, Macau casting ($6-8)
5. white cab with chrome exhausts, white plastic rocket, Macau casting ($5-7)
6. white cab with gray exhausts, white plastic rocket, Macau casting ($5-7)
7. white cab with gray exhausts, white plastic rocket, Thailand casting ($5-7)
8. white cab with gray exhausts, chrome plastic rocket, Thailand casting ($4-5)

CY-3-A PETERBILT DOUBLE CONTAINER, issued 1982 (continued)

8. MB43-D cab in red, clear windows, brown containers, black trailer, England casting, "Uniroyal" labels ($10-15)
9. MB45-C cab in white, amber windows, beige containers, black trailer, England casting, "Pepsi" labels ($350-500)
10. MB43-D cab in black, clear windows, cream containers, black trailer, England casting "Pepsi" labels ($350-500)
11. MB43-D cab in black, clear windows, cream containers, black trailer, England casting, "Federal Express" labels ($15-18)
12. MB43-D cab in white, clear windows, beige containers, black trailer, England casting "Federal Express" labels ($15-18)
13. MB43-D cab in white , clear windows, white containers, black trailer, England casting, "Federal Express" tempa ($6-8)
14. MB41-D cab in red, clear windows, light tan containers, white trailer, England casting, "Mayflower" labels ($1000+)
15. MB43-D cab in white with gray exhausts, clear windows, white containers, black trailer, Macau casting, "Federal Express" tempa ($6-8)
16. MB43-D cab in white with gray exhausts, clear windows, white containers, black trailer, Thailand casting, "Federal Express" tempa ($4-5)

CY-3-B KENWORTH BOX TRUCK, issued 1985

1. red cab, red container, yellow trailer casting, "Linfox" tempa ($12-15)(AU)

CY-4-A KENWORTH BOAT TRANSPORTER, issued 1982 (continued)

7. dark orange cab, boat with dark orange hull & green windows, pearly silver trailer, Macau casting ($6-8)
8. very light orange cab, boat with dark orange hull & green windows, pearly silver trailer, Macau casting ($6-8)

CY-4-B SCANIA BOX TRUCK, issued 1985

1. white cab, white container, black trailer base, "Ansett" (towards front) labels ($12-15)(AU)
2. white cab, white container, black trailer base, "Ansett" (toward rear) labels ($15-20)(AU)

CY-5-A PETERBILT COVERED TRUCK, issued 1982 (continued)

3. MB41-D cab in green, green cover, white trailer, England casting, "Interstate Trucking" labels ($18-25)

4. MB43-D cab in green with chrome exhausts, green cover, white trailer, Macau casting, "Interstate Trucking" tempa ($12-15)
5. MB43-D cab in yellow with chrome exhausts, yellow cover, pearly silver trailer, Macau casting, "Michelin" tempa ($8-12)
6. MB43-D cab in orange with chrome exhausts, light gray cover, pearly silver trailer, Macau casting, "Walt's Farm Fresh Produce" tempa ($6-8)
7. MB43-D cab in orange with gray pipes light gray cover, pearly silver trailer, Macau casting, "Walt's Farm Fresh Produce" tempa ($6-8)
8. MB43-D cab in orange with gray pipes, light gray cover, pearly silver trailer, Thailand casting, "Walt's Farm Fresh Produce" tempa ($4-6)

CY-6-A KENWORTH HORSE BOX, issued 1982 (continued)

5. green cab with white & black stripes, beige box, green door, pearly silver trailer, Macau casting, "Blue Grass Farms" tempa ($8-12)
6. green cab with white & black stripes, beige box, white door, pearly silver trailer, Macau casting, "Blue Grass Farms" tempa ($8-12)

CY-7-A PETERBILT GAS TANKER, issued 1982 (continued)

7. MB43-D cab in white with no tempa, light amber windows, white tank, black trailer base, England casting, "Supergas" labels ($1000+)
8. MB43-D cab in black with yellow & red with "Z" tempa, clear windows, orange-yellow tank, black trailer base, Macau casting, "Supergas" tempa ($8-12)
9. MB43-D cab in red with "Getty" tempa, clear windows, chrome tank, pearly silver trailer base, Macau casting, "Getty" tempa ($6-8)

CY-8-A KENWORTH BOX TRUCK, issued 1982 (continued)

4. MB41-D cab in red with black/white tempa, red container with white roof & doors, England casting, "Redcap" labels ($45-60)
5. MB41-D cab in red with black/white tempa, red container with white roof & black doors, England casting, "Redcap" labels ($45-60)
6. MB41-D cab in red with black/white tempa, red container with black roof & doors, England casting, "Redcap" labels ($150-200)
7. MB41-D cab in red with black/white tempa, red container with black roof & white doors, England casting, "Redcap" labels ($150-200)
8. MB45-C cab in white with brown/blue tempa, red container with white roof & doors, England casting, "Ski Fruit Yogurt" labels ($100-150)(AU)(C2)
9. MB45-C cab in white with orange/yellow tempa, red container with white roof & doors, England casting, "Ski Fruit Yogurt" labels ($100-150)(AU)(C2)
10. MB45-C cab in white with orange/yellow tempa, red container with white roof & red doors, England casting, "Redcap" labels ($45-60)
11. MB45-C cab in silver-gray & blue with yellow/red tempa, silver-gray container with blue roof & doors, Macau casting,

"Matchbox Showliner" labels ($300-375)
12. MB45-C cab in white with red band tempa, white container with white roof & doors, Macau casting, "K-Line tempa ($300-450)(US)
13. MB45-C cab in white with yellow/red/blue tempa, white container with white roof & doors, Macau casting, "Matchbox" tempa ($12-15)
14. MB45-C cab in white with yellow/red/blue tempa, white container with white roof & doors, Macau casting, "Matchbox", "This Truck Delivers 1988" roof label ($40-60)(UK)
15. MB45-C cab in white with yellow/red/blue tempa, white container with white roof & doors, Macau casting, "Matchbox", "This Truck Delivers 1989" roof label ($350-500)(HK)
16. MB45-C cab in white with red band tempa, red container with red roof & doors, Macau casting, "K-Line" tempa ($125-175)(US)
17. MB45-C cab in black with "Harley Davidson" tempa, black container with black roof & doors, Macau casting, "Harley Davidson" labels ($5-8)(HD)

CY-9-A KENWORTH BOX TRUCK, issued 1982 (continued)

3. MB45-C cab in black, black container, England casting, "Midnight X-Press" tempa ($40-60)
4. MB41-D cab in black, black container, Macau casting, "Midnight X-Press" tempa ($6-8)
5. MB41-D cab in black, black container, Macau casting, "Moving In New Directions" tempa ($125-175)(AU)
6. MB41-D cab in black, black container,

Macau casting, "Moving In New Directions" tempa, "Personal Contact Is Barry Oxford" roof label ($250-300)(AU)
7. MB41-D cab in black, black container, Macau casting, "Moving In New Directions" tempa, "Personal Contact Is Anita Jones" roof label ($250-300)(AU)
8. MB41-D cab in black, black container, Macau casting, "Moving In New Directions" tempa, "Personal Contact Is Keith Mottram" roof label ($250-300)(AU)
9. MB41-D cab in black, black container, Macau casting, "Moving In New Directions" tempa, "Personal Contact Is Terry Blyton" roof label ($250-300)(AU)
10. MB41-D cab in black, black container, Macau casting, "Moving In New Directions" tempa, "Personal Contact is Jenny Brindley" roof label ($250-300)(AU)
11. MB41-D cab in black, yellow container, Macau casting, "Stanley" tempa ($12-15)(US)(OP)
12. DAF cab in yellow, yellow container, Macau casting, "IPEC" tempa ($12-15)(AU)
13. MB41-D cab in white, white container, Macau casting, "Paul Arpin Van Lines" tempa ($01-15)(US)
14. MB41-D cab in white, white container, Macau casting, "Matchbox Compliments Macau Diecast Co. Ltd." tempa ($1000+)(Macau)
15. MB41-D cab in white, white container, Macau casting, "Matchbox In Celebration of Universal Group's 20th Anniversary" tempa ($1000+)(HK)
16. MB41-D cab in white, white container, Macau casting, "Canadian Tire" tempa

17. MB41-D cab in white, white container, Macau casting, "Merry Christmas 1988 MICA Members" with roof label ($25-40)
18. MB41-D cab in blue, blue container, Macau casting, "Mitre 10" tempa ($10-15)(AU)
19. MB41-D cab in blue, blue container, Macau casting, "Spaulding" tempa ($25-40)(US)
20. MB41-D cab in white, white container, Macau casting "Merry Christmas MICA Members 1990" labels ($25-40)(C2)
21. MB41-D cab in black, black container, Thailand casting, "Midnight X-Press" tempa ($4-6)
22. MB41-D cab in black, black container, Thailand casting, "Cool Paint Co." labels ($4-6)

CY-10-A RACING TRANSPORTER, issued 1983
1. white body, "Tyrone Malone" tempa, MB66-D Super Boss with green windows ($20-25)
2. white body, "Tyrone Malone" tempa, MB66-d Super Boss with red windows ($25-35)

CY-11-A KENWORTH HELICOPTER TRANSPORTER, issued 1983
NOTE; helicopter comes with clear or amber windows; orange or black base; gray or black interior for additional variants.
1. silver-gray cab with "Ace Hire"; MB75-D Helicopter in silver-gray with "600" tempa, England casting ($8-12)
2. pearly silver cab with "Ace Hire"; MB75-D Helicopter in pearly silver with "600" tempa, Macau casting ($7-10)
3. black cab with chrome exhausts & "Air Car" tempa; MB75-D Helicopter in black with "Aircar" tempa, Macau casting ($7-10)
4. black cab with gray exhausts & "Air Car" tempa; MB75-D Helicopter in black with "Air Car" tempa, Macau casting ($6-8)
5. dark blue cab with chrome exhausts & white/gold stripes tempa; MB75-D Helicopter in white with "Rescue" tempa ($8-12)(MC)
6. dark blue cab with gray exhausts & white/gold stripes tempa; MB75-D Helicopter in white with "Rescue" tempa ($8-12)(MC)
7. black cab with gray exhausts & "Air Car" tempa; MB75-D Helicopter in black in "Air Car" tempa, Thailand casting ($5-7)

CY-12-A KENWORTH BOAT TRANSPORTER, issued 1984
1. white cab with blue & dark green tempa, silver-gray trailer, blue plane with "Darts" tempa, England casting ($12-15)
2. white cab with blue & brown tempa, silver-gray trailer, blue plane with "Darts" tempa, England casting ($12-15)
3. white cab with 2 tone blue tempa, pearly silver trailer, blue plane with "Darts" tempa, Macau casting ($8-12)

CY-13-A PETERBILT FIRE ENGINE, issued 1984
1. MB43-D cab in red, red trailer with white lettered "8" & "Fire Dept.", white ladder, England casting ($15-18)
2. MB45-C cab in red with "Denver" label, red trailer with white lettered "8" & "Fire Dept.", white ladder, England casting ($1000+)
3. Peterbilt with dome lights in red, red trailer with white lettered "8" & "Fire Dept.", white ladder, Macau casting ($7-10)
4. Peterbilt with dome lights in red, red trailer with yellow lettered "8" & "Fire Dept.", white ladder, Macau casting ($7-10)
5. Peterbilt with dome lights in red, red trailer with white lettered "8" & white lettered "Fire Dept.", white ladder, Macau casting ($7-10)
6. Peterbilt with dome lights in red, red trailer with white lettered "8" & "Fire Dept.", white ladder, Thailand casting ($4-6)
7. Peterbilt with dome lights in fluorescent orange, fluorescent orange trailer with "City Fire Dept. 15" & checkers, white ladder, Thailand casting ($4-6)(EM)

CY-14-A KENWORTH BOAT TRANSPORTER, issued 1985
1. white cab, white boat, pearly silver trailer with brown cradle ($8-12)

CY-15-A PETERBILT TRACKING VEHICLE, issued 1985
1. white cab with chrome exhausts & "NASA" tempa, white container with pearly silver trailer, Macau casting, "NASA" tempa ($6-8)
2. yellow cab with chrome exhausts & no tempa, yellow container with pearly silver trailer, Macau casting, "British Telecom" tempa ($8-12)(CY)
3. powder blue cab with chrome exhausts & "Satellite" & TV bolt, powder blue container with pearly silver trailer, Macau casting, "MB TV News" tempa ($12-15)
4. powder blue cab with chrome exhausts & "Peterbilt" & TV bolt, powder blue container with pearly silver trailer, Macau casting, "MB TV News" tempa ($6-8)
5. powder blue cab with gray exhausts & "Peterbilt" & TV bolt, powder blue container with pearly silver trailer, Macau casting, "MB TV News" tempa ($6-8)
6. olive cab with black exhausts/base & "Strike Team" tempa, olive container with olive trailer, Macau casting, "LS2009" tempa ($20-25)(CM)
7. dark blue cab with gray exhausts & "Peterbilt" & TV bolt, dark blue container with pearly silver trailer, Macau casting, "MB TV News" tempa ($6-8)
8. dark blue cab with gray exhausts & "Peterbilt" & TV bolt, dark blue container with pearly silver trailer, Thailand casting, "MB TV News" tempa ($6-8)
9. powder blue cab with gray exhausts & "Peterbilt" & TV bolt, powder blue container with pearly silver trailer, Thailand casting, "MB TV News" tempa ($6-8)
10. white cab with gray exhausts & "Sky TV", white container with pearly silver trailer, Thailand casting, "Sky Satellite TV" tempa ($4-5)
11. white cab with chrome exhausts & "Sky TV", white container with pearly silver trailer, Thailand casting, "Sky Satellite TV" tempa ($4-5)

CY-16-A SCANIA BOX TRUCK, issued 1985
1. white cab/green chassis & chrome base,

white container with black trailer, Macau casting, "7 Up" (towards rear) labels ($12-15)(US)

2. white cab/green chassis & chrome base, white container with black trailer, Macau casting, "7 Up" (towards front) labels ($12-15)(US)

3. white cab/green chassis & chrome base, white container with black trailer, Macau casting, "7 Up" (upside down) labels ($20-35)(US)

4. white cab/dk. blue chassis & chrome base, dark blue container with white trailer, Macau casting, "Duckham's Oils" tempa ($10-15)

5. purple cab/chassis & chrome base, white container with purple trailer, Macau casting, "Edwin Shirley" tempa ($12-15)(UK)

6. white cab/black chassis & chrome base, white container with black trailer, Macau casting, "Wimpey" tempa ($12-15)(UK)

7. red cab/white chassis & chrome base, red container with white trailer, Macau casting, "Kentucky Fried Chicken" tempa ($12-15)(UK)

8. white cab/blue chassis & chrome base, white container with blue trailer, Macau casting, "Signal Toothpaste" tempa ($12-15)(UK)

9. white cab/red chassis & chrome base, red container with red trailer, Macau casting, "Heinz Tomato Ketchup Squeezable" labels ($12-15)(UK)

10. yellow cab/white chassis & chrome base, white container with red trailer, Macau casting, "Weetabix" labels ($12-15)(UK)

11. blue cab/chassis & chrome base, blue container with blue trailer, Macau

casting, "Matey Bubble Bath" tempa ($12-15)(UK)

12. white cab/black chassis & chrome base, white container with black trailer, Macau casting, "Golden Wonder Potato Crisps" tempa ($12-15)(UK)

13. white cab/red chassis & chrome base, white container with red trailer, Macau casting, "Merchant Tire & Auto Centers" tempa ($10-15)(US)

14. white cab/red chassis & chrome base, white container with red trailer, Macau casting, "Merry Christmas 1988 MICA Members" with roof label ($25-40)(C2)

15. yellow cab/white chassis & black base, yellow container with yellow trailer, Macau casting, "Weetabix" tempa ($12-15)(UK)

16. purple cab/red chassis & black base, purple container with red trailer, Macau casting, "Ribena" tempa ($12-15)(UK)

17. red cab/white chassis & black base, red container with white trailer, Macau casting, "Kentucky Fried Chicken" tempa ($12-15)(UK)

18. white cab/green chassis & chrome base, white container with black trailer, Macau casting, "Merry Christmas 1989 MICA Members ($25-40)(C2)

19. white cab/blue chassis & black base, white container with black trailer, Thailand casting, "Goodyear Vector" tempa ($6-8)

CY-17-A SCANIA PETROL TANKER, issued 1985

1. white cab/blue chassis & chrome base, white tank with blue base, Macau casting, "Amoco" tempa ($6-8)

2. red cab/red chassis & chrome base, red

tank with red base, Macau casting "Tizer" tempa ($12-15)(UK)

3. white cab/green chassis & chrome base, white tank with green base, Macau casting, "Diet 7 Up" tempa ($12-15)(UK)

4. orange cab/chassis & chrome base, orange tank with white base, Macau casting, "Cadbury's Fudge" tempa ($12-15)(UK)

5. white cab/dark gray chassis & chrome base, chrome tank with dark gray base, Macau casting, "Shell" tempa ($6-8)

6. white cab/dark gray chassis & black base, chrome tank with dark gray base, Macau casting, "Shell" tempa ($6-8)

7. white cab/dark gray chassis & black base, chrome tank with dark gray base, Thailand casting, "Shell" tempa ($6-8)

8. white cab/dark gray chassis & black base, white tank with dark gray base, Thailand casting, "Feoso" tempa ($35-45)(CH)

CY-18-A SCANIA DOUBLE CONTAINER, issued 1986

1. blue cab with black interior, dark blue containers, yellow trailer base, "Varta Batteries" tempa ($30-45)

2. blue cab with gray interior, dark blue containers, yellow trailer base, "Varta Batteries" tempa ($10-15)

3. white cab, white containers, dark blue trailer base, "Wall's Ice Cream" tempa ($12-15)(UK)

4. red cab, red containers, red trailer base, "Kit Kat" tempa ($12-15)(UK)

5. orange cab, orange containers, black trailer base, "Breakaway" tempa ($12-15)(UK)

6. white cab, white containers, green trailer base, "7 Up" tempa ($12-15)(UK)

7. red cab, red containers, black trailer base, "Beefeater Steak Houses" tempa ($12-15)(UK)

CY19-A PETERBILT BOX TRUCK, issued 1987

1. white cab, white container with pearly silver trailer base, "Ansett Wridgeays" tempa ($15-20)

CY-20-A KENWORTH ARTICULATED DUMPER TRUCK, issued 1987

1. MB45C cab in yellow with chrome exhausts, yellow trailer with black base, Macau casting, "Taylor Woodrow" tempa ($6-8)

2. MB8E cab in pink with chrome base, pink trailer with black base, Macau casting, "Readymix" tempa ($15-18)(AU)

3. MB45-C cab in yellow with chrome exhausts, yellow trailer with black base, Macau casting, "Eurobran" tempa ($6-8)

4. MB45-C cab in yellow with chrome exhausts, yellow trailer with black base, Macau casting, black & white road design tempa ($6-8)

5. MB45-C cab in green with chrome exhausts, yellow trailer with black base, Macau casting, "Eurobran" tempa ($6-8)(MC)

6. MB45-C cab in green with gray exhausts, yellow trailer with black base, Macau casting, "Eurobran" tempa ($6-8)(MC)

7. MB45-C cab in yellow with gray exhausts, yellow trailer with black base, Macau casting, "Taylor Woodrow" tempa ($6-8)

8. MB8-E cab in pink with black base, pink trailer with black base, Macau casting, "Readymix" tempa ($18-25)(AU)

9. MB45-C cab in green with gray exhausts, yellow trailer with black base, Thailand casting, "Eurobran" tempa ($5-7)
10. MB45-C cab in red with gray exhausts, yellow trailer with black base, Thailand casting, red design tempa ($4-5)(CS)

CY-21-A DAF AIRCRAFT TRANSPORTER, issued 1987

1. white cab/blue chassis with "Space Cab" tempa, dark blue trailer, orange plane with "Airtrainer" tempa, Macau casting ($7-10)
2. white cab/dark blue chassis with "Red Rebels" tempa, white trailer, red plane with "Red Rebels" tempa, Macau casting ($6-8)(MC)
3. black cab/gray chassis with "AC102" tempa, black trailer, black plane with "AC102" tempa ($25-40)(CM)
4. white cab/gray chassis with no tempa, white trailer, white plane with no tempa ($15-25)(GF)

CY-22-A DAF BOAT TRANSPORTER, issued 1987

1. white cab/blue chassis with "Lakeside" tempa, dark blue trailer, boat with white deck, red hull & "Shark" tempa, Macau casting ($7-10)
2. white cab/blue chassis with "P&G" tempa, dark blue trailer, boat with white deck, orange-brown hull & "CG22" tempa, Macau casting ($7-10)
3. white cab/black chassis with "Coast Guard" tempa, black trailer, boat with gray deck, white hull & "Coast Guard" tempa, Macau casting ($6-8)
4. white cab/black chassis with "Coast Guard" tempa, black trailer, boat with

gray deck, white hull & "Coast Guard" tempa, Thailand casting ($5-7)
5. white/fluorescent orange chassis & orange trailer, boat with fluorescent orange deck, white hull & "Marine Rescue" tempa ($4-5)

CY-23-A SCANIA COVERED TRUCK, issued 1988

1. yellow cab/blue chassis & chrome base, yellow canopy with blue trailer sides on pearly silver base, "Michelin" tempa ($7-10)
2. yellow cab/blue chassis & black base, yellow canopy with blue trailer sides on pearly silver base, "Michelin" tempa ($7-10)

CY-24-A DAF BOX TRUCK, issued 1988

1. red cab with black chassis, red container with pearly silver base, Macau casting, "Ferrari" tempa ($6-8)
2. blue cab with black chassis, blue container with black base, Macau casting, "Pickfords" tempa ($6-8)
3. white cab with black chassis, white container with pearly silver base, Macau casting, "Porsche" labels ($5-7)
4. white cab with black chassis, white container with pearly silver base, Thailand casting, "Porsche" labels ($5-7)
5. red cab with black chassis, dark red container with pearly silver base, Thailand casting, "Ferrari" tempa ($5-7)
6. white cab with red chassis, white container with red base, Thailand casting, "Circus Circus" labels ($5-7)(MC)

CY-25-A DAF CONTAINER TRUCK, issued 1989

1. yellow cab/chassis, yellow container with yellow base, Macau casting, "IPEC" tempa ($12-15)(AU-CY9)
2. blue cab/chassis, blue container with blue base, Macau casting, "Crookes Healthcare" tempa ($12-15)(UK)
3. white & orange cab/black chassis, white container with black base, Macau casting, "Unigate" tempa ($12-15)(UK)
4. red cab/black chassis, red container with black base, Macau casting, "Royal Mail Parcels" tempa ($12-15)(UK)
5. blue cab/chassis, blue container with blue base, Macau casting, "Comma Performance Oil" tempa ($12-15)(UK)
6. white cab/bright blue chassis, white container with bright blue base, Macau casting, "Leisure World" tempa ($12-15)(UK)
7. white cab/blue chassis, white container with black base, Macau casting, "Pepsi Team Suzuki" tempa ($12-15)
8. white & orange cab/orange chassis, white container with orange base, Macau casting, "TNT IPEC" tempa ($8-12)(TC)
9. white cab/blue chassis, white container with blue base, Macau casting "Pioneer" tempa ($12-15)(UK)
10. met. gold cab/chassis, black container "Duracell" tempa ($12-15)(UK)
11. yellow cab/black chassis, yellow container with black base, Macau casting, "Zweifel Pomy Chips" tempa ($12-15)(SW)
12. green cab/chassis, white container with black base, Macau casting, "M" & orange stripe tempa ($12-15)(SW)
13. white cab/red chassis, white container

with red base, Macau casting, "Toblerone" tempa ($7-10)
14. white cab/black chassis, white container with black base, Macau casting, "Pirelli Gripping Stuff" tempa ($7-10)(TC)
15. white cab/black chassis, white container with black base, Macau casting, "XP" tempa ($7-10)(TC)
16. white cab/red chassis, white container with red base, Thailand casting "Toblerone" tempa ($6-8)
17. white cab/yellow chassis, white container with yellow base, Thailand casting, "HB Racing" tempa ($25-40)
18. white cab/green chassis, white container with green base, Thailand casting, "Garden Festival Wales" tempa ($12-15)(WL)(GS)
19. brown cab/black chassis, light gray container with black chassis, Thailand casting, "United Parcel Service" tempa ($20-25)(UK)

CY-26-A DAF DOUBLE CONTAINER TRUCK, issued 1989

1. powder blue cab, black chassis, dark blue containers with black trailer, Macau casting, "P & O" tempa ($10-12)
2. powder blue cab, black chassis, dark blue containers with black trailer, Thailand casting, "P & O" tempa ($10-12)

CY-27-A MACK CONTAINER TRUCK, issued 1990

1. white cab/black chassis, white container with black trailer, "A Great Name In Trucks-Mack" labels ($12-15)
2. chrome cab/black chassis, black container with black trailer, "Celebrating A Decade of Matchbox Conventions 1991" labels ($30-45)(C2)

CY-28-A MACK DOUBLE CONTAINER, issued 1990

1. white cab/black chassis, white containers with black trailer, "Big Top Circus" tempa ($10-15)
2. white cab/blue chassis, white containers with blue trailer, "Big top Circus" tempa ($6-8)
3. white cab/black chassis, white containers with black trailer, "DHL Worldwide Express" tempa ($5-7)(TC)
4. red cab/black chassis, white containers with red trailer, "Big Top Circus" tempa ($4-5)

CY-29-A MACK AIRCRAFT TRANSPORTER, issued 1991

1. red cab/black chassis, white carriage with black trailer, red plane with "Red Rebels" tempa ($4-6)

CY-30-A GROVE CRANE, issued 1992

1. orange-yellow body, red crane cab, yellow boom, "AT1100 Grove" tempa ($10-15)

CY-31-A MACK PIPE TRUCK, issued 1992

1. red cab/black chassis, red trailer with silver-gray sides with black base, yellow plastic pipes ($8-12)

CY-32-A MACK SHOVEL TRANSPORTER, issued 1992

1. orange-yellow cab/red chassis, red trailer, MB29-C Tractor Shovel in yellow with red scoop ($8-12)

CY-33-A MACK HELICOPTER TRANSPORTER, issued 1992

1. white cab/blue chassis, white carriage with blue base, "Police" tempa; MB46-F Mission Helicopter in white & black ($7-10)
2. white cab/green chassis, green carriage with silver-gray base, "Polizei" tempa; MB46-F Mission Helicopter in green & white ($7-10)
3. white cab/black chassis, white carriage with black base, "Police" tempa; MB46-F Mission Helicopter in black & white ($7-10)

CY-34 PETERBILT EMERGENCY CENTER, issued 1992

1. fluorescent orange cab, fluorescent orange trailer with gray roof & white boom, "Rescue" with checkers tempa ($8-12)

CY-35-A MACK TANKER, issued 1992

1. white & fluorescent green with chrome base, chrome tank with fluorescent green base, "Orange Juice" tempa ($7-10)
2. white & fluorescent green with gray base, chrome tank with fluorescent green base, "Orange Juice" tempa ($6-8)

CY-36-A KENWORTH TRANSPORTER, issued 1992

1. white cab, white container with black base, "Charitoys" labels ($25-35)(US)
2. orange cab, black container with black base, "Trick Truckin" labels ($4-6)
NOTE: due to the popularity of Nascar racing in the 1990s please consult various Nascar diecast magazines for the most current prices on Nascar related Convoys. Prices fluctuate tremendously!

CY-104-A KENWORTH SUPERSTAR TRANSPORTER, issued 1989

1. white cab with "STP" logo, white container with black base, "Richard Petty/STP" labels ($85-125)(WR)
2. white cab, white container with black base, "Neil Bonnett/Citgo" labels ($100-125)(WR)
3. white cab, white container with black base, "Hardee's Racing" labels ($100-125)(WR)
4. black cab, black container with silver-gray base, "Goodwrench Racing Team" labels ($100-175)(WR)
5. white cab, dark blue container with black base, "Goodyear Racing" labels ($75-100)(WR)
NOTE: all models listed below have black trailer bases.
6. black cab, black container, "Exxon 51" labels ($25-40)(DT)
7. black/green cab, black container, "Mello Yello 51" labels ($25-40)(DT)
8. orange cab, orange container, "Hardees 18" labels ($25-40)(DT)
9. pink cab, white container, "Superflo" labels ($25-40)(DT)
10. white cab, white container, "City Chevrolet" labels ($25-40)(DT)
11. white cab with red/blue tempa, white container with black base, "Richard Petty/STP labels ($90-125)(WR)
NOTE: above models with Macau base, below models with Thailand base.
12. black cab, black container, "Goodwrench Racing Team" labels ($60-85)(WR)
13. gold cab without "6" on doors, gold container, "Folgers" labels ($50-75)(WR)
14. white cab, white container, "Trop Artic" with "Dick Trickle" labels ($35-50)(WR)
15. white & blue cab, white container, "Valvoline" labels ($7-10)(IN)
16. yellow cab, yellow container, "Pennzoil 4" labels ($7-10)(IN)
17. gold cab with "6" on doors, gold container, "Folgers" labels ($25-35)(WR)
18. black cab, black container, "Goodwrench Racing Team" with team car depicted on sides labels ($20-35)(WR)
19. dark blue cab, dark blue container, "94 Sunoco" without "Sterling Marlin" on labels ($250-300)(WR)
20. dark blue cab, dark blue container, "94 Sunoco" with "Sterling Marlin" on labels ($250-300)(WR)
21. white cab, white container, "Crown Moroso" labels ($35-50)(WR)
22. black cab, black container, "Texaco Havoline/Davey Allison" labels ($35-50)(WR)
23. white cab, white container, "Richard Petty" with portrait labels ($25-40)(WR)
24. dark blue cab, white container, "Penn State 1855-1991" labels ($12-15)(WR)
25. white cab, white container, "Trop Artic" with "Lake Speed" labels ($18-25)(WR)
26. blue cab, blue container, "Maxwell House Racing" labels ($7-10)(WR)
27. white cab, white container, "Ken Schraeder 25" labels ($10-15)(WR)
28. orange-yellow cab, orange-yellow container, "Kodak Racing" labels ($7-10)(WR)
29. white cab, white container, "Purolator"

with red car labels ($60-75)(WR)

30. white cab, white container, "Purolator" with orange car labels ($18-20)(WR)

31. white cab, white container, "Western Auto 17" labels ($10-15)(WR)

32. white cab, white container, "Country Time" labels ($10-15)(WR)

33. black cab, black container, "MAC Tools" labels ($20-35)(WR)

34. black cab, black container, "Mello Yello 42" labels ($10-15)(WR)

35. black cab, black container, "Alliance" labels ($15-20)(WR)

36. yellow cab, yellow container, "Pennzoil 2" labels ($7-10)(IN)

37. white cab, white container, "Panasonic" labels ($7-10)(IN)

38. white cab, white container, "K-Mart/ Havoline" labels ($10-15)(IN)

39. yellow cab, yellow container, "Pennzoil (Waltrip)" labels ($7-10)(WR)

40. yellow cab, dark yellow container, "Pennzoil (Waltrip)" labels ($7-10)(WR)

41. white cab, white container, "STP-Richard Petty Fan Appreciation Tour 1992" labels ($18-25)(WR)

42. white cab, white container, "Baby Ruth Racing" labels ($7-10)(WR)

43. black body, black container, "Goodwrench Racing Team" with checkered flags labels ($15-20)(WR)

44. metallic blue body, blue container, "Raybestos" labels ($7-10)(WR)

45. black cab, red container, "Slim Jim Racing Team" labels ($10-15)(WR)

46. white/green cab, white container, "Quaker State" labels ($7-10)(WR)

47. blue/white cab, blue container, "Evinrude 89" labels ($7-10)(WR)

48. white cab, white container, "Jasper Engines 55" labels ($7-10)(WR)

49. black cab, black container, "MAC Tools Racing (Harry Gant)" labels ($15-18)(WR)

50. black cab, black container, "Martin Birrane-Team Ireland" labels ($7-10)(WR)

51. white cab, white container, "Penn State Nittany Lions-Happy Valley Express" labels ($10-15)(WR)

CY-105-A KENWORTH GAS TANKER, issued 1989

1. white cab, white tank with black base, Macau casting, gold & black stripes tempa ($15-20)(GS)(JB)

2. white cab, white tank with matt gray base, Thailand casting, "Shell" tempa ($12-15)(GS)

CY-106-A PETERBILT ARTICULATED TIPPER, issued 1990

1. pink cab, gray tipper with black base, "Readymix" tempa ($12-15)(AU)

CY-107-A MACK SUPERSTAR TRANSPORTER, issued 1990

1. orange cab, black container, "Baltimore Orioles" labels ($15-20)(WR)

2. red cab, white container, "Melling Racing 9" labels ($50-75)(WR)

3. red/white cab, maroon container, "Nascar-America's Ultimate Motorsport" labels ($75-90)(WR)

4. blue cab, white container, "Bill Elliot 9" labels ($18-25)(WR)

5. white/blue cab, white container, "ADAP/ Auto Palace" labels ($18-25)(WR)

6. white cab, white container, "Ferree Chevrolet 49" labels ($7-10)(WR)

7. black cab, black container, "Interstate Racing Batteries" labels ($7-10)(WR)

CY-108-A DAF AIRCRAFT TRANS-PORTER, issued 1992

1. red cab, red carriage with red trailer base, red SB-37-A Hawk with rondels & white stripe livery ($8-12)

CY-109-A FORD AEROMAX SUPERSTAR TRANSPORTER, issued 1991

1. gold cab, gold container, "Folgers" labels ($15-20)(WR)

2. blue cab, white container, "Bill Elliot 9" labels ($15-20)(WR)

3. white cab, white container, "Hooters Racing" labels ($25-40)(WR)

4. black cab, black container, "Texaco Havoline/Davey Allison" labels ($25-40)(WR)

5. white cab, dark blue container, "Goodyear Racing" labels ($10-15)(WR)

6. red cab, red container, "Melling Performance 9" labels ($10-15)

7. red cab, red container, "Motorcraft" labels ($10-15)

8. white cab, dark blue container, "Penn State Nittany Lions" labels ($10-15)(WR)

9. red-brown cab, red-brown container, "Washington Redskins Super Bowl Champions" labels ($50-75)(WR)

10. white cab, white container, "Snickers Racing Team" labels ($10-15)(WR)

11. black cab, black container, "Stanley Mechanic Tools 92" labels ($7-10)(WR)

12. green cab, red container, "Merry Christmas White Rose Collectibles 1992" labels ($250-300)(WR)

CY-110-A KENWORTH SUPERSTAR TRANSPORTER, issued 1992

1. black cab, black container, "Rusty Wallace-Pontiac" labels ($10-15)(WR)

CY-203-A CONSTRUCTION LOWLOADER, issued 1989

1. yellow cab, pearly silver trailer, Macau casting, MB32 Excavator ($8-12)(MC)

2. yellow cab, pearly silver trailer, Thailand casting, MB32 Excavator ($8-12)(MC)

CY-803 SCANIA LOWLOADER WITH DODGE TRUCK, issued 1992

1. red cab with silver-gray trailer; MB72-E Dodge Truck with' Wigwam" tempa ($20-25)(DU)

TEAM MATCHBOX & TEAM CONVOY

Team Matchbox was first introduced in 1985 as a secondary Convoy line which included team transporters with two miniature vehicles. By 1988, the line was renamed Team Convoy. Team Convoy included some of the Team Matchbox items but also included a combination of one Convoy plus one miniature. Several models were never numbered, however late in 1992, a CY-111 was issued for Australia which includes a Team Matchbox type issued. For the sake of cataloging, all unnumbered sets, especially those from the Indy and Days of Thunder line will be cataloged under "CY-111". White Rose Collectibles also introduced Team Convoy sets in 1990 with numbers starting at TC-54.

TM-1-A Pepsi Team
TM-2-A Superstar Team
TM-3-A Dr. Pepper Team
TM-4-A Brut Team
TM-5-A 7 Up Team
TM-6-A Duckham's Team
TM-X-A Son of Gun & STP Teams

TC-1-A Fire Set
TC-2-A Tanker Set
TC-3-A Construction Set
TC-4-A Cargo Set
TC-5-A NASA Set
TC-6-A Rescue Set
TC-7-A Pepsi Team (was TM-1)
TC-8-A 7 Up Team (was TM-5)
TC-9-A Duckham's Team (was TM-6)
TC-10-A Fuji Team
TC-11-A Pirelli Team
TC-12-A Tizer Team
TC-13-A TV News Set
TC-14-A Ferrari Set
TC-15-A Pirelli Set
TC-16-A Coast Guard Set
TC-17-A Farm Set
TC-18-A Transport Set

TC-54-A Goodwrench Racing Team
TC-55-A not issued
TC-56-A Purolator Racing Team
TC-57-A Kodak Racing Team
TC-58-A not issued
TC-59-A Schraeder Racing Team
TC-60-A Pennzoil Racing Team
TC-61-A STP (Petty) Racing Team
TC-62-A Mello Yello Racing Team
TC-63-A J.D. McDuffie Racing Team
TC-64-A Pontiac Excitement Racing Team
TC-65-A Bill Elliot Racing Team

CY-111-A Team Sets

TM-1-A PEPSI TEAM, issued 1985 (reissued as TC-7 in 1988)
1. yellow transporter with ram clips toward rear; includes MB34-C Prostocker & MB68 Chevy Van all in "Pepsi" liveries ($18-25)
2. yellow transporter with ramp clips toward front; includes MB34-C Prostocker & MB68 Chevy Van all in "Pepsi" liveries ($18-25)

TM-2-A SUPERSTAR TEAM, issued 1985
1. white Transporter with ramp clips toward rear; includes MB34-C Prostocker & MB58-D Ruff Trek in "Superstar" liveries ($18-25)
2. white Transporter with ram clips toward front; includes MB34-C Prostocker & MB58-D Ruff Trek in "Superstar" liveries ($18-25)

TM-3-A DR. PEPPER TEAM, issued 1985
1. white Transporter with ramp clip towards rear; includes MB17-D AMX Prostocker & MB68-C Chevy Van in "Dr. Pepper" liveries ($30-50)
2. white Transporter with ram clip towards front; includes MB17-D AMX Prostocker & MB68-C Chevy Van in "Dr. Pepper" liveries ($30-50)

TM-4-A BRUT TEAM, issued 1985
1. white Transporter with ramp clip towards rear; includes MB62-D Corvette & MB58-D Ruff Trek in "Brut" liveries ($20-25)
2. white Transporter with ramp clip towards front; includes MB62-D Corvette &

MB58-D Ruff Trek in "Brut" liveries ($20-25)

TM-5-A 7 UP TEAM, issued 1985 (reissued in 1988 as TC-8)
1. green Transporter with ramp clip casting toward rear; includes MB34-C Chevy Prostocker & MB58-D Ruff Trek in "7 Up" liveries ($25-40)
2. green Transporter with ramp clip casting toward front; includes MB34-C Chevy Prostocker & MB58-D Ruff Trek in "7 Up" liveries ($25-40)

TM-6-A DUCKHAMS TEAM, issued 1985 (reissued in 1988 as TC-9)
1. dark blue Transporter with ramp clip casting toward rear; includes MB15-D Ford Sierra & MB6-H Ford Supervan II in "Duckhams QXR" liveries ($18-25)

TM-X-A STP TEAMS, issued 1985 (no numbers issued)
1. dark blue Transporter; includes MB65-D Indy Racer & MB58-D Ruff Trek in "STP 20" liveries ($175-225)(US)
2. yellow Transporter; includes MB60-E Firebird Racer & MB68-C Chevy Van in "STP Son of Gun" liveries ($175-225)(US)

TC-1-A FIRE SET, issued 1988
1. includes CY-13-A in red with MB54-E Command Vehicle in red ($10-15)

TC-2-A TANKER SET, issued 1988
1. includes CY-17-A in "Shell" livery with MB56-D Tanker in "Shell" livery ($10-15)

TC-3-A CONSTRUCTION SET, issued 1988
1. includes CY-20-A in yellow with MB29-C Tractor Shovel in yellow ($8-12)

TC-4-A CARGO SET, issued 1988
1. includes CY-25-A DAF Box Truck & MB20-D Volvo Container Truck in "TNT Ipec" liveries ($15-20)
2. includes CY-25-A DAF Box Truck & MB72-E Dodge Delivery Truck in "XP Parcels" liveries ($15-20)

TC-5-A NASA SET, issued 1988
1. includes CY-2-A Rocket Transporter & MB54-E Command Vehicle in "NASA" liveries ($8-12)

TC-6-A RESCUE SET, issued 1988
1. includes CY-22-A Boat Transporter in "Rescue" livery with MB75-D Helicopter in "Rescue" livery ($8-12)

TC-7-A PEPSI TEAM, reissued 1988 (see TM-1-A above)
TC-8-A 7 UP TEAM, reissued 1988 (see TM-5-A above)
TC-9-A DUCKHAMS TEAM, reissued 1988 (see TM-6-A above)

TC-10 FUJI TEAM, issued 1988
1. white & green Transporter with chrome exhausts; includes MB24-E Nissan 300ZX & MB6-H Ford Supervan II in "Fuji Racing Team" liveries ($20-25)
2. white & green Transporter with gray exhausts; includes MB24-E Nissan 300ZX & MB6-H Ford Supervan II in "Fuji Racing Team" liveries ($20-25)

TC-11-A PIRELLI TEAM, issued 1989
1. white Transporter with chrome exhausts; includes MB7-F Porsche 959 & MB72-E Dodge Van in "Pirelli Gripping Stuff" liveries ($18-25)
2. white Transporter with gray exhausts; includes MB7-F Porsche 959 & MB72-E Dodge Van in "Pirelli Gripping Stuff" liveries ($18-25)

TC-12-A TIZER TEAM, issued 1989
1. red Transporter with chrome exhausts; includes MB15-D Ford Sierra & MB6-H Ford Supervan II in "Tizer" liveries ($18-25)
2. red Transporter with gray exhausts; includes MB15-D Ford Sierra & MB6-H ford Supervan II in "Tizer" liveries ($18-25)

TC-13-A TV NEWS SET, issued 1990
1. includes CY-15-A Tracking Vehicle & MB68-G TV Van in dark blue with "MB TV News" liveries ($8-12)
2. includes CY-15-A Tracking Vehicle & MB68-G TV Van in white with "Sky Satellite TV" liveries ($8-12)

TC-14-A FERRARI SET, issued 1990
1. includes CY-24-A in red & MB74-H Racer in red with "Ferrari" liveries ($7-10)
2. includes CY-24-A in red & MB70-D Ferrari in red with "Ferrari" liveries ($7-10)

TC-15-A PIRELLI SET, issued 1990
1. includes CY-25-A DAF Container & MB7-F Porsche 959 in white with "Pirelli Gripping Stuff" liveries ($12-15)

TC-16-A COAST GUARD SET, issued 1990
1. includes CY-22-A DAF Boat Transporter & MB39-E Ford Bronco II in "Coast Guard" liveries ($7-10)

TC-17-A FARM SET, issued 1991
1. includes CY-20-A Tipper Truck with green cab/yellow dumper with "Eurobran" livery with MB73-E Mercedes Tractor with yellow & green ($7-10)

TC-18-A TRANSPORTER SET, issued 1991
1. includes CY-28-A Mack Double Container in white with "DHL" livery with MB28-H Forklift in white with red stripes ($10-15)

TC-54-A GOODWRENCH RACING TEAM, issued 1990
1. black Transporter with "GM" roof tempa & no door tempa; MB54-H Chevy Lumina in black with Goodyear slicks, plain trunk & no "Western Steer emblem' MB68-C Chevy Van in black-all with "Goodwrench Racing", Macau casting ($125-150)(WR)
2. black Transporter with "GM" roof tempa & "1990 Champion" on door, Thailand casting; MB54-H Chevy Lumina in black with Goodyear Slicks, with trunk design & no "Western Steer", China casting; MB68-C Chevy Van in black, Thailand casting-all with "Goodwrench Racing" ($25-40)(WR)
3. black Transporter with "GM" roof tempa & "1990 Champion" on door, Thailand casting; MB54-H Chevy Lumina in black

with black disc & rubber tires, with trunk design & with "Western Steer", China casting; MB68-C Chevy Van in black, Thailand casting-all with "Goodwrench Racing" ($20-35)(WR)
4. black Transporter with "GM" roof tempa & "1990 Champion" on door, Thailand casting; MB54-H Chevy Lumina in black with gray disc & rubber tires, with trunk design & with "Western Steer" China casting; MB68-C Chevy Van in black, Thailand casting-all with "Goodwrench Racing" ($20-35)(WR)
5. black Transporter with "Dale Earnhardt" roof tempa & "1990 Champion" on door, Thailand casting; MB54-H Chevy Lumina in black with black disc & rubber tires, with trunk design & with "Western Steer"; MB68-C Chevy Van in black, Thailand casting-all with "Goodwrench Racing" ($15-20)(WR)
6. black Transporter with "Dale Earnhardt" roof tempa & "1991 Champion" on door, Thailand casting; MB54-H Chevy Lumina in black with black disc & rubber tires, with trunk design & with "Western Steer"; MB68-C Chevy Van in black with "5 Time Champion" tempa-all with "Goodwrench Racing" ($15-18)(WR)

TC-56-A PUROLATOR RACING TEAM, issued 1991
1. white & fluorescent orange Transporter, Thailand casting; MB54-H Chevy Lumina in white & fluorescent orange, China casting; MB68-C Chevy Lumina in white & fluorescent orange, Thailand casting-all with "Purolator 10 Racing Team" liveries ($18-25)(WR)

TC-57-A KODAK RACING TEAM, issued 1991
1. orange-yellow Transporter, Thailand casting; MB54-H Chevy Lumina in orange-yellow, China casting; MB68-C Chevy Van in orange-yellow, Thailand casting-all with "Kodak Racing" liveries ($15-20)(WR)

TC-59-A SCHRAEDER RACING TEAM, issued 1991
1. white/green Transporter, gold hubs, Thailand casting; MB54-H Chevy Lumina in white/green, gold disc wheels, China casting; MB68-C Chevy Van in white/green, gold hubs, Thailand casting-all with "Schraeder 25" liveries ($15-20)(WR)

TC-60-A PENNZOIL RACING TEAM, issued 1992
1. yellow Transporter, Thailand casting; MB54-H Chevy Lumina in yellow, China casting; MB68-C Chevy Van in yellow, Thailand casting-all with "Pennzoil" liveries ($15-20)(WR)

TC-61-A STP (PETTY) RACING TEAM, issued 1992
1. blue Transporter, Thailand casting; MB54-H Chevy Lumina in blue, blue disc wheels, China casting; MB68-C Chevy Van in blue, Thailand casting-all with "STP Oil Treatment" liveries ($15-20)(WR)

TC-62-A MELLO YELLO RACING TEAM, issued 1992
1. black/green Transporter, Thailand casting; MB54-H Chevy Lumina in

green/black, China casting; MB68-C Chevy Van in black/green, Thailand casting-all with "Mello Yello 42" liveries ($15-20)(WR)

TC-63-A J.D. McDUFFIE RACING TEAM, issued 1992

1. blue Transporter, Thailand casting; MB216 Pontiac Grand Prix in blue, China casting; MB34-C Chevy Prostocker in blue/white, Thailand casting-all with "Rumple 70", "Son's or "J.D. McDuffie" liveries ($25-40)(WR)

TC-64-A PONTIAC EXCITEMENT RACING TEAM, issued 1992

1. black Transporter, Thailand casting; MB216 Pontiac Grand Prix in black, China casting; MB68-C Chevy Van in black, Thailand casting-all with "Pontiac Excitement" & "Rusty Wallace" liveries ($15-20)(WR)

TC-65-A BILL ELLIOT RACING TEAM, issued 1992

1. red Transporter, Thailand casting; MB212 Ford Thunderbird in red, China casting; MB53-D Flareside Pickup in red, "Thailand casting; all with "Bill Eliot 11 Racing" liveries ($15-20)(WR)

CY-111-A TEAM TRANSPORTERS, issued 1989-1992

1. white Transporter with gray exhausts with "Pioneer Racing Team" livery; includes MB6-D F.1 in white with "Matchbox" tempa ($12-15)(MC)
2. orange Transporter with chrome exhausts; includes MB54-G Chevy Lumina in orange with "Hardees 18" livery ($15-20)(DT)

3. orange Transporter with chrome exhausts; includes MB54-H Chevy Lumina in orange with "Hardees" livery ($15-20)(DT)
4. black/green Transporter with chrome exhausts; includes MB54-G Chevy Lumina in black/green with "Mello Yello 51" livery ($15-20)(DT)
5. black/green Transporter with chrome exhausts; includes MB54-H Chevy Lumina in black/green with "Mello Yello 51" livery ($15-20)(DT)
6. blue/white Transporter with chrome exhausts; includes MB65-D F.1 Racer in blue/white with "Valvoline" livery ($10-15)(IN)
7. yellow Transporter with chrome exhausts; includes MB74-H G.P. Racer in yellow with "Pennzoil 4" livery ($10-15)(IN)
8. pink/white/blue Transporter with chrome exhausts; includes MB65-D F.1 Racer in pink/white/blue with "Amway" livery ($25-30)(IN)
9. white/black Transporter with chrome exhausts; includes MB74-H G.P. Racer in white/black with "Havoline" livery ($10-15)(IN)
10. lemon Transporter with chrome exhausts; includes MB74-H G.P. Racer in lemon with "Pennzoil 2" livery ($10-15)(IN)
11. lavender/orange/white Transporter with chrome exhausts; includes MB74-H G.P. racer in lavender/orange/white in "Indy 1" livery ($20-25)(IN)
12. white Transporter with chrome exhausts; includes MB54-G Chevy Lumina in white with no livery ($15-20)(GF)
13. blue/white Transporter with chrome exhausts; includes MB65-D F.1 Racer in

blue/white with "Mitre 10" livery ($15-20)(AU)

SUPERKINGS

The Superkings range was diversified in 1991 to a point that the line has been divided into different categories-Emergency (EM), Construction (CS) and Farming (FM) and redesignated with new prefixes and numbers. These models will be listed under their original number release. If the model is new and never received an original "K" number these will be listed under the new prefix numbers. Only the new models are added to the basic listing below, although earlier models were still available and will be listed in the variation list. In 1988, a special series of 4 Superkings under the Roadblaster line were released as these are noted here with their respective "RB" number. Many models in the Superking range retained Lesney and England baseplates for several years, although released years after Lesney's demise!

K-8-E Ferrari F40	1989
K-98-B Porsche 944	1983
K-99-B Dodge Polizei Van	1983
K-100-A Ford Sierra XR4i	1983
K-101-B Racing Porsche	1983
K-102-B Race Support Set	1983
K-103-B Peterbilt Tanker	1983
K-104-B Rancho Rescue Set	1983
K-105-B Peterbilt Tanker	1984
K-106-B Kenworth Aircraft Transporter	1984
K-107-B Powerboat Transporter	1984
K-108-B Peterbilt Digger Transporter	1984
K-109-B Iveco Tanker	1984

K-110-B Fire Engine	1985
K-111-B Peterbilt Refuse Truck	1985
K-112-B Fire Spotter Transporter	1985
K-113-B Gargae Transporter	1985
K-114-B Mobile Crane	1985
K-115-B Mercedes Benz 190E	1985
K-116-B Racing Car Transporter	1985
K-117-B Scania Bulldozer Transporter	1985
K-118-B Road Construction Set	1985
K-119-A Fire Rescue Set	1985
K-120-A Car Transporter	1986
K-121-A Peterbilt Wrecker	1986
K-122-A DAF Road Train	1986
K-123-A Leyland Cement Truck	1986
K-124-A Mercedes Container Truck	1986
K-125-A no model issued at this number!	
K-126-A DAF Helicopter Transporter	1986
K-127-A Peterbilt Tanker	1986
K-128-A DAF Aircraft Transporter	1986
K-129-A Powerboat Transporter	1986
K-130-A Peterbilt Digger Transporter	1986
K-131-A Iveco Petrol Tanker	1986
K-132-A Fire Engine	1986
K-133-A Iveco Refuse Truck	1986
K-134-A Fire Spotter Transporter	1986
K-135-A Garage Transporter	1986
K-136-A Racing Car Transporter	1986
K-137-A Road Construction set	1986
K-138-A Fire Rescue Set	1986
K-139-A Iveco Tipper Truck	1987
K-140-A Car Recovery Vehicle	1987
K-141-A Skip Truck	1987
K-142-A BMW Police Car	1987
K-143-A Emergency Van	1987
K-145-A Iveco Tipper	1988
K-146-A Jaguar XJ6	1988
K-147-A BMW 750iL	1988
K-148-A Crane Truck	1988
K-149-A Ferrari Testarossa	1988
K-150-A Leyland Truck	1988

K-151-A Skip Truck 1988
K-152-A no model issued at this number!
K-153-A Jaguar XJ6 Police 1988
K-154-A BMW 750iL Police 1988
K-155-A Ferrari Testarossa Rallye 1988
K-156-A Porsche Turbo Rallye 1988
K-157-A Porsche 944 Rallye 1988
K-158-A no model issued at this number!
K-159-A Racing Car Transporter 1988
K-160-A Racing Car Transporter 1988
K-161-A Rolls Royce Silver Spirit 1989
K-162-A Sierra RS500 Cosworth 1989
K-163-A Unimog Snow Plow 1989
K-164-A Range Rover 1989
K-165-A Range Rover Police Car 1989
K-166-A Mercedes Benz 190E Taxi 1989
K-167-A Ford Transit 1990
K-168-A Porsche 911 Carrera 1990
K-169-A Ford Transit Ambulance 1990
K-170-A JCB Excavator (reissue K-41)
1991
K-171-A Toyota Hi-Lux 1990
K-172-A Mercedes Benz 500SL 1991
K-173-A Lamborghini Diablo 1992

CS-5-A Unimog Tar Sprayer 1991
FM-3-A Shovel Tractor 1992
FM-5-A Muir Tractor & Trailer 1992
EM-13-A Helicopter 1992

RB-2521 A.L.T.R.A.C. 1988
RB-2522 V.A.R.M.I.T. 1988
RB-2531 T.R.A.P.P.E.R. 1988
RB-2532 M.O.R.G. 1988

32630 Speedway Team Transporter 1991

K-8-E FERRARI F40, issued 1989
1. red body, clear windows, small logo
tempa, China casting ($6-8)

2. white, amber windows, no tempa, China
casting ($6-8)
3. white, clear windows, "Ferrari" tempa,
China casting ($6-8)(alarm car)
NOTE: although technically a "Special",
this model was never given an "SP"
number and has always been known with
"K" number.

K-15-B THE LONDONER, issued 1973
(continued)
14. powder blue upper deck, white & dark
blue lower deck, black interior, "1234-
1984 Parish Church 750th Anniversary"
labels, England casting ($20-25)(UK)
15. powder blue upper deck, white lower
deck, orange-yellow interior, "Macleans
Toothpaste" labels, England casting
($20-25)(UK)
16. blue upper & lower deck, gray interior,
"Alton Towers" labels, England casting
($20-25)(UK)
17. dark cream upper deck, blue lower deck,
blue interior, "Telegraph & Argus" labels,
England casting ($20-25)(UK)
18. red upper & lower deck, orange-yellow
interior, "Nestles Milkybar" labels,
England casting ($15-18)
19. white upper deck, red lower deck,
orange-yellow interior, "Nestles
Milkybar' labels, England casting
($15-18)
20. red upper & lower deck, orange-yellow
interior, "Petticoat Lane" labels, England
casting ($10-15)
21. yellow upper deck, brown lower deck,
yellow interior, "Butterkist Butter Toffee
Popcorn" labels, England casting
($25-40)(OP)(UK)
22. red upper & lower deck, orange-yellow

interior, "London Wide Tour Bus" labels,
England casting ($10-15)
23. red upper & lower deck, yellow interior,
"The Planetarium" labels, Macau casting
($10-15)
24. red upper & lower deck, yellow interior,
"Besuchen Sie Berlin Hauptstadt Der
DDR" labels, Macau casting
($25-40)(GR)
25. red upper & lower deck, yellow interior,
Chinese lettered side labels with
"Matchbox" logo label on roof, China
casting ($250-400)(CH)
26. red upper & lower deck, yellow interior,
"Around London Tour Bus" labels, China
casting ($10-12)

K-21-D FORD TRANSCONTINENTAL,
issued 1979 (continued)
12. blue cab, white chassis, orange (solid)
tarp, white roof mount, mag wheels,
"Sunkist" livery, England casting
($35-50)

K-25-B DIGGER AND PLOW, issued
1977 (continued)
3. yellow body & base, red plow, amber
windows, yellow backhoe with red scoop,
red wheels, England casting ($15-20)
4. yellow body, base & plow, amber
windows, yellow backhoe with yellow
scoop, yellow wheels, no origin cast
($15-20)
5. yellow body & base, red plow, amber
windows, yellow backhoe with red scoop,
red wheels, China casting ($15-20)
6. green body, yellow base, no plow, amber
windows, silver-gray backhoe with green
scoop, yellow wheels, China casting
($15-20)(FM)

K-35-B MASSEY FERGUSON
TRACTOR & TRAILER, issued 1979
(continued)
2. red tractor with silver-gray base & white
canopy; red trailer with silver-gray base,
Macau casting ($12-15)(Early Learning
Centre)
3. red tractor with silver-gray base & white
canopy; red trailer with silver-gray base,
China casting ($12-15)(FM)

K-42-B TRAXCAVATOR ROAD
RIPPER, issued 1979 (continued)
2. yellow body & base, yellow shovel &
rear ripper, red arms, black plastic roof,
England casting ($15-20)(K-117-A
component)
3. yellow body & base, red shovel and rear
ripper, yellow arms, red plastic roof,
China casting ($12-15)(CS)

K-43-B LOG TRANSPORTER, issued
1981 (continued)
2. yellow body, red chassis, red boom, 3
gray plastic pipes ($15-18)(CS)

K-69-A JAGUAR/VOLVO & EUROPA
CARAVAN, issued 1980 (continued)
8. Jaguar with light brown body, brown
base, white interior, amber windows, mag
wheels, England casting; Caravan with
white body, brown base, unpainted gas
cover, 5 arch wheels, England casting
($25-35)
9. Volvo with red body, brown interior,
amber windows, England casting;
Caravan with white body, red base,
unpainted gas cover, 5 arch wheels,
England casting ($25-35)

K-70-A PORSCHE TURBO, issued 1979 (continued)

3. black body, black base, red interior, clear windows, black roof label and "Turbo" on sides, England casting ($8-12)
4. black body, black base, red interior, clear windows, "Turbo" on sides, Macau casting ($8-12)
5. red body, black base, black interior, clear windows, no tempa, Macau casting ($8-12)
6. metallic red body, black base, cream interior, clear windows, "Turbo" tempa, Macau casting ($8-12)
7. red body, black base, red interior, clear windows, "Dragon Racing Team" tempa, Macau casting ($15-18)(HK)

K-78-A GRAN FURY POLICE CAR, issued 1980 (continued)

12. white body, blue interior, blue dome lights on white frame, "City Police" tempa, Macau casting ($8-12)
13. white body, blue interior, blue dome lights on white frame, "City Police" tempa, China casting ($8-12)
14. maroon body, blue interior, blue dome lights on white frame, "Fire Dept./Fire Chief" tempa, Macau casting ($8-12)
15. maroon body, blue interior, blue dome lights on white frame, "Fire Dept./Fire Chief" tempa, China casting ($8-12)
16. red body, black interior, red dome lights on white frame, "IAAFC" decals, China casting ($12-15)(WR)
17. orange body, black interior, blue dome lights on white frame, "Lindberg City Police" tempa, China casting ($8-12)(EM)

K-83-A HARLEY DAVIDSON MOTORCYCLE, issued 1981 (continued)

2. white body & frame, tan gas tank, blue & black driver with painted features, "Police" labels, Macau casting ($8-12)
3. red body & frame, red gas tank, no driver, "Harley Davidson" tempa, Thailand casting ($6-8)(HD)
4. dark blue body & frame, blue gas tank, no driver, "Harley Davidson" tempa, Thailand casting ($6-8)(HD)
5. black body & frame, black gas tank, no driver, "Harley Davidson" tempa, Thailand casting ($6-8)(HD)(GS)

K-87-A MASSEY FERGUSON TRACTOR & ROTARY RAKE, issued 1981 (continued)

2. green tractor with silver-gray base, green rake frame with yellow rakes ($12-15)(FM)

K-88-A MONEY BOX, issued 1981 (continued)

6. lime cab, black base, white container with lime roof, "De Speelboom Spaaravto" labels, K-19 China casting ($25-35)(DU)

K-90-A MATRA RANCHO, issued 1982 (continued)

3. white body, white base, black interior & side molding, "Coast Guard Patrol 70" tempa, England casting ($8-12)(K-104 component)
4. red body, red base, black interior & side molding, "2 Fire Control Unit" tempa, Macau casting ($8-12)
5. red body, red base, black interior & side molding, "2 Fire Control Unit" tempa, China casting ($7-10)

K-95-A AUDI QUATTRO, issued 1982 (continued)

5. white body, "Pirelli/Duckhams 10" & assorted liveries tempa, England casting ($7-10)
6. metallic blue body, "0000" tempa, Macau casting ($7-10)
7. white body, "Pirelli/Duckhams 10" & assorted liveries tempa, China casting ($7-10)

K-98-B PORSCHE 944, issued 1983

1. silver/gray body, red interior, 2 tone blue with "Porsche 944" tempa, Macau casting ($7-10)
2. brown body, red interior, orange/brown with "Porsche 944" tempa, Macau casting ($7-10)
3. red body, black interior, "Porsche 944" tempa, Macau casting ($7-10)
4. silver-gray body, black interior, small logo tempa, Macau casting ($7-10)
5. black body, tan interior, small logo tempa, Macau casting ($8-12)
6. pearly white body, black interior, small logo tempa, China casting ($6-8)

K-99-B DODGE POLIZEI VAN, issued 1983

1. white body, green plastic roof, "Polizei" labels, England base ($12-15)

K-100-A FORD SIERRA XR4i, issued 1983

1. white body, gray base, gray interior, no tempa, England casting ($8-12)
2. silver-gray body, gray base, gray interior, no tempa, England casting ($7-10)
3. powder blue body, gray base, gray interior, no tempa, England casting ($12-15)

4. red body, gray base, gray interior, small "XR4i" tempa, England casting ($12-15)
5. plum body, gray base, gray interior, no tempa, England casting ($12-15)
6. black body, gray base, gray interior, red stripe tempa, Macau casting ($7-10)
7. black body, gray base, gray interior, red stripe with "XR4i" tempa, Macau casting ($7-10)

K-101-B RACING PROSCHE, issued 1983

1. metallic beige body, red dash, black interior, "TC Racing" & "Sunoco 16" tempa, England casting ($20-25)
2. red body, red dash & interior, "TC Racing" & "Sunoco 16" tempa, England casting ($7-10)
3. red body, red dash, black interior, "TC Racing" & "Sunoco 16" tempa, England casting ($7-10)
4. white body, red dash, black interior, "Team Porsche" & "9 Esso" with dark blue sides tempa, England casting ($7-10)
5. white body, red dash, black interior, "Team Porsche" & "9 Esso" with plain sides tempa, England casting ($7-10)(K-102 component)

K-102-B RACE SUPPORT SET, issued 1983

1. Dodge van in yellow with red interior, "Team Porsche" labels includes K-101-A-5 Racing Porsche in white, England castings ($18-25)

K-103-B PETERBILT TANKER, issued 1983

1. silver-gray cab with solid blue stripe label, blue cab & trailer chassis, white

plastic tank, "Comet" labels, England casting ($45-60)
2. white cab with solid blue stripe label, blue cab & trailer chassis, white plastic tank, "Comet" labels, England casting ($20-25)
3. white cab with two stripe tempa, blue cab & trailer chassis, white plastic tank, "Comet" labels, England casting ($20-25)

K-104-B RANCHO RESCUE SET, issued 1983
1. K-90-A Matra Rancho in white with "Coast Guard Patrol 70" tempa; K-25-A Boat with white deck, blue hull, tan seats, "370" on sides only, white plastic cradle on orange trailer ($18-25)
2. K-90-A Matra Rancho in white with "Coast Guard Patrol 70" tempa; K-25-A Boat with white deck, blue hull, tan seats, "370" on sides & deck, white plastic cradle on orange trailer ($18-25)
3. K-90-A Matra Rancho in white with "Coast Guard Patrol 70" tempa; K-25-A Boat with white deck, blue hull, tan seats, "370" on sides & deck, black plastic cradle on orange trailer ($18-25)

K-105-B PETERBILT TIPPER, issued 1984
1. white body, black chassis, red plastic dump with red stripes, England casting ($12-15)
2. yellow body, black chassis, yellow plastic dump with "Taylor Woodrow" tempa, England casting ($12-15)

K-106-B KENWORTH AIRCRAFT TRANSPORTER, issued 1984
1. red cab with silver-gray chassis, silver-gray trailer, red plastic airplace with "Aces" on wings, England casting ($15-20)

K-107-B POWERBOAT TRANSPORTER, issued 1984
1. white cab with blue chassis, no dome lights cast; white trailer with blue base; boat with white deck & hull, "Spearhead" labels, England casting ($20-25)

K-108-B PETERBILT DIGGER TRANSPORTER, issued 1984
1. yellow cab with black chassis, "Avro" tempa, black trailer; K-25-B in yellow with yellow base, amber windows, yellow backhoe with red scoop, red wheels, England casting ($25-30)
2. yellow cab with black chassis, "Avro" tempa, black trailer; K-25-B in yellow with yellow base, green windows, yellow backhoe with red scoop, red wheels, England castings ($25-30)

K-109-B IVECO TANKER, issued 1984
1. yellow cab with white front fenders, no dome lights cast, white tank, "Shell" labels, England casting ($8-12)
2. yellow cab with black front fenders, no dome lights cast, white tank, "Shell" labels, England casting ($8-12)

K-110-B FIRE ENGINE, issued 1985
1. red body, gray ladder, "008" tempa, England casting ($8-12)

K-111-B PETERBILT REFUSE TRUCK, issued 1985
1. orange body, gray chassis, orange dump, "Waste Beater" tempa, England casting ($12-15)

K-112-B FIRE SPOTTER TRANSPORTER, issued 1985
1. red cab & chassis with "Fire 23" tempa, red trailer, red plastic plane with chrome plated wings with "Fire" on black cradle, cab features nonsteerable wheels, England casting ($20-25)

K-113-B GARAGE TRANSPORTER, issued 1985
1. yellow cab with red chassis, no dome lights cast, red plastic trailer with black base, gray plastic garage, orange-yellow tarp with "Shell", England casting ($20-25)
2. yellow cab with red chassis, no dome lights cast, red plastic trailer with black base, gray plastic garage, white tarp with "Shell", England casting ($20-25)

K-114-B CRANE TRUCK, issued 1985
1. yellow body, yellow crane cab with yellow boom & black extending crane, no roof tempa, "Taylor Woodrow" tempa on crane cab, black base, Macau casting ($15-20)
2. yellow body, yellow crane cab with yellow boom & black extending crane, "113" roof tempa, "Cotras" tempa on crane cab, black base, Macau casting ($15-20)
3. yellow body, red crane cab with yellow boom & red extending crane, "113" roof tempa, red/white stripes tempa on crane cab, red base, China casting ($15-20)

K-115-B MERCEDES BENZ 190E, issued 1985
1. white body, black interior, silver wheels, England casting ($7-10)

2. silver blue body, black interior, silver wheels, Macau casting ($6-8)
3. black body, tan interior, silver wheels, Macau casting ($6-8)
4. gunmetal gray body, red interior, silver wheels, Macau casting ($6-8)
5. white body, black interior, red wheels, "Fuji" tempa, Macau casting ($6-8)
6. white body, black interior, red wheels, "Fuji" tempa, China casting ($6-8)
7. lt. silver green body, black interior, silver wheels, China casting ($6-8)

K-116-B RACING CAR TRANSPORTER, issued 1985
1. red cab, black interior & front hubs, white chassis, no dome lights cast; trailer with red body, red base, "Ferrari" labels; includes 2X MB16-D in red ($20-25)
2. red cab, white interior & front hubs, white chassis, no dome lights cast; trailer with red body, red base, "Ferrari" labels; includes 2X MB16-D in red ($20-25)

K-117-B SCANIA BULLDOZER TRANSPORTER, issued 1985
1. yellow cab with brown chassis, "Taylor Woodrow" roof tempa, brown trailer; with K-42-B3 Traxcavator Road Ripper, England castings ($25-30)

K-118-B ROAD CONSTRUCTION SET, issued 1985
1. Ford Truck with yellow cab & body with black chassis & white roof, small design tempa, England casting; support trailer in yellow, small design tempa, Macau casting; Unimog in yellow with black chassis, "115" & small design labels, Macau casting; compressor trailer in

yellow with yellow top & small design tempa, Macau casting ($35-45)

K-119-A FIRE RESCUE SET, issued 1985

1. Unimog in red, red interior with small "Fire" label on hood, "8 Fire Dept." labels on white canopy, "Fire Dept." label on sides, England base; K-110-A1 version Fire Truck; Compressor trailer in red, England casting ($25-35)

K-120-A CAR TRANSPORTER, issued 1986

1. white cab with black chassis, white trailer with pearly silver ramp, "Carrier" labels, Macau casting ($15-20)
2. white cab with black chassis, white trailer with white ramp, "Carrier" labels, Macau casting ($15-20)

K-121-A PETERBILT WRECKER, issued 1986

1. dark blue cab, white chassis & rear bed, red dome lights, black booms, "Highway patrol City Police" tempa, Macau casting ($15-18)
2. dark blue cab, white chassis & rear bed, red dome lights, black booms, "Highway Patrol City Police" tempa, China casting ($15-18)
NOTE: this is a reissue of K-20-C, but a new feature includes steerable front wheels, otherwise identical!

K-122-A DAF ROAD TRAIN, issued 1986

1. white cab, black chassis, white bed, white canopies, amber dome lights, "Eurotrans" tempa, Macau casting ($18-25)

2. white cab, black chassis, red bed, white canopies, amber dome lights, "Toblerone" tempa, Macau casting ($25-35)(SW)
3. cream cab, dark gray chassis, cream bed, gray canopies, amber dome lights, "Marti" tempa, Macau casting ($25-35)(SW)
4. blue cab, black chassis, blue bed, blue canopies, amber dome lights, "Henniez" tempa, Macau casting ($25-35)(SW)
5. yellow cab, black chassis, yellow bed, yellow canopies, amber dome lights, "Zweifel Pomy Chips" tempa, Macau casting ($25-35)(SW)

K-123-A LEYLAND CEMENT TRUCK, issued 1986

1. yellow-orange cab & body, red chassis gray barrel with black stripes, road design tempa, red hubs, Macau casting ($7-10)
2. yellow-orange cab & body, red chassis, gray barrel with red stripes on white, red/white stripes tempa, red hubs, China casting ($7-10)(CS)

K-124-A MERCEDES CONTAINER TRUCK, issued 1986

1. white cab, red interior & front wheels hub, dark gray chassis; white container with dark gray chassis, red wheels, "7 Up" labels, England casting ($15-20)
2. pearly silver cab, black interior & front wheel hub, black chassis; silver-gray container with black chassis, silver wheels, "Taglisch Frisch" labels ($30-45)(GR)

K-126-A DAF HELICOPTER TRANSPORTER, issued 1986

1. blue cab with blue chassis, "Royal Navy"

tempa, blue trailer; helicopter with dark blue body, bright orange pontoons, white interior, "RN Royal Navy" & "Rescue" tempa, Macau casting ($15-20)
2. white cab with powder blue chassis, "Coast Guard" tempa, powder blue trailer; helicopter with white body, orange pontoons, white interior, "Coast Guard" tempa, China casting ($15-20)

K-127-A PETERBILT TANKER, issued 1986

1. white Peterbilt cab with red chassis, white plastic tank with red chassis, red wheels, "Total" labels, Macau casting ($15-20)
2. red Peterbilt cab with red chassis, white plastic tank with light gray chassis, red wheels, "Getty" labels, Macau casting ($15-20)
NOTE: this is a reissue of K-103-B with the only difference being the addition of steerable front axle.

K-127-B IVECO TANKER, issued 1989

1. blue Iveco cab with white chassis, blue plastic tank with white chassis, white wheels, "British Farm Produce-Milk" tempa, China casting ($25-40)(UK)(Early Learning Centre)

K-128-A DAF AIRCRAFT TRANSPORTER, issued 1986

1. red cab with red chassis, red trailer with gray cradle; light brown plastic jet plane with olive/white/red tempa, Macau casting ($15-18)
2. red cab with red chassis, red trailer with white cradle; silver-gray plastic spotter plane with red pontoons & wings, no tempa, China casting ($35-50)(UK) (released in Woolworth's)

K-129-A POWERBOAT TRANSPORTER, issued 1986

1. white cab with blue chassis, blue stripes tempa, amber dome lights, cream plastic trailer with blue base, chrome wheels; boat with white deck & hull, "Spearhead" in blue letters with blue side stripes labels, England casting ($15-20)
2. blue cab with powder blue chassis, "Kruger" tempa, amber dome lights, gray plastic trailer with powder blue base, chrome wheels; boat with white deck & hull, "Spearhead" in blue letters with brown side stripes labels, England casting ($15-20)
3. blue cab with powder blue chassis, "Kruger" tempa, amber dome lights, gray plastic trailer with powder blue base, chrome wheels; boat with white deck & tan hull, "Spearhead" in brown letters with brown side stripes labels, England casting ($15-20)
4. blue cab with powder blue chassis, "Kruger" tempa, amber dome lights, gray plastic trailer with dark blue base, red wheels; boat with white deck & tan hull, "Spearhead" in brown letter with brown side stripes labels, England casting ($15-20)
5. white cab with black chassis, "Coast Guard" tempa, amber dome lights, gray plastic trailer with black base, chrome wheels; boat with light gray deck & white & black hull, "4439 Coast Guard" labels, Macau casting ($15-20)
6. white cab with powder blue chassis, "Coast Guard" tempa, amber dome lights, gray plastic trailer with powder blue base, white wheels; boat with light gray &

fluorescent orange deck & white & black hull, "4439 Coast Guard" labels, China casting ($15-20)(EM)
NOTE: this is a reissued of the K-107-B in which the addition of dome lights and steerable front axle have been added.

K-130-A PETERBILT DIGGER TRANSPORTER, issued 1986
1. yellow cab with brown chassis, "Plant Hire' & "B1" tempa, brown trailer; K-25-B in yellow with yellow base, amber windows, yellow backhoe with yellow scoop, yellow wheels, Macau casting ($20-25)
NOTE: this is a reissue of K-108-B in which the addition of a steerable front axle has been incorporated otherwise no different!

K-131-A IVECO PETROL TANKER, issued 1986
1. yellow cab, amber dome lights, chrome wheels, white tank, "Shell" labels, England casting ($7-10)
2. red cab, amber dome lights, white wheels, red tank, "Texaco" labels, Macau casting ($12-15)
3. white cab, amber dome lights, white wheels, white tank, "Texaco" labels, Macau casting ($15-20)
NOTE: this is a reissue of the K-109-B in which the addition of dome lights and steerable front axle have been added.

K-132-A FIRE ENGINE, issued 1986
1. red body, gray ladder, white interior, chrome wheels, "008" tempa, England casting ($8-12)
2. red body, white ladder, white interior, chrome wheels, "Fire 201" tempa,

England casting ($8-12)
3. red body, white ladder, black interior, red wheels, "Fire 201" tempa, Macau casting ($8-12)
NOTE: this is a reissue of K-110-B in which the addition of a steerable front axle has been added.

K-133-A IVECO REFUSE TRUCK, issued 1986
1. maroon cab & dump, maroon wheels, "Refuse City Corp." tempa, England casting ($10-15)
2. maroon cab & dump, red wheels, "Refuse City Corp." tempa, Macau casting ($10-15)
3. white cab & dump, white wheels, "Recycle" tempa, Macau casting ($10-15)
4. white cab & dump, red wheels, caricature tempa, Macau casting ($10-12)

K-134-A FIRE SPOTTER TRANS-PORTER, issued 1986
1. red cab with white chassis, "Patrol Unit 12 Fire" tempa, red trailer with brown cradle; plane in red with red pontoons & chrome wings, "12" on plane tail & "C-802 Fire" with red stripe on wings, Macau casting ($15-20)
2. red cab with white chassis, "Patrol Unit 12 Fire" tempa, red trailer with brown cradle; plane in red with red pontoons & chrome wings, "12" on plane tail & "C-802 Fire" with red stripe on wings, China casting ($15-20)
3. red cab with white chassis, "Patrol Unit 12 Fire" tempa, red trailer with white cradle; plane in silver-gray with red pontoons & wings, "12" on plane tail & "Fire" on wings, China casting ($15-20)(EM)

NOTE: this is a reissue of K-112-B with the addition of a steerable front axle, otherwise no other difference.

K-135-A GARAGE TRANSPORTER, issued 1986
1. yellow cab with red chassis, red interior; dull red plastic trailer with black base, gray plastic garage, orange-yellow tarp with "Shell", England casting ($18-25)
2. yellow cab with red chassis, black interior; red plastic trailer with black base, gray plastic garage, yellow tarp with "Shell", Macau casting ($18-25)
3. red cab with red chassis, black interior; gray plastic trailer with red base, gray plastic garage, red tarp with "Texaco", Macau casting ($18-25)
NOTE: this is a reissue of K-113-B with the addition of dome lights and a steerable front axle.

K-136-A RACING CAR TRANS-PORTER, issued 1986
1. red cab with white chassis, black interior, red container with dark gray base, white wheels, "Ferrari" labels; includes 2X MB16-D in red, England casting ($18-25)
NOTE: this is a reissue of K-116-B with the addition of dome lights and a steerable front axle.

K-137-A ROAD CONSTRUCTION SET, issued 1986
1. DAF truck with orange-yellow body, black chassis, red stripe & small "1104" tempa, Macau casting; support trailer in yellow with black chassis, red stripe with "4172" tempa, England casting; Unimog

in yellow with black chassis, red stripe with "126" tempa, England casting; compressor trailer in yellow with yellow canopy, red stripe with "1127" tempa, England casting ($30-45)

K-138-A FIRE RESCUE SET, issued 1986
1. Unimog with red body, black interior, "Fire" hood tempa, "9 Foam Pump" tempa on white canopy, "Fire Dept." on sides tempa, Macau casting; K-132-A1 Fire Engine; compressor trailer in red with red canopy, Macau casting ($25-30)

K-139-A IVECO TIPPER TRUCK, issued 1987
1. orange-yellow body, black chassis, orange-yellow dump, "Wimpey" tempa, Macau casting ($10-15)
2. orange-yellow body, black chassis, orange-yellow dump, "Department of Highways" tempa, Macau casting ($10-15)
3. red body, black chassis, red dump, silver stripe tempa, Macau casting ($10-15)
4. orange-yellow body, red chassis, red dump, red/white stripes tempa, China casting ($8-12)(CS)

K-140-A CAR RECOVERY VEHICLE, issued 1987
1. white cab, blue chassis, white bed with white drop ramp, blue/red tempa, Macau casting ($8-12)

K-141-A SKIP TRUCK, issued 1987
1. red cab, black chassis, gray dump, "ECD" tempa, Macau casting ($10-15)
2. red cab, red chassis, orange-yellow dump, no tempa, Macau casting ($10-15)

K-142-A BMW POLICE CAR, issued 1987
1. white body, black interior, "Polizei" with green hood & doors tempa, Macau casting ($8-12)
2. white body, black interior, "Police" with checkers tempa, Macau casting ($8-12)
3. white body, black interior, "Fina Pace Car" tempa, Macau casting ($8-12)

K-143-A EMERGENCY VAN, issued 1987
1. white body, blue windows, pink cross & stripe tempa, Macau casting ($8-12)

K-144-A LAND ROVER, issued 1987
1. yellow body & roof, blue dome lights, chrome wheels, "Frankfurt Flughafen" tempa, Macau casting ($8-12)
2. orange-yellow body & roof, amber dome lights, red wheels, "Heathrow Airport" tempa, Macau casting ($8-12)(UK)
3. orange-yellow body & roof, amber dome lights, red wheels, "Road Maintenance" tempa, Macau casting ($8-12)
4. green body, cream roof, amber dome lights, white wheels, "Veterinary Surgeon" tempa, China casting ($15-18)(UK)(Early Learning Centre)

K-145-A IVECO TIPPER, issued 1988
1. orange cab, black chassis, "S&G" tempa, pearly silver twin tippers with black chassis, orange stripes tempa, Macau casting ($20-25)

K-146-A JAGUAR XJ6, issued 1988
1. white body, black interior, Macau casting ($10-15)(GS)
2. metallic red body, gray interior, Macau casting ($6-8)
3. metallic red body, gray interior, China

casting ($6-8)
4. dark green body, brown interior, China casting, display stand ($15-18)(UC)

K-147-A BMW 750iL, issued 1988
1. silver blue body, gray interior, Macau casting ($6-8)
2. silver blue body, gray interior, China casting ($6-8)
3. red body, gray interior, China casting ($25-40)(AU)
4. silver-gray body, white & black interior, China casting, display stand ($15-18)(UC)

K-148-A CRANE TRUCK, issued 1988
1. white cab, bed & crane post, red chassis, black plastic boom, blue stripe & "PEX" tempa, Macau casting ($15-18)

K-149-A FERRARI TESTAROSSA, issued 1988
1. red body, black & cream interior, red base, Macau casting ($6-8)
2. red body, black & cream interior, red base, detailed trim, display stand, China casting ($15-18)(UC)

K-150-A LEYLAND TRUCK, issued 1988
1. blue cab, blue chassis with black base; includes three removeable rear components-plastic container in blue with "SMF" tempa, white tipper bed with "SMF" tempa, powder blue flatbed with "SMF" tempa, China casting ($15-20)

K-151-A SKIP TRUCK, issued 1988
1. orange-yellow cab, orange-yellow bed, black chassis, orange-yellow dump, orange skip arms, road design tempa, Macau casting ($6-8)

2. orange-yellow cab, orange-yellow bed, black chassis, orange-yellow dump, orange skip arms, road design tempa, China casting ($6-8)
3. orange-yellow cab, yellow bed, black chassis, yellow dump, orange skip arms, road design tempa, China casting ($6-8)
4. orange-yellow cab, yellow bed, red chassis, red dump, red skip arms, white/red stripes tempa, China casting ($6-8)(CS)

K-153-A JAGUAR XJ6 POLICE CAR, issued 1988
1. white body, black interior, blue dome lights, "Police" with checkers tempa, Macau casting ($7-10)

K-154-A BMW 750iL POLICE CAR, issued 1988
1. white body, black interior, blue dome lights, "Polizei" with green doors, hood & trunk tempa, Macau casting ($7-10)
2. white body, black interior, blue dome lights, "Police" with orange & black stripes tempa, China casting ($7-10)

K-155-A FERRARI TESTAROSSA RALLYE, issued 1988
1. yellow body & base, black & cream interior, gold & black stripes with "61" tempa, Macau casting ($6-8)

K-156-A PORSCHE TURBO RALLYE, issued 1988
1. red body, white interior, black base, white with "18 Elf" & "Pioneer" tempa, Macau casting ($6-8)

K-157-A PORSCHE 944 RALLYE, issued 1988
1. orange-yellow body, black interior, "Turbo Porsche 944" with red/blue stripes tempa, Macau casting ($6-8)

K-159-A RACING CAR TRANSPORTER, issued 1988
1. white cab with white chassis, black interior, white container with black chassis, "Porsche" tempa; includes 2X MB7-F Porsche 959 in white, Macau casting ($20-25)

K-160-A RACING CAR TRANSPORTER, issued 1988
1. white cab with blue chassis, black interior, white container with blue chassis, "Matchbox Formula Racing Team" labels; includes 2X MB16-D in white, Macau casting ($20-25)

K-161-A ROLLS ROYCE SILVER SPIRIT, issued 1989
1. silver-gray body, gray interior, black base, Macau casting ($6-8)
2. metallic red body, gray interior, black base, Macau casting ($10-12)(GS)

K-162-A SIERRA RS500 COSWORTH, issued 1989
1. white body, white lower body, dark gray interior with white steering wheel, white spoiler, white wheels, small "Cosworth" tempa, Macau casting ($6-8)
2. white body, dark gray lower body, dark gray interior with black steering wheel, dark blue spoiler, white wheels, "Gemini" & "British Open Rally Championship" tempa, Macau casting ($6-8)

3. black body, dark gray lower body, light gray interior with black steering wheel, black spoiler, chrome wheels, "Texaco 6" tempa, Macau casting ($6-8)
4. white body, white lower half, black interior with white steering wheel, black spoiler, white wheels, "Caltex Bond" tempa, Macau casting ($20-35)(AU)

K-163-A UNIMOG SNOW PLOW, issued 1989
1. orange body, black chassis, blue canopy on bed, orange plow with "Schmidt" tempa, Macau casting ($10-15)

K-164-A RANGE ROVER, issued 1989
1. dark blue body, gray interior, "Range Rover" on hood tempa, Macau casting ($6-8)
2. white body, gray interior, "Africa Safari Tours" & black zebra stripes tempa, China casting ($6-8)
3. beige body, gray interior, brown stripes tempa, China casting ($10-15)(UK)(Early Learning Centre)
4. green body, gray interior, yellow/white stripes tempa, China casting ($6-8)(FM)

K-165-A RANGE ROVER POLICE, issued 1989
1. white body, black interior, blue stripe with yellow "Police" tempa, Macau casting ($6-8)
2. white body, black interior, orange band with blue "Police" tempa, China casting ($6-8)

K-166-A MERCEDES BENZ 190E TAXI, issued 1989
1. dark cream body, black interior, yellow "Taxi" roof sign, Macau casting ($6-8)

K-167-A FORD TRANSIT, issued 1990
1. lavender body with white roof, black interior, clear windows, chrome wheels, "Milka" labels, Macau casting ($12-15)(UK)
2. powder blue body with white roof, black interior, blue windows, chrome wheels, "Surf N Sun" labels, China casting ($7-10)
3. orange-yellow body with red roof, black interior, clear windows, red wheels, "Miller Construction Company" labels, China casting ($7-10)(CS)

K-168-A PORSCHE 911 CARRERA, issued 1990
1. red body, black interior, black base, "Carrera 4 Porsche" tempa, Macau casting ($6-8)
2. pearly white body, black interior, black base, "Carrera" tempa, with display stand, China base ($15-18)(UC)

K-169-A FORD TRANSIT AMBU-LANCE, issued 1990
1. white body with cream roof, "Air Ambulance" with yellow band & orange stripe tempa, Macau casting ($7-10)
2. white body, with cream roof, "Air Ambulance" with light orange band & blue stripe tempa, China casting ($7-10)

K-170-A JCB EXCAVATOR, issued 1991
1. orange-yellow body, chassis and crane booms, red cab & shovel, "JCB", "808" & red/white stripes tempa, China casting ($18-25)(CS)
NOTE: this is a reissue of K-41-C.

K-171-A TOYOTA HI-LUX, issued 1990
1. white body, black interior, white bed, "Runnin Brave" & Peacock feather design tempa; includes MB23-F Honda ATC in fluorescent green/orange, Macau casting ($15-20)
2. red body, gray interior, white bed, yellow stripes tempa; includes hay bale, 2 milk cans & plastic cow ($8-12)(FM)

K-172-A MERCEDES BENZ 500SL, issued 1991
1. silver-gray body, blue interior, gray base, China casting ($6-8)
2. red body, dark gray interior, red base, display stand, China casting ($15-18)(UC)

K-173-A LAMBORGHINI DIABLO, issued 1992
1. yellow body, black interior, yellow base, "Diablo" tempa, China casting ($6-8)

CS-5-A UNIMOG TAR SPRAYER, issued 1991
1. orange-yellow body, red chassis, black interior, red tank with white stripes tempa, red plastic tar sprayer units, China casting ($10-15)

FM-3-A SHOVEL TRACTOR, issued 1992
1. red body, silver-gray base, white shovel arms with red shovel, white plastic roof, red wheels, base cast K-35, China casting ($8-12)

FM-5-A MUIR HILL TRACTOR & TRAILER, issued 1992
1. K-25-C Muir Hill Tractor with green body, yellow chassis, yellow wheels; trailer with green body & yellow chassis, 3 plastic logs ($12-15)

EM-13-A HELICOPTER, issued 1992
1. red body, white base, yellow pontoons with red stripes tempa, "Patrol Unit" & "4 Fire" tempa, no origin cast ($10-12)

RB-2521-A A.L.T.R.A.C., issued 1988
1. pea green cab with dark blue chassis, dark blue trailer with pea green ramp; helicopter in pea green with brown base, "89" with red design tempa, plastic battle armament attachments, Macau casting ($45-60)(RB)

RB-2522 V.A.R.M.I.T., issued 1988
1. plum body and crane cab, yellow wheels, blue & yellow tempa, plastic battle armament attachments, Macau casting ($35-50)(RB)

RB-2531 T.R.A.P.P.E.R., issued 1988
1. brown body with black chassis, black trailer with large gray & dark blue plastic container attachment, green/blue/yellow tempa, plastic battle armament attachments, Macau casting ($45-60)(RB)

RB-2532 M.O.R.G., issued 1988
1. purple cab with purple chassis, yellow tempa, black plastic trailer with purple chassis, large plastic accessories as load, plastic battle armament attachments, Macau casting ($45-60)(RB)

32630 SPEEDWAY TEAM TRANS-PORTER, issued 1991
1. black cab & chassis, black container, pink lettered "Indy" with "75th Indianapolis 500" labels, black chassis; includes MB65-D in black/yellow & MB74-H in orange/white ($18-25)
2. black cab & chassis, black container,

yellow lettered "Indy" with "Indianapolis 500-the Seventy Sixth" labels; includes 2X MB74-H Racer-one in orange/lavender/white & one in white/blue ($18-25)

SPECIALS & TURBO SPECIALS

First making their mark in 1984, Specials were a series of race cars in approximately 1/43 scale. Although still available today, this series has yet to find a real niche in the Matchbox series. Although all models were and are cast "Specials" the model series underwent many different "names" including L.A. Wheels, Super GT Sport, Superkings, Muscle Cars and Alarm Cars. Some models were also introduced into the Graffic Traffic (GF) series. A secondary series called "Turbo Specials" feature pullback action with two speed motor. Specials designated as Muscle Cars (MU), Alarm Cars (AL), LA Wheels (LA), Turbo Specials (TS) or King-size (KS) only are denoted below in the variation text.

SP-1/2-A KREMER PORSCHE, issued 1984

1. white body, brown interior, clear windows, "22 Grand Prix" tempa, Macau base, chromed wheels ($6-8)(SP1)
2. white body, brown interior, clear windows, "35 Porsche" tempa, Macau base, chromed wheels ($6-8)(SP1)
3. white body, brown interior, clear windows, "35 Porsche" tempa, Macau base, chromed wheels ($6-8)(TS3)
4. white body, brown interior, clear windows, "35 Porsche" tempa, China base, unchromed wheels ($6-8)(SP1)

5. pearly silver body, brown interior, clear windows, stripes & "19" tempa, Macau base, chromed wheels ($6-8)(SP2)
6. maroon body, brown interior, clear windows, "Michelin 15" tempa, Macau base, chromed wheels ($6-8)(TS3)
7. white body, brown interior, clear windows, green & yellow with "2" tempa, Macau base, chromed wheels ($6-8)(SP2)
8. white body, brown interior, clear windows, green & yellow with "2" tempa, China base, unchromed wheels ($6-8)(SP2)
9. white body, brown interior, clear windows, red & yellow with "2" tempa, China base, unchromed wheels ($6-8)(SP2)
10. black body, brown interior, clear windows, "35 Porsche" tempa, China base, unchromed wheels ($10-15)(LA)
11. white body, white interior, pink windows, no tempa, white China base, unchromed wheels ($10-15)(GF)
12. white body, brown interior, clear windows, "Lloyds 1" tempa, black China base, unchromed wheels ($20-30)(UK)

SP-3/4-A FERRARI 512B, issued 1984

1. blue body, black interior, clear windows, "Pioneer 11" tempa, Macau base, chromed wheels ($6-8)(SP3)
2. red body, black interior, clear windows, "European University 11" tempa, Macau base, chromed wheels ($6-8)(SP4)
3. black body, gray interior, clear windows, "Michelin 88" tempa, Macau base, chromed wheels ($6-8)(SP4)
4. white body, black interior, clear windows, "Pioneer 11" tempa, Macau base,

chromed wheels ($8-10)(TS6)
5. orange body, gray interior, clear windows, "147" tempa, Macau base, chromed wheels ($6-8)(SP3)
6. white body, gray interior, clear windows, "147" tempa, China base, unchromed wheels ($6-8)(LA)
7. lime body, gray interior, clear windows, "Michelin 88" tempa, China base, unchromed wheels ($6-8)(LA)
8. white body, white interior, amber windows, no tempa, white China base, unchromed wheels ($10-15)(GF)

SP-5/6-A LANCIA RALLYE, issued 1984

1. yellow body, dark gray interior, clear windows, "102" tempa, Macau base, chromed wheels ($6-8)(SP5)
2. white body, dark gray interior, clear windows, "Martini Racing 1" tempa, Macau base, chromed wheels ($6-8)(SP6)
3. green body, light gray interior, clear windows, "Pirelli 116" tempa, Macau base, chromed wheels ($6-8)(SP5)
4. white body, dark gray interior, clear windows, "102" tempa, Macau base, chromed wheels ($6-8)(TS5)
5. dark blue body, light gray interior, clear windows, "Pirelli 116" tempa, China base, unchromed wheels ($6-8)(LA)
6. white body, white interior, blue windows, no tempa, white China base, unchromed wheels ($10-15)(GF)

SP-7/8-A ZAKSPEED MUSTANG, issued 1984

1. white body, black interior, clear windows, "Ford 16" tempa, Macau base, chromed wheels ($6-8)(SP7)
2. pearly silver body, black interior, clear

windows, "Motul 28" tempa, Macau base, chromed wheels ($6-8)(SP8)
3. yellow body, gray interior, clear windows, blue & red with "20" tempa, Macau base, unchromed wheels ($6-8)(TS2)
4. black body, maroon interior, clear windows, "83" tempa, Macau base, chromed wheels ($6-8)(SP8)
5. blue body, black interior, clear windows, "QXR Duckhams" tempa, Macau base, chromed wheels ($6-8)(SP7)
6. black body, black interior, clear windows, "83" tempa, Macau base, unchromed wheels ($6-8)(TS2)
7. orange body, maroon interior, clear windows, blue & red with "20" tempa, China base, unchromed wheels ($6-8)(KS1)
8. blue body, black interior, clear windows, "QXR Duckhams" tempa, China base, unchromed wheels ($6-8)(KS6)
9. white body, black interior, clear windows, blue stripes tempa, China base, chromed wheels ($6-8)(MU)
10. orange body, black interior, clear windows, black stripes tempa, China base, chromed wheels ($6-8)(MU)

SP9/10-A CHEVY PROSTOCKER, issued 1984

1. lemon body, brown interior, clear windows, "NGK 12" tempa, black Macau base, chromed wheels ($6-8)(SP9)
2. metallic blue body, brown interior, clear windows, "Heuer 111" tempa, black Macau base, chromed wheels ($6-8)(SP10)
3. green body, brown interior, clear windows, "Heuer 9" tempa, black Macau

base, chromed wheels ($6-8)(SP10)

4. green body, gray interior, clear windows, "Heuer 9" tempa, gray Macau base, unchromed wheels ($6-8)(TS4)
5. white body, gray interior, clear windows, "Momo 5" tempa, gray Macau base, unchromed wheels ($8-12)(TS4)
6. yellow body, brown interior, clear windows, "Heuer 9" tempa, black China base, unchromed wheels ($6-8)(LA)
7. orange-yellow body, brown interior, clear windows, "Heuer 9" tempa, black China base, unchromed wheels ($6-8)(KS)
8. dark blue body, gray interior, clear windows, white stripes & "SS454" tempa, black China base, chromed wheels ($6-8)(MU)
9. red body, gray interior, clear windows, black stripes tempa, black China base, chromed wheels ($6-8)(MU)

SP11/12-A CHEVROLET CAMARO, issued 1984

1. white body, black interior, clear windows, "Goodyear 18" tempa, Macau base, chromed wheels ($6-8)(SP11)
2. red body, black interior, clear windows, "56" tempa, Macau Base, chromed wheels ($6-8)(SP12)
3. white body, black interior, clear windows, "7 Total" tempa, Macau base, chromed wheels ($6-8)(SP12)
4. white body, brown interior, clear windows, "Firestone 4" tempa, Macau base, unchromed wheels ($6-8)(TS1)
5. white body, brown interior, clear windows, "7 Total" tempa, Macau base, unchromed wheels ($6-8)(TS1)
6. white body, black interior, clear windows, "Michelin 3" tempa, Macau base,

chromed wheels ($6-8)(SP11)

7. white body, black interior, clear windows, "Michelin 3" tempa, China base, unchromed wheels ($6-8)(LA)(KS)
8. yellow body, black interior, clear windows, "7 Total" tempa, China base, unchromed wheels ($6-8)(LA)(KS)
9. white body, black interior, clear windows, orange stripes tempa, China base, chromed wheels ($6-8)(MU)
10. black body, black interior, clear windows, orange & black stripes tempa, China base, chromed wheels ($6-8)(MU)

SP13/14-A PORSCHE 959, issued 1986

1. red body, gray interior, clear windows, "Michelin 44" tempa, Macau base, chromed wheels ($6-8)(SP13)
2. white body, gray interior, clear windows, "53" with stripes tempa, Macau base, chromed wheels ($6-8)(SP14)
3. black body, gray interior, clear windows, "Michelin 44" tempa, China base, unchromed wheels ($6-8)(LA)
4. yellow body, gray interior, clear windows, "53" with stripes tempa, China base, unchromed wheels ($6-8)(LA)
5. silver/gray body, tan interior, clear windows, "27" with stripes tempa, China base, unchromed wheels ($8-12)(TS8)
6. silver/gray body, gray interior, clear windows, "3" with stripes tempa, China base, unchromed wheels ($8-12)(KS)
7. white body, gray interior, blue windows, "Porsche 959" tempa, China base, chromed wheels ($8-12)(KS)
8. white body, white interior, blue windows, no tempa, China base, unchromed wheels ($10-15)(GF)
9. dark blue body, no interior, black

windows, "Turbo" tempa, China base, chromed wheels ($15-18)(AL)

10. black body, no interior, black windows, "Turbo" tempa, China base, chromed wheels ($15-18)(AL)

YESTERYEARS

Yesteryears have been produced since 1956, when the Y-1-A Allchin Traction Engine was introduced. Yesteryears were produced in England through 1985 and many of these models retained their "Lesney" baseplates until the dies were moved to the Orient, although produced years after the company's demise. New additions to the line are noted in the list below. "LE" denotes Limited Editions.

Model	Name	Year
Y-2-D	1930 Bentley 2-1/2 Litre	1985
Y-3-E	1912 Model T Tanker	1985
Y-5-E	1929 Leyland Titan	1989
Y-6-E	1932 Mercedes Benz Lorry	1988
Y-7-D	1930 Model A Wrecker	1985
Y-8-E	1917 Yorkshire Steam Wagon	1987
Y-9-C	1920 Leyland 3 Ton Lorry	1985
Y-9-D	1936 Leyland Cub Fire Engine	1989
Y-10-D	1957 Maserati 250F	1986
Y-10-E	1931 Diddler Trolley Bus	1988
Y-11-D	1932 Bugatti Type 51	1986
Y-12-D	1912 Model T Pickup	1986
Y-12-E	Stephenson's Rocket	1987
Y-12-F	1937 GMC Van	1988
Y-14-D	1935 E.R.A.	1986
Y-15-C	Preston Type Tramcar	1987
Y-16-C	1960 Ferrari Dino 246/V12	1986
Y-16-D	1922 Scania Vabis Postbus	1988
Y-16-E	Scammel 100 Ton Transporter with Class 2-4-0 Loco	1989
Y-18-B	1918 Atkinson Steam Lorry	1985
Y-18-C	1918 Atkinson Steam Lorry	1986
Y-18-D	1918 Atkinson D-Type Steam Lorry	1987
Y-19-B	Fowler B6 Showman's Engine	1986
Y-19-C	1929 Morris Light Van	1988
Y-21-B	Aveling-Porter Steam Roller	1987
Y-21-C	1957 BMW 507	1988
Y-21-D	Ford Model TT Van	1989
Y-23-A	1922 AEC 'S' Type Bus	1982
Y-23-B	Mack Petrol Tanker	1989
Y-24-A	1928 Bugatti Type 44	1983
Y-25-A	1910 Renault Type AG Van	1983
Y-25-B	1910 Renault Type AG Ambulance	1986
Y-26-A	1913 Crossley Beer Lorry	1983
Y-27-A	1922 Foden C Type Steam Wagon	1985
Y-27-B	1922 Foden Steam Wagon & Trailer	1986
Y-28-A	1907 Unic Taxi	1984
Y-29-A	1919 Walker Electric Van	1985
Y-30-A	1920 Model AC Mack Truck	1985
Y-30-B	1920 Mack Canvasback Truck	1986
Y-31-A	1931 Morris Pantechnicon	1990
Y-32-A	1917 Yorkshire Steam Wagon	1990
Y-33-A	1920 Mack Truck	1990
Y-34-A	1933 Cadillac V-16	1990
Y-35-A	1930 Model A Ford Pickup	1990
Y-36-A	1926 Rolls Royce Phantom I	1990
Y-37-A	1931 Garrett Steam Wagon	1990
Y-38-A	Rolls Royce Armoured Car	1990
Y-39-A	1820 English Coach	1990
Y-40-A	1931 Mercedes Benz Type 770	1991
Y-41-A	1932 Mercedes Benz Lorry	1991
Y-42-A	1938 Albion CX27	1991
Y-43-A	1905 Busch Steam Fire Engine	1991
Y-44-A	1910 Renault Bus	1991
Y-45-A	1930 1930 Bugatti Royale	1991
Y-46-A	1868 Merryweather Fire Engine	1991

Y-47-A 1928 Morris Van   1991
(no numbers issued for Y-48 to Y-60)
YY-61-A 1933 Cadillac Fire Engine 1992
Y-62-A 1932 Ford AA Truck   1992
Y-63-A 1939 Bedford Truck   1992
Y-64-A 1938 Lincoln Zephyr   1992
Y-65-A Austin 7/BMW/Rosengart   1992
Y-66-A Gold State Coach   1992
Y-901 1936 Jaguar SS100 (pewter) 1991

Y-1-B MODEL FORD, issued 1964 (continued)

9. black body & chassis, black roof, tan seats, black plastic sterring wheels, England casting ($15-25)(LE) (Connoisseur Set)

Y-1-C 1936 JAGUAR SS, issued 1977 (continued)

5. light yellow body & chassis, black seats, whitewall tires, England casting ($75-100)
6. dark yellow body & chassis, black seats, whitewall tires, issued with 2 piece plastic diorama, Macau casting ($12-15)
7. red body & chassis, black seats, blackwall tires, China casting ($15-18)

Y-2-D 1930 BENTLEY 2-1/2 LITRE, issued 1985

1. dark green body & base, green fenders, reddish brown interior, green wheels, door flag tempa, England casting ($12-15)
2. dark green body & base, dark green fenders, reddish brown interior, green wheels, door flag tempa, Macau casting ($12-15)
3. dark blue body & base, dark blue fenders, black interior, blue wheels, door flag

tempa, Macau casting ($12-15)
4. dark blue body & base, dark blue fenders, black interior, blue wheels, door flag tempa, China casting ($25-35)
5. purple body & base, black fenders, brown interior, chrome wheels, no tempa, China casting ($15-18)

Y-3-B 1910 BENZ LIMOUSINE, issued 1966 (continued)

23. black body with blue sides tempa, black roof, tan seats, brown grille, black plastic steering wheels, England casting ($15-25)(LE)(Connoisseur set)

Y-3-D 1912 MODEL T FORD TANKER, issued 1981 (continued)

4. blue body, black chassis, blue tank, white roof, gold 12 spoke wheels, "Express Dairy" tempa, England casting ($12-15)
5. blue body, black chassis, blue tank, white roof, red 12 spoke wheels, "Express Dairy" tempa, England casting ($25-40)
6. cream body, maroon chassis, maroon tank, white roof, red 12 spoke wheels with white walls, "Carnation Farm Products" tempa, England casting ($12-15)
NOTE: above models have BLACK seats, models below with TAN seats.
7. blue & red body, black chassis, blue tank, blue roof, red 12 spoke wheels, "Mobiloil" tempa, England casting ($15-18)
8. blue & red body, black chassis, blue tank, blue roof, red 24 spoke wheels, "Mobiloil" tempa, England casting ($25-40)
9. dark green body, black chassis, dark green tank, white roof, red 12 spoke

wheels, "Castrol" tempa, England casting ($25-35)
10. dark green body, black chassis, dark green tank, white roof, gold 12 spoke wheels, "Castrol" tempa, England casting (Int'l)($12-15)
11. dark green body, brown chassis, dark green tank, white roof, red 12 spoke wheels, "Castrol" tempa, England casting (Int'l)($150-250)
12. yellow body, black chassis, yellow tank, white roof, black seat, red 12 spoke wheels, "Shell" tempa, Macau casting ($12-15)

Y-3-E 1912 MODEL T TANKER, issued 1985

1. red body, black chassis, red tank, red roof, gold 12 spoke wheels, "Red Crown Gasoline" tempa ($18-25)(LE)

Y-4-C 1909 OPEL COUPE, issued 1967 (continued)

9. red body, dark red chassis, red seats, black grille, 12 spoke brass wheels, textured tan roof ($15-25)(LE)(Connoisseur set)

Y-4-D 1930 MODEL "J" DUESENBERG, issued 1976 (continued)

9. brown/beige body, brown chassis, light tan roof & seats, disc wheels, England casting ($15-18)
10. brown/beige body, brown chassis, light tan roof & seats, 24 spoke wheels, England casting ($15-18)
11. brown/beige body, brown chassis, dark tan roof & seats, 24 spoke wheels, England casting ($15-18)
NOTE: above models with "Lesney" base,

below models with "International" base.
12. brown/beige body, brown chassis, rust brown roof & seats, 24 spoke wheels, England casting ($15-18)
13. silver/blue body, blue chassis, black roof & seats, 24 spoke wheels, England casting ($15-18)
14. silver/blue body, blue chassis, black roof & seats, 24 spoke light blue wheels, Macau casting ($15-18)
15. powder blue body, blue chassis, creamy tan roof & seats, 24 spoke chrome wheels, Macau casting ($15-18)
16. silver/blue body, blue chassis, black roof & seats, 24 spoke blue wheels, China casting ($25-40)

Y-5-D 1927 TALBOT VAN, issued 1978 (continued)

29. powder blue body, black chassis, black roof, black seat, 12 spoke red wheels, "Nestles Milk" decals, England casting ($75-100)
30. blue body, black chassis, white roof, black seat, 12 spoke chrome wheels, "EverReady" tempa, England casting ($12-15)
31. blue body, black chassis, white roof, tan seat, 12 spoke chrome wheels, "EverReady" tempa, England casting ($12-15)
32. black body & chassis, yellow roof, tan seat, 12 spoke yellow wheels, "Dunlop" tempa, England casting ($15-18)
33. black body & chassis, yellow roof, black seat, 12 spoke yellow wheels, "Dunlop" tempa, England casting ($12-15)
34. cream body, green chassis & seat, black seat, 12 spoke olive wheels, "Rose's Lime Juice" tempa, England casting ($15-18)

35. cream body, dark green chassis & seat, black seat, 12 spoke olive wheels, "Rose's Lime Juice" tempa, Macau casting ($15-18)
36. green body, black chassis, white roof, tan seat, 12 spoke gold wheels, "Lyle's Golden" tempa, Macau casting ($12-15)

Y-5-E 1929 LEYLAND TITAN, issued 1989

1. lime green body, dark green fenders & base, cream roof, tan seats, "Robin The New Starch" tempa, Macau casting ($15-18)
2. blue body, black fenders & base, cream roof, tan seats, "Swan Fountpens". This model is issued in a framed cabinet with one model assembled and one model disassembled, Macau casting ($75-100)
3. maroon body, black fenders & base, cream roof, tan seats, "Newcastle Brown Ale" tempa, China casting ($15-18)
4. maroon body, black fenders & base, cream roof, tan seats, "MICA for collectors of Matchbox" labels, China casting ($35-50)(C2)

Y-6-D ROLLS ROYCE FIRE ENGINE, issued 1977

7. red body & chassis, white ladder, black seat, 12 spoke gold wheels, England casting ($10-15)
8. red body, black chassis, white ladder, black seat, 12 spoke gold wheels, England casting ($10-15)

Y-6-E 1932 MERCEDES BENZ LORRY, issued 1988

1. cream body, black fenders, dark gray base, cream canopy, red wheels,

"Gtuttgarter Hofbrau" tempa, Macau casting ($12-15)

Y-7-D 1930 MODEL A FORD WRECK TRUCK, issued 1985

1. orange body, black chassis, brown interior, dark green boom, orange wheels, "Barlow Motor Sales" tempa, England casting ($10-15)
2. orange body, black chassis, brown interior, dark green boom, orange wheels, "Barlow Motor Sales" tempa, Macau casting ($20-25)
3. yellow body, black chassis, brown interior, gray boom, red wheels, "Shell" tempa, Macau casting ($10-15)

Y-8-D 1945 M.G. TD, issued 1978 (continued)

9. blue body & chassis, tan roof, black interior, 24 spoke chrome wheels, England casting ($12-15)
10. blue body & chassis, tan roof & interior, 24 spoke chrome wheels, England casting ($12-15)
11. blue body & chassis, rust roof, tan interior, 24 spoke wheels, England casting ($12-15)
12. cream body, brown chassis, tan roof & interior, 24 spoke wheels, England casting ($12-15)
13. cream body, brown chassis, rust roof, tan interior, 24 spoke wheels, England casting ($12-15)
14. cream body, brown chassis, rust roof, light tan interior, 24 spoke wheels, England casting ($12-15)

Y-8-E 1917 YORKSHIRE STEAM WAGON, issued 1987

1. plum body, black chassis, gray cab roof, tan canopy, red wheels, "Johnnie Walker Whisky" tempa, Macau casting ($15-18)
2. green body, black chassis, white cab roof & canopy, green wheels, "Samuel Smith" tempa-this model is in a framed cabinet unit. The disassembled model has metal parts in chrome plate and the canopy has no tempa, Macau casting ($125-150)
3. dark blue body, black chassis, gray cab roof, white sack load, cream wheels, "William Prichard Millenium Flour" tempa, Macau casting ($15-18)
4. yellow body, black chassis, beige cab roof & canopy, yellow wheels, "Fyffes" tempa, China casting ($15-18)

Y-9-B 1912 SIMPLEX, issued 1968 (continued)

12. yellow body, black chassis, black roof, tan seats, black grille, gold 12 spoke wheels, England casting ($12-15)
13. yellow body, black chassis, yellow roof, tan seats, black grille, gold 12 spoke wheels, includes 2 piece plastic diorama, England casting ($25-40)
14. yellow body, black chassis, black roof, tan seats, black grille, gold 12 spoke wheels, includes 2 piece plastic diorama, England casting ($12-15)

Y-9-C 1920 LEYLAND 3 TON LORRY, issued 1985

1. dark green body, red chassis, black base, tan seats, green wheels, "A. Luff & Sons" tempa, Macau casting ($25-40)(LE)
2. dark green body, red chassis, charcoal base, tan seats, green wheels, "A. Luff &

Sons" tempa, Macau casting ($25-40)(LE)

Y-9-D 1936 LEYLAND CUB FIRE ENGINE, issued 1989

1. red body & fenders, black roof, brown ladder ,red wheels, "Works Fire Service" tempa, Macau casting ($200-275)(LE)

Y-10-D 1957 MASERATI 250F, issued 1986

1. red body & base, black seat, chrome wheels, Macau casting ($15-18)
2. red body & base, black seat, silver painted wheels, Macau casting ($15-18)

Y-10-E 1931 "DIDDLER" TROLLEY BUS, issued 1988

1. red body, brown interior, gray roof, black roof, "Ronuk" & "Jeyes' Kills" labels, Macau casting ($25-35)(LE)

Y-11-B 1912 PACKARD LAUNDALET, issued 1964 (continued)

6. beige & brown body, brown interior, plastic steering wheel, brass grille & wheels, England casting ($15-25)(LE) (Connoisseur set)

Y-11-C 1938 LAGONDA DROPHEAD COUPE, issued 1972 (continued)

13. plum body & chassis, black interior & grille, chrome 24 spoke wheels, England casting ($20-25)(GS)
14. plum body & chassis, maroon interior & grille, chrome 24 spoke wheels, England casting ($35-50)(GS)

Y-11-D 1932 BUGATTI TYPE 51, issued 1986

1. blue body, blue base, brown interior with gray dash, silver painted wheels, "4" tempa, England casting ($10-15)
2. blue body, blue base, black interior with cream dash, chrome wheels, "6" tempa, Macau casting ($10-15)

Y-12-C 1912 MODEL T FORD VAN, issued 1978 (continued)

23. yellow body, black chassis & roof, black seat, red wheels, "Sunlight Seife" labels, England casting ($150-200)(GR)
24. red body, black chassis & roof, black seat, red wheels, "Royal Mail" tempa, England casting ($12-15)
25. black body & chassis, white roof, tan seat, gold wheels, "Captain Morgan" decals, England casting ($12-15)
26. black body & chassis, white roof, tan seat, gold wheels, "Captain Morgan" labels, England casting ($12-15)
27. blue body, black chassis, white roof, black seat, gold wheels, "Hoover" tempa, England casting ($1000+)
28. orange body, black chassis & roof, black seat, solid disc chrome wheels, "Hoover" tempa, England casting ($25-40)
29. orange body, black chassis & roof, black seat, black wheels, "Hoover" tempa, England casting ($12-15)
30. tan body, brown chassis & roof, black seat, gold wheels, "Motor 100" tempa, England casting ($25-40)
31. tan body, brown chassis & roof, black seat, red wheels, "Motor 100" tempa, England casting ($12-15)
32. white body, blue chassis, red roof, black seat, red wheels, "Pepsi Cola" tempa, England casting ($25-40)
33. white body, blue chassis, red roof, black seat, chrome wheels, "Pepsi Cola" tempa, England casting ($12-15)
34. white body, blue chassis, red roof, tan seat, chrome wheels, "Pepsi Cola" tempa, England casting ($12-15)
35. light gray body, green chassis, gray roof with pickle cast, black seat, red wheels, "Heinz 57" tempa, England casting ($15-25)
36. blue body, black chassis, yellow roof, black seat, gold wheels, "Rosella" tempa, England casting ($15-20)

Y-12-D 1912 MODEL T FORD PICKUP, issued 1986

1. blue body, base & roof, tan seats, gold wheels, "Imbach" livery, Macau casting ($18-25)(LE)
2. blue body, base & roof, tan seats, red wheels, "Imbach" livery, Macau casting ($20-35)(LE)

Y-12-E STEPHENSON'S ROCKET, issued 1987

1. yellow engine body with yellow wheels & white stack; tender in yellow with yellow barrel, Macau casting ($20-35)(LE)

Y-12-F 1937 GMC VAN, issued 1988

1. black body with gray painted roof, black chassis, tan interior, chrome wheels, "Goblin" tempa, Macau casting ($50-65)
2. black body, black chassis, tan interior, chrome wheels, "Goblin" tempa, Macau casting ($12-15)
3. cream body, green chassis, black interior, chrome wheels, "Baxter's" tempa, Macau casting ($12-15)

4. dark blue, black chassis, tan interior, red wheels, "Goanna" tempa, China casting ($)

Y-13-B 1911 DAIMLER, issued 1966 (continued)

7. blue body with light blue pinstripe tempa, powder blue chassis, brown interior, black grille, chrome 24 spoke wheels, England casting ($15-25)(LE)(Connoisseur set)

Y-13-C 1918 CROSSLEY, issued 1974 (continued)

13. cream body, black chassis, maroon seat, green roof, grille & rear canopy, 12 spoke chrome wheels, "Carlsberg" labels, England casting ($12-15)
14. cream body, black chassis, maroon seat, green roof, grille & rear canopy, 12 spoke gold wheels, "Carlsberg" labels, England casting ($12-15)
15. dark green body, black chassis, maroon seat, cream roof & canopy, black grille, 12 spoke gold wheels, "Warings" label with cream background, England casting ($15-20)
16. dark green body, black chassis, maroon seat, white roof & canopy, black grille, 12 spoke gold wheels, "Warings" label with background, England casting ($18-25)
17. dark green body, black chassis, maroon seat, white roof & canopy, black grille, 12 spoke gold wheels, "Warings" label with white background, England casting ($15-20)
18. yellow body, black chassis, black seat, roof, grille & rear load, 12 spoke black wheels, "Kohle & Koks" tempa, Macau casting ($15-18)

Y-14-B 1911 MAXWELL ROADSTER, issued 1965 (continued)

6. beige body, green chassis, black grille & interior, black roof, copper gas tank, England casting ($15-18)(LE) (Connoisseur set)

Y-14-C 1931 STUTZ BEARCAT, issued 1975 (continued)

9. blue body, dark gray chassis, tan seats, black grille, chrome 12 spoke wheels, England casting ($15-18)
10. dark blue & cream body, blue chassis, red seats, black grille, chrome 12 spoke wheels, Macau casting ($18-25)

Y-14-D 1935 E.R.A., issued 1986

1. black body, black base, silver-gray exhaust/grille/steering, brown interior, chrome plated wheels, "7" tempa, England casting ($10-15)
2. blue body, yellow base, dk. silver-gray exhaust/grille/steering, black interior, yellow wheels, "4" tempa, England casting ($10-15)
3. blue body, yellow base, dk. silver-gray exhaust/grille/steering, black interior, yellow wheels, "4" tempa, China casting ($25-35)

Y-15-B 1930 PACKARD VICTORIA, issued 1969 (continued)

12. beige body, brown chassis, white roof, maroon grille & seats, black trunk, red 12 spoke wheels, England casting ($50-65)
13. beige body, brown chassis, tan roof & seats, black grille & trunk, red 12 spoke wheels, England casting ($15-18)
14. beige body, brown chassis, tan roof & seats, black grille & trunk, red 24 spoke

wheels, England casting ($25-40)

15. beige body, brown chassis, rust roof, tan seats, black grille & trunk, red 12 spoke wheels, England casting ($25-40)

Y-15-C PRESTON TYPE TRAMCAR, issued 1987

1. red body, white window area, dark gray roof, dark gray base, brown interior, "Swan Vestas" tempa, Macau casting ($12-15)
2. red body, white window area, dark gray roof, dark gray base, brown interior, "Swan Vestas" tempa-this model comes in a framed cabinet unit with a disassembled version in chrome plate with no tempa, Macau casting ($75-100)
3. blue body, cream window area, blue roof, dark gray base, brown interior, "Swan Soap" tempa, Macau casting ($12-15)
4. orange-red body, cream window area, light gray roof, black base, brown interior, "Golden Shred" tempa, Macau casting ($12-15)
5. brown body, cream window area, dark gray roof, black base, black interior, "Zebra Grate" tempa, China casting ($12-15)

Y-16-B 1928 MERCEDES BENZ SS, issued 1972 (continued)

17. red body, silver-gray chassis, no roof, black trunk & seats, red 24 spoke wheels, England casting ($18-25)
18. lavender gray body, black chassis, black roof, grille, trunk & seats, chrome 24 spoke wheels, Macau casting ($20-25)

Y-16-C 1960 FERRARI DINO 246/V16, issued 1986

1. red body & base, black interior, gray exhaust, chrome wheels, "17" tempa, Macau casting ($15-18)
2. red body & base, black interior, gray exhaust, silver painted wheels, "17" tempa, Macau casting ($15-18)

Y-16-D 1922 SCANIA VABIS POSTBUS, issued 1988

1. yellow body, black chassis, gray roof, reddish brown interior, gold wheels, yellow skis, Macau casting ($25-40)(LE)

Y-16-E SCAMMEL 100 TON TRANSPORTER WITH CLASS 2-4-0 LOCO, issued 1989

1. dark blue cab with red chassis, blue trailer, red wheels, "Pickfords" livery; Locomotive in black, Macau casting ($85-110)(LE)

Y-17-A 1938 HISPANO SUIZA, issued 1975 (continued)

5. green body, dark green chassis, black interior, green grille, gold 24 spoke wheels, includes 2 piece plastic diorama, England casting ($15-20)
6. green body, dark green chassis, black interior, black grille, gold 24 spoke wheels, includes 2 piece plastic diorama, England casting ($15-20)
7. green body, dark green chassis, black interior, black grille, gold 24 spoke wheels, includes 2 piece plastic diorama, Macau casting ($15-20)
8. green body with lime sides, green chassis, cream interior, black grille, chrome 24 spoke wheels with white walls, Macau casting ($15-20)

Y-18-A 1937 CORD, issued 1979 (continued)

4. plum body & chassis, black base, white roof & interior, chrome 24 spoke wheels, England casting ($25-40)
5. plum body & chassis, black base, white roof & interior, chrome disc wheels, England casting ($25-40)
6. pale yellow body & chassis, black base, tan roof & interior, chrome 24 spoke wheels, Macau casting ($20-25)

Y-18-B 1918 ATKINSON STEAM LORRY, issued 1985

1. green body, red base, green wheels, "Sand & Gravel" tempa, England casting ($18-25)(LE)

Y-18-C 1918 ATKINSON STEAM LORRY, issued 1986

1. yellow body, black base, yellow wheels, no load, "Blue Circle Portland Cement", tempa, England casting ($15-18)
2. red body, black base, red wheels, beige sack load, "Burghfield Mills Reading" tempa, Macau casting ($15-18)

Y-18-D 1918 ATKINSON D-TYPE STEAM LORRY, issued 1987

1. blue body, black base, red wheels, brown barrels with gold chain, "Bass & Co." tempa, Macau casting ($18-25)(LE)

Y-19-A 1935 AUBURN 851 SUPER-CHARGED ROADSTER, issued 1980 (continued)

6. cream body, black chassis, red seats, red disc wheels, England casting ($12-15)
7. white with blue side tempa body, white chassis, blue seats, 24 spoke blue wheels,

England casting ($18-25)

8. beige & cream body, beige chassis, tan seats, 24 spoke chrome wheels, Macau casting ($18-25)

Y-19-B FOWLER B6 SHOWMAN'S ENGINE, issued 1986

1. blue body, cream roof, black boiler, red wheels, gray tires, Macau casting ($20-35)(LE)
2. blue body, white roof, black boiler, red wheels, gray tires, Macau casting ($20-25)(LE)

Y-19-C 1929 MORRIS LIGHT VAN, issued 1988

1. blue body, white roof, brown interior, red 12 spoke wheels, "Brasso" tempa, Macau casting ($15-18)
2. blue body, yellow roof, brown interior, chrome 12 spoke wheels, "Michelin" tempa, Macau casting ($15-18)
3. brown body, white roof, brown interior, chrome 12 spoke wheels, "Sainsbury" tempa, Macau casting ($15-18)
4. blue body, white roof, brown interior, red 12 spoke wheels, "Brasso" tempa, China casting ($25-35)

Y-20-A 1937 MERCEDES -BENZ 540K, issued 1979 (continued)

6. white body & chassis, red seats, red 24 spoke wheels with white wall tires, England casting ($12-15)
7. white body & chassis, red seats, red 24 spoke wheels with black wall tires, England casting ($12-15)
8. red body & chassis, red seats, red 24 spoke wheels, includes 2 piece plastic diorama, Macau casting ($15-20)

9. black body & chassis, red seats, chrome 24 spoke wheels, Macau casting ($15-18)

Y-21-A 1930 MODEL A FORD "WOODY WAGON", issued 1981 (continued)

4. metallic bronze hood, brown rear body, brown chassis, cream interior, "A&J Box" tempa, England casting ($10-15)
5. orange-brown hood, brown rear body, brown chassis, cream interior, "A&J Box" tempa, England casting (Lesney or Int'l)($10-15)
6. blue hood, dark cream rear body, black chassis, cream interior, "Carter's Seeds" tempa, England casting ($25-40)
7. blue hood, dark cream rear body, black chassis, brown interior, "Carter's Seeds" tempa, England casting ($10-15)
8. blue hood, pale cream rear body, black chassis, brown interior, "Carter's Seeds" tempa, England casting ($10-15)

Y-21-B 1920 AVELING -PORTER STEAM ROLLER, issued 1987

1. green body, green wheels with gray tires, black stack, gray roof without inscription underneath ($250-350)(LE)
2. green body, green wheels with gray tires, black stack, gray roof with inscription underneath ($20-30)(LE)

Y-21-C 1957 BMW 507, issued 1988

1. blue body, black roof, red interior, clear windshield, chrome wheels, black base, Macau casting ($15-20)(LE)

Y-21-D FORD MODEL TT VAN , issued 1989

1. green body, black chassis, red roof, gray

interior, red wheels, silver grille, "O for an Osram" tempa, Macau casting ($15-18)
2. green body, black chassis, red roof, gray interior, red wheels, black grille, "O for an Osram" tempa, Macau casting ($15-18)
3. beige body, black chassis, beige roof, gray-brown interior, red wheels, "My Bread" tempa, Macau casting ($12-15)
4. black body, black chassis, black roof, gray interior, gold wheels, "Drambuie" tempa, China casting ($12-15)
5. black body, black chassis, black roof, gray interior, gold wheels, "Antiques Road Show-Next Generation 1992" labels, China casting ($1500+)(UK)
6. dark blue body, dark blue chassis & roof, gray interior, gold wheels, "Jenny Kee" & 'Waratah' painting tempa, China casting (AU)
7. dark green body, dark green chassis & roof, gray interior, gold wheels, "Pro Hart" & 'Lunchtime' painting tempa, China casting (AU)

NOTE: #6 & 7 issued as a boxed pair only! ($1500+)

Y-22-A 1930 MODEL A FORD VAN, issued 1982 (continued)

2. yellow body, black chassis, red roof, chrome 24 spoke wheels, "Maggi's" tempa, England casting ($10-15)
3. beige body, brown chassis, red roof, chrome 24 spoke wheels, "Toblerone" tempa, England casting ($10-15)
4. red body, black chassis, black roof, black 24 spoke wheels, "Canada Poste" tempa, England casting ($12-15)
5. cream body, red chassis, red roof, gold 24

spoke wheels, "Walter's Palm Toffee" tempa, England casting ($12-15)
6. reddish brown body, brown chassis, white roof, chrome 24 spoke wheels, "Spratt's" tempa, England casting ($12-15)
7. blue body, black chassis, white roof, chrome 12 spoke wheels, "Lyon's Tea" tempa, England casting ($12-15)
8. white body, black chassis, black roof, red 12 spoke wheels, "Cherry Blossom" tempa, Macau casting ($15-18)
9. white body, black chassis, black roof, red 12 spoke wheels, "Cherry Blossom" tempa, China casting ($25-35)
10. blue body, black chassis, white roof, chrome 12 spoke wheels, "Lyon's Tea" tempa, China casting ($25-35)
11. white body, black chassis, black roof, orange disc wheels, "Pratt's" tempa, China casting ($15-18)

Y-23-A 1922 AEC 'S' TYPE BUS, issued 1982

1. red body, red upper deck, chocolate interior, red wheels, "Schweppes Tonic Water" labels, England casting ($18-25)
2. red body, red upper deck, light tan interior, red wheels, "Schweppes Tonic Water" labels, England casting ($15-18)
3. red body, red upper deck, tan interior, red wheels, "Schweppes" labels, England casting ($15-18)
4. red body, red upper deck, tan interior, red wheels, "Maples Furniture" labels, England casting ($15-20)(GS)
5. red body, red upper deck, tan interior, red wheels, "The RAC" labels, England casting ($12-15)
6. brown body, cream upper deck, chocolate interior, dark red wheels, "Haig" labels,

England casting ($12-15)
7. red body, red upper deck, tan interior, red wheels, "Rice Krispies" labels, Macau casting ($12-15)
8. blue body, cream upper deck, chocolate interior, cream wheels, "Lifebuoy Soap" labels, Macau casting ($12-15)

Y-23-B MACK PETROL TANKER, issued 1989

1. red body, fenders & roof, red tank, "Texaco" tempa, Macau casting ($12-15)
2. red body, fenders & roof, white tank, "Conoco" tempa, Macau casting ($10-15)

Y-24-A 1928 BUGATTI TYPE 44, issued 1983

1. black body & chassis, tan interior, chrome 24 spoke wheels, full yellow tempa, England casting ($12-15)
2. black body & chassis, tan interior, chrome 24 spoke wheels, pinstriped lemon yellow tempa, England casting ($12-15)
3. black body & chassis, black interior, chrome 24 spoke wheels, pinstriped lemon yellow tempa, England casting ($18-25)
4. gray body, plum chassis, tan interior, chrome disc wheels, pinstriped plum tempa, includes 2 piece plastic diorama, Macau casting ($15-18)
5. black body & chassis, tan interior, chrome 24 spoke wheels, pinstriped red tempa, Macau casting ($15-18)

Y-25-A 1910 RENUALT TYPE AG VAN, issued 1983

1. green body, dark green chassis, white interior, gold 12 spoke wheels, "Perrier"

tempa, England casting ($12-15)
2. green body, dark green chassis, white interior, red 12 spoke wheels, "Perrier" tempa, England casting ($25-40)
3. yellow body, blue chassis, black interior, yellow 12 spoke wheels, "James Neale & Sons" tempa, England casting ($15-20)
4. yellow body, dark blue chassis, black interior, yellow 12 spoke wheels, "James Neale & Sons" tempa, England casting ($35-50)
5. yellow body, dark blue chassis, black interior, red 12 spoke wheels, "James Neale & Sons" tempa, England casting ($45-60)
6. silver-gray body & chassis, maroon interior, red 12 spoke wheels, "Duckham's Oils" tempa, England casting ($15-18)
7. powder blue body, dark blue chassis, maroon interior, gold 12 spoke wheels, "Eagle Pencils" tempa, England casting ($15-18)
8. red body, black chassis, white interior, black 12 spoke wheels, "Tunnock" tempa, England casting ($15-18)
9. green body, black chassis, maroon interior, gold 12 spoke wheels, "Delhaize" tempa, Macau casting ($15-18)
10. lavender body, black chassis, brown interior, gold 12 spoke wheels, "Suchard Chocolate" tempa, Macau casting ($15-18)

Y-25-B 1910 RENAULT TYPE AG AMBULANCE, issued 1986

1. olive body, black chassis, olive roof, brown interior, 12 spoke olive wheels, red cross & "British Red Cross Society-St. John Ambulance Assn" tempa, England casting ($18-25)(LE)

Y-26-A 1913 CROSSLEY BEER LORRY, issued 1983

1. light blue body, black chassis, tan canopy, brown barrels, maroon seat, black grille, "Lowenbrau" tempa, England casting ($12-15)
2. light blue body, black chassis, light tan canopy, brown barrels, maroon seat, black grille, "Lowenbrau" tempa, England casting ($12-15)
3. light blue body, black chassis, light tan canopy, gray-brown barrels, maroon seat, black grille, "Lowenbrau" tempa, England casting ($12-15)
4. black body, reddish brown chassis, black canopy, dark brown barrels, brown seat, black grille, "Romford Brewery" tempa, England casting ($12-15)
5. white body, red chassis, maroon canopy, dark brown barrels, maroon seat & grille, "Gonzales Byass" tempa, England casting ($12-15)

Y-27-A 1922 FODEN STEAM WAGON, issued 1985

1. blue body, red chassis, gray roof, gray rear canopy, red wheels, without tow hook cast, "Pickfords" tempa, England casting ($15-18)
2. blue body, red chassis, gray roof, gray rear canopy, red wheels, with tow hook cast, "Pickfords" tempa, England casting ($25-40)
3. brown body, black chassis, tan roof, tan rear canopy, tan wheels, "Hovis" tempa, England casting ($12-15)
4. light brown body, black chassis, black roof, black rear canopy, red wheels, "Tate & Lyle" tempa, England casting ($12-15)
5. cream body, dark green chassis, greenish black roof, beige sack load, red wheels, "Spillers" tempa, England casting ($12-15)
6. dark blue body, black chassis, black roof, brown barrel load, red wheels, "Guinness" tempa, Macau casting ($15-18)
7. dark green body, brown chassis, brown roof, cream sack load, dark green wheels, "Joseph Rank" tempa, Macau casting ($15-18)
8. black body, red chassis, cream roof, brown barrels load, red wheels, "McMullen" tempa, China casting ($15-18)

Y-27-B 1922 FODEN STEAM WAGON & TRAILER, issued 1986

1. dark green body with white roof, red chassis & white canopies on both wagon & trailer, "Frasers" tempa, England casting ($25-40)(LE)

Y-28-A 1907 UNIC TAXI, issued 1984

1. maroon body, black chassis & roof, with window frame tempa, red wheels, plastic meter, England casting ($10-15)
2. maroon body, black chassis & roof, without window frame tempa, red wheels, plastic meter, England casting ($10-15)
3. maroon body, black chassis & roof, without window frame tempa, red wheels, metal meter, England casting ($10-15)
4. maroon body, black chassis & roof, with window frame tempa, red wheels, metal meter, England casting ($10-15)
5. maroon body, black chassis & roof, with window frame tempa, maroon wheels, metal meter, England casting ($10-15)
6. maroon body, black chassis & roof, without window frame tempa, maroon wheels, metal meter, England casting ($10-15)
7. blue body, black chassis & roof, without window frame tempa, maroon wheels, metal meter, England casting ($10-15)
8. blue body, black chassis & roof, without window frame tempa, red wheels, metal meter, England casting ($10-15)
9. white body, black chassis & roof, without window frame tempa, gold wheels, Macau casting ($10-15)

Y-29-A WALKER ELECTRIC VAN, issued 1985

1. olive body, beige canopy, olive wheels, tan interior, "Harrods Ltd." tempa, England casting ($15-18)(UK)
2. green body, green canopy, red wheels, tan interior, "Joseph Lucas" tempa, England casting ($10-15)
3. dark blue body, gray canopy, dark blue wheels, tan interior, "His Master's Voice" tempa, Macau casting ($10-15)
4. olive body, dark olive canopy, olive wheels, red-brown interior, "Harrod's Special Bread" tempa, Macau casting ($12-15)

Y-30-A 1920 MODEL AC MACK TRUCK, issued 1985

1. light blue body, dark gray cab roof, dark blue fenders, brown wheels, dark blue chassis, light blue container with dark gray roof, "Acorn Storage" tempa, England casting ($15-18)
2. light blue body, dark gray cab roof &

fenders, brown wheels, dark blue chassis, light blue container with dark gray roof, "Acorn Storage" tempa, England casting ($15-18)

3. cream body & cab roof, dark green fenders, cream wheels, dark green chassis, cream container with tan roof, "Artic Ice Cream" tempa, Macau casting ($12-15)

4. red body, cab roof & fenders, tan wheels, black chassis, red container with tan roof, "Kiwi Boot Polish" tempa, Macau casting ($12-15)

Y-30-B 1920 MACK CANVASBACK TRUCK, issued 1985

1. yellow body & cab roof, dark brown fenders, red wheels, black chassis, tan canopy on yellow bed, "Consolidated Transport" tempa, England casting ($15-18)(LE)

Y-31-A 1931 MORRIS PANTECHNINCON, issued 1990

1. red cab & container, white container roof, red wheels, black base, "Kemp's Biscuits" tempa, Macau casting ($15-18)

2. orange-yellow cab, container & container roof, red wheels, black base, "Weetabix" tempa, China casting ($15-18)

Y-32-A 1917 YORKSHIRE STEAM WAGON, issued 1990

1. plum body, black chassis, cream cab roof & wheels, "Samuel Smith", Macau casting (base cast Y-8 or Y-32)($15-18)

Y-33-A 1920 MACK TRUCK, issued 1990

1. blue body, gray cab roof, blue container

with gray roof, blue wheels, dark blue chassis, "Goodyear" tempa, Macau casting (reissued of Y-30 but base is cast Y-23!)($15-18)

Y-34-A 1933 CADILLAC V-16

1. dark blue body & chassis, brown interior, cream roof, Macau casting ($15-18)

2. white body, dark blue chassis, black interior, black roof, China casting ($15-18)

Y-35-A 1930 MODEL A FORD PICKUP, issued 1990

1. cream body, black roof & chassis, brown interior, yellow wheels, cream bumper, "W. Clifford & Sons" & "Fresh Farm Milk" tempa, Macau casting ($15-18)

2. blue body, cream roof & chassis, brown interior, cream wheels, blue bumper, "From Our Devon Creamery-Ambrosia" tempa, China casting ($15-18)

Y-36-A 1926 ROLLS ROYCE PHANTOM I, issued 1990

1. plum body, black roof & chassis, brown interior, China casting ($15-18)

2. blue body, black roof & chassis, brown interior, China casting ($15-18)

Y-37-A 1931 GARRETT STEAM WAGON, issued 1990

1. powder blue body with white roof, powder blue container with white roof, dark blue chassis, blue wheels, "Chubb's Safe Deposits" tempa, Macau casting ($15-18)

2. dark blue cab & cab roof, dark cream container with white roof, dark blue chassis, red wheels, "Milkmaid Brand Milk" tempa, China casting ($15-18)

Y-38-A ROLLS ROYCE ARMOURED CAR, issued 1990

1. khaki tan body, chassis & turret, khaki wheels, small red/white/blue roundels tempa, Macau casting ($25-30)

Y-39-A 1820 ENGLISH STAGE COACH, issued 1990

1. black body with maroon door, red chassis, hitch & wheels, 4 horses with 1 each in brown, white, red-brown and dark brown, five painted figures, China casting ($45-60)(LE)

Y-40-A 1931 MERCEDES BENZ TYPE 770, issued 1991

1. dark gray body & chassis, dark blue roof, maroon interior, chrome wheels, China casting ($15-20)

Y-41-A 1932 MERCEDES TRUCK, issued 1991

1. dark green body with dark gray roof, black fenders, gray base, red wheels, light gray casting load with brown bed, "Howaldtswerke A.G. Kiel" tempa, China casting (base cast Y-6)($15-20)

Y-42-A ALBION CX27, issued 1991

1. white cab, powder blue bed, dark blue chassis, black interior, blue wheels, chrome milk can load, "Libby's" tempa, China casting ($15-20)

Y-43-A 1905 BUSCH STEAM FIRE ENGINE, issued 1991

1. dark green body & chassis, bluish boiler, red wheels with black tire rims, 4 dark blue painted figures, small lettered tempa, China casting ($50-65)(LE)

Y-44-A 1910 RENAULT BUS, issued 1991

1. orange-yellow body, black chassis, red roof, brown seats, red wheels, "Wesserling -Bassang" tempa, China casting ($150-200)

2. orange-yellow body, black chassis, black roof, brown seats, red wheels, "Wesserling-Bassang" tempa, China casting ($15-20)

Y-45-A 1930 BUGATTI ROYALE, issued 1991

1. black body with blue flash tempa, black chassis, dark blue interior, chrome wheels, China casting ($15-20)

Y-46-A 1869 MERRYWEATHER FIRE ENGINE, issued 1991

1. red body & chassis, red wheels & hitch, black rear base section, 2 white horses, 4 dark blue fireman, "Tehidy House" tempa, China casting ($45-60)(LE)

Y-47-A 1929 MORRIS VAN, issued 1991

1. black body & chassis, yellow roof, chrome wheels, brown interior, "Chocolat Lindt" labels, China casting ($15-20)

2. black body & chassis, yellow roof, chrome wheels, brown interior, "Antiques Road Show 1991" labels, China casting ($1500+)(UK)

NOTE: reissue of Y-19 but without small side window cast. Bases cast Y-19.

Y-61-A 1933 CADILLAC FIRE ENGINE, issued 1992

1. red body & chassis, brown ladder with red roof rack, red dome lights, black interior, chrome wheels, "Feurwehr Aarau" tempa, China casting ($18-25)

Y-62-A 1932 FORD AA PICKUP, issued 1992

1. lime body with gray roof, black chassis, red wheels, brown sack load, China casting ($20-30)

Y-63-A 1939 BEDFORD TRUCK, issued 1992

1. red cab, brown bed, black chassis, black interior, chrome wheels, real stone load, "George Farrar Yorkshire Stone" tempa, China casting ($20-30)

Y-64-A 1938 LINCOLN ZEPHYR, issued 1992

1. cream body & chassis, brown interior, yellowish cream tonneau, China casting ($35-50)

Y-65-A AUSTIN 7/BMW/ROSENGART SET, issued 1992

1. three model set includes Austin 7 in red with black chassis, brown interior, chrome wire wheels, "Castrol" tempa; BMW in white with black chassis, brown interior, chrome wire wheels; Rosengart in blue with black chassis & roof, brown interior, chrome disc wheels, China casting ($45-50)(LE)

Y-66-A GOLD STAGE COACH, issued 1992

1. gold plate body & horsebar, opaque blue windows, white horses with painted features, China casting ($15-25)(LE)

Y-901-A JAGUAR SS100, issued 1991

1. pewter casting with free-standing wooden plinth ($75-100)(LE)

DINKY

The Dinky name has been around since the 1930's. Meccano owned the trademark until their bankruptcy in 1979. Nearly a decade later, Matchbox obtained the Dinky trademark to market a new line of 1/43 vehicles. This was in 1988. But in 1988, no molds were ready to release these models. In order not to lose the trademark, Matchbox Toys scrambled to release some sort of Dinky Toy. The original idea was to release their new Commando miniatures under the Dinky name. In fact, some preproductions of Commando are cast "Dinky"! It was finally decided to release six miniature cars in new colors and place them in a red and yellow blistercards denoting "Dinky". It is also of note that the MB6-B Mercedes that was released with Bulgarian cast base features the "Dinky" name but reasons unknown. The miniature Dinkys are cataloged under the miniatures and denoted with (DY).

DY-1A 1967 Series 1-1/2 E Type Jaguar
DY-2A 1957 Chevrolet Belair Sports Coupe
DY-3A 1965 MGB-GT
DY-4A 1950 Ford E.83.W Van
DY-5A 1949 Ford V-8 Pilot
DY-6A 1951 Volkswagen Deluxe Sedan
DY-7A 1959 Cadillac Coupe De Ville
DY-8A 1948 Commer 8 CWT Van
DY-9A 1949 Land Rover Series 1.80
DY-10A 1950 Mercedes Benz Konferenz Type Omnibus 0-3500
DY-11A 1948 Tucker Torpedo
DY-12A 1955 Mercedes 300SL
DY-13A 1955 Bentley 'R' Type Continental
DY-14A 1946 Delahaye 145 Chapron

DY-15A 1952 Austin A40 GV4 10-CWT Van
DY-16A 1967 Ford Mustang Fastback 2 + 2
DY-17A 1939 Triumph Dolomite
DY-18A 1967 Series 1-1/2 E Type Jaguar Convertible
DY-19A 1973 MGB-GT V8
DY-20A 1965 Triumph TR4A-IRS
DY-21A 1964 Austin Mini Cooper 'S'
DY-22A 1952 Citroen 15CV 6CYL
DY-23A 1956 Chevrolet Corvette
DY-24A 1973 Ferrari Dino 246 GTS
DY-25A 1958 Porsche 356A Coupe
DY-26A 1958 Studebaker Golden Hawk
DY-27A 1957 Chevrolet Belair Sports Coupe Convertible
DY-28A 1969 Triumph Stag
DY-29A 1953 Buick Skylark
DY-30A 1956 Austin Healey
DY-31A 1955 Ford Thunderbird
DY-32A 1957 Citroen 2CV
DY-902 3 piece set with DY-12, DY-24 & DY-25
DY-903 3 piece set with DY-18, DY-20, & DY-30
DY-921 1967 Series 1-1/2 E Type Jaguar (Pewter)

DY-1A 1967 SERIES 1-1/2 E TYPE JAGUAR, issued 1989

1. dark green body, black roof, tan interior, green Macau base ($10-15)
2. yellow body, black roof, black interior, yellow China base ($17-20)

DY-2A 1957 CHEVROLET BELAIR SPORTS COUPE, issued 1989

1. red body, white roof, red interior, pearly silver Macau base ($17-20)

DY-3A 1965 MGB-GT, issued 1989

1. blue body with black painted roof, red interior, black Macau base ($10-15)
2. orange body, black interior, black China base ($10-15)

DY-4A 1950 FORD E.83.W VAN, issued 1989

1. yellow-orange body, red interior, yellow-orange hubcaps, black China base, "Heinz 57 Varieties" tempa ($10-15)
2. olive body, red interior, silver hub caps, olive China base, "Radio Times' tempa ($10-15)

DY-5A 1949 FORD V-8 PILOT, issued 1989

1. black body, tan interior, black Macau base ($10-15)
2. silver-gray body with black painted roof, maroon interior, silver-gray China base ($10-15)
3. tan body with black painted roof, maroon interior, tan China base ($10-15)

DY-6A 1951 VOLKSWAGEN DELUXE SEDAN, issued 1989

1. powder blue body with gray painted roof, light gray interior, powder blue wheels, black Macau base ($18-25)
2. black body with gray painted roof, light gray interior, black wheels, black China base ($17-20)
3. red body with gray painted roof, light gray interior, red wheels, black China base ($17-20)

DY-7A 1959 CADILLAC COUPE DE VILLE, issued 1989

1. metallic red body, cream roof, tan interior

with orange steering wheel, black China base ($17-20)
2. pink body, cream roof, cream interior with red steering wheel, black China base ($17-20)

DY-8A 1948 COMMER 8 CWT VAN, issued 1989
1. red body, red-brown interior, black Macau base, "Sharp's Toffee" tempa ($10-15)
2. dark blue body, gray interior, black China base, "His Master's Voice" tempa ($10-15)
3. dark blue body, gray interior, black China base, "Motorfair 91" labels ($50-65)(C2)

DY-9A 1949 LAND ROVER SERIES 1.80, issued 1989
1. green body & windscreen, green wheels, black interior, tan canopy, black Macau base ($10-15)
2. yellow body & windscreen, black wheels, black interior, tan canopy, black China base, "AA Road Service" tempa ($17-20)

DY-10A 1950 MERCEDES BENZ KONFERENZ TYPE OMNIBUS 0-3500, issued 1989
1. cream body with blue hood & fenders, red interior, clear windows, black China base, "Reiseburo Ruoff Stuttgart" tempa ($50-75)(LE)

DY-11A 1948 TUCKER TORPEDO, issued 1990
1. metallic red body, brown interior, black Macau base ($17-20)
2. metallic blue body, white interior, black China base ($17-20)

DY-12A 1955 MERCEDES 300SL, issued 1990
1. cream body, red interior, cream China base ($10-15)
2. silver-gray body, red interior, silver-gray China base (see DY902)
3. black body, cream interior, black China base ($10-15)

DY-13A 1955 BENTLEY 'R' TYPE CONTINENTAL, issued 1990
1. silver blue body, black interior, black China base ($10-15)
2. dark blue body, tan interior, black China base ($17-20)

DY-14A 1946 DELAHAYE 145 CHAPRON, issued 1990
1. dark metallic blue body, red interior, black China base ($10-15)
2. metallic red body, tank interior, black China base ($10-15)

DY-15A 1952 AUSTIN A40 GV4 10-CWT VAN, issued 1990
1. red body, tan interior, black China base, "Brooke Bond Tea" tempa ($10-15)
2. yellow body, black interior, black China base, "Dinky Toys" tempa ($10-15)

DY-16A 1967 FORD MUSTANG FASTBACK 2 + 2, issued 1990
1. dark green body, black interior, black China base ($10-15)

DY-17A 1939 TRIUMPH DOLOMITE, issued 1990
1. red body, black roof & interior, maroon wheels, red China base ($15-25)(LE)

DY-18A 1967 SERIES 1-1/2 E TYPE JAGUAR CONVERTIBLE, issued 1990
1. red body, brown interior, clear windshield, red Macau base ($10-15)
2. cream body, red interior, clear windshield, cream China base (see DY903)

DY-19A 1973 MGB-GT V8, issued 1990
1. brownish body, black interior, black Macau base ($10-15)

DY-20A 1965 TRIUMPH TR4A-IRS, issued 1991
1. white body, black interior & tonneau, white China base ($10-15)
2. red body, black interior & tonneau, red China base (see DY903)

DY-21A 1964 AUSTIN MINI COOPER 'S', issued 1991
1. cream body with black roof, red interior, cream China base ($10-15)
2. cream body with black roof, red interior, cream China base, small side labels depicting steering wheel & flags ($85-110)(C2)

DY-22A 1952 CITROEN 15CV 6 CYL, issued 1991
1. black body, tan interior, cream wheels, black China base ($10-15)
2. cream body, brown interior, cream wheels, cream China base ($10-15)

DY-23A 1956 CHEVROLET CORVETTE, issued 1991
1. red body with cream side flash, red roof, red interior, red China base ($10-15)
2. metallic copper body with cream side flash, met. copper roof, cream interior, metallic copper China base ($17-20)

DY-24A 1973 FERRARI DINO 246 GTS, issued 1991
1. red body, black roof & interior, red China base ($17-20)
2. metallic blue body, black roof, cream interior, metallic blue China base (see DY902)

DY-25A 1958 PORSCHE 356A COUPE, issued 1991
1. silver-gray body, red interior, cream wheels, black China base ($10-15)
2. red body, cream interior, cream wheels, black China base (see DY902)

DY-26A 1958 STUDEBAKER GOLDEN HAWK, issued 1991
1. metallic brown with cream flash, red interior, black China base ($25-40)

DY-27A 1957 CHEVROLET BEL AIR COUPE CONVERTIBLE, issued 1991
1. powder blue body, brown & blue interior, silver-gray China base ($90-125)
2. powder blue body, cream & blue interior, silver-gray China base ($17-20)

DY-28A 1969 TRIUMPH STAG, issued 1992
1. white body, red interior, white China base ($10-15)
2. white body, red interior, white China base, small label depicting steering wheel & flag with "1992" on hood & trunk ($50-75)(C2)

DY-29-A 1953 BUICK SKYLARK, issued 1992
1. powder blue body, white & blue interior, white tonneau, black China base ($17-20)

BY-30A 1956 AUSTIN HEALEY, issued 1992

1. dark green body, black seats, black tonneau, black China base ($17-20)
2. silver blue body, cream seats, black tonneau, black China base (see DY903)

DY-31A 1955 FORD THUNDERBIRD, issued 1992

1. red body, red & white interior, black China base ($17-20)

DY-32A 1957 CITROEN 2CV, issued 1992

1. gray body with dark gray painted roof, gray & white interior, cream wheels, gray China base ($17-20)

DY902 THREE PIECE DINKY SET, issued 1991

1. wooden plinth with DY-12A3, DY-24A2, DY-25A2 ($60-75)

DY903 THREE PIECE DINKY SET, issued 1992

1. wooden plinth with DY-18A2, DY-20A2, DY-30A2 ($60-75)

DY-921 1967 SERIES 1-1/2 E TYPE JAGUAR, issued 1992

1. pewter casting mounted to wooden plinth ($35-50)

SKYBUSTERS

The list below is a continuation from book 2 and denotes new introductions.

SB-28-A A300 Airbus	1983
SB-29-A Lockhead SR-71 Blackbird	1990
SB-30-A Grumman F-14 Tomcat	1990
SB-31-A Boeing 747-400	1990
SB-32-A Fairchild A1O Thunderbolt	1990
SB-33-A Bell Jet Ranger	1990
SB-34-A Lockheed A130 Hercules	1990
SB-35-A MIL Mi Hind-D	1990
SB-36-A Lockhead F-117A (Stealth)	1990
SB-37-A Hawk	1992
SB-38-A B.Ae 146	1992
SB-39-A Boeing Stearman	1992
SB-40-A Boeing 737.300	1992

NOTE: all Skybusters now have thick cast axles unless otherwise noted.

SB-1-A LEAR JET, issued 1973 (continued)

5. lemon fuselage, white undercarriage, "D-IDLE" tempa, clear window, Macau casting ($6-8)
6. red fuselage, red undercarriage, "Datapost" tempa, clear window, Macau casting ($6-8)
7. purple fuselage, white undercarriage, "Federal Express" tempa, clear window, Macau casting ($6-8)(GS)
8. white fuselage, white undercarriage, "G-JCB" tempa, clear window, Macau casting ($6-8)(GS)
9. white fuselage, white undercarriage, "U.S. Air Force" tempa, clear window, Macau casting ($4-6)(US)
10. purple fuselage, purple undercarriage, "U.S. Air Force" tempa, clear window, Macau casting ($6-8)(SC)
11. white fuselage, white & orange undercarriage, "QX-Press Freight Delivery Service" tempa, clear window, Macau casting ($4-6)
12. white fuselage, white undercarriage, "DHL" tempa, clear window, Thailand casting ($3-4)
13. white fuselage, white undercarriage, "U.S. Air Force" tempa, clear window, Thailand casting ($3-4)(US)

SB-2-A CORSAIR A7D, issued 1973 (continued)

6. khaki tan body, white base, clear window, brown & green camouflage tempa, Macau casting ($6-8)(US)
7. orange body, white base, clear window, olive & brown camouflage tempa, Macau casting ($6-8)(SC)

SB-3-B NASA SPACE SHUTTLE, issued 1980 (continued)

2. white body, pearly silver base-Macau casting, "United States NASA" labels ($3-5)
3. white body, pearly silver base-Thailand casting. "United States NASA" labels ($3-5)
4. white body, white base-Thailand casting, no labels ($6-8)(GF)

SB-4-A MIRAGE F1, issued 1973 (continued)

5. light orange body, brown base, clear window, "122-18" tempa, Macau casting ($4-6)
6. dark orange body, brown base, clear window, "122-18" tempa, Macau casting ($4-6)
7. yellow body & base, clear window, blue stripes on red painted wings tempa, Macau casting ($3-5)(US)
8. white body & base, clear window, "ZE-164" with blue wings tempa, Macau casting ($3-5)(UK)
9. pink body & base, clear window, blue stripes on red painted wings tempa, Macau casting ($6-8)(SC)
10. blue body & base, clear window, white & red stripes tempa, Thailand casting ($3-5)

SB-6-A MIG 21, issued 1973 (continued)

6. silver-gray flecked body & base, clear window, star wing & tail tempa, Macau casting ($4-6)(US)
7. black body & base, clear window, lightning bolts on wings & scorpion & "3" on tail tempa, Macau casting ($4-6)(US)
8. light purple body & base, clear window, lightning bolts on wings & scorpion & "3" on tail tempa, Macau casting ($6-8)(SC)
9. black body & base, clear window, lightning bolts on wings & scorpion & "3" on tail tempa, Thailand casting ($3-5)(US)
10. pearly silver body & base, clear window, star wing & tail tempa, Thailand casting ($3-5)(US)

SB-7-A JUNKERS, issued 1973 (continued)

4. black body, beige & brown base & wings, cross label on wings, cross tail label, Macau casting ($7-10)

SB-8-A SPITFIRE, issued 1973 (continued)

9. light brown body, khaki base & wings, clear windows, light brown camouflage tempa & bullseye labels, Macau casting ($7-10)
10. khaki body, base & wings, clear

window, green camouflage tempa, Thailand casting ($20-25)(0P)

SB-9-A CESSNA 402, issued 1973 (continued)

11. dark green body, white base & wings, clear window, large "N7873Q" tempa, Macau casting ($7-10)
12. brown body, beige base & wings, clear window, "N402CW" tempa, Macau casting ($6-8)
13. white body, red base & wings, clear window, "DHL World-Wide Express" tempa, Macau casting ($4-6)
14. blue body, yellow base & wings, clear window, "S7-402", etc. tempa, Macau casting ($4-6)
15. white body, red base & wings, clear window, "DHL World-Wide Express" tempa, Thailand casting ($3-5)

SB-10-A BOEING 747, issued 1973 (continued)

10. white fuselage, dark blue undercarriage, "British" tempa, Macau casting ($15-20)
11. white fuselage, pearly silver undercarriage, "Cathay Pacific" tempa, Macau casting ($6-8)
12. white fuselage, pearly silver undercarriage, "British Caledonia" tempa, Macau casting ($8-12)
13. white fuselage, pearly silver undercarriage, "Lufthansa" tempa, Macau casting ($6-8)
14. white fuselage, pearly silver undercarriage, "Pan Am" tempa, Macau casting ($12-15)
15. white fuselage, white undercarriage, "Virgin" tempa, Macau casting ($6-8)
16. powder blue fuselage, pearly silver

undercarriage, "KLM" tempa, Macau casting ($6-8)
17. white body, pearly silver undercarriage, "Air Nippon" tempa, Macau casting ($12-15)(JP)(GS)
18. lime body, white undercarriage with pearly silver wings, "Aer Lingus" tempa, Macau casting ($6-8)
19. powder blue fuselage, pearly silver undercarriage, "KLM" tempa, Thailand casting ($3-5)
20. white body, pearly silver undercarriage, "Pan Am" tempa, Thailand casting ($7-10)
21. white body, pearly silver undercarriage, "Lufthansa" tempa, Thailand casting ($3-5)
22. white body, white undercarriage, "South African Airways" tempa, Thailand casting ($10-15)(GR)
23. white body, white undercarriage, "Virgin" tempa, Thailand casting ($3-5)
24. white body, pearly silver undercarriage, "Saudi" tempa, Thailand casting ($3-5)
25. white body, pearly silver undercarriage, "Olympic" tempa, Thailand casting ($3-5)

SB-11-A ALPHA JET, issued 1973 (continued)

4. blue body, blue base, clear window, "162" on wings tempa, Macau casting ($7-10)
5. white body, red base, clear window, "AT39" on wings tempa, Macau casting ($7-10)

SB-12-B PITTS SPECIAL, issued 1980 (continued)

2. dark green body, white upper wings, red

flares tempa, cream driver, Macau casting ($7-10)
3. blue body, white upper wings, "Matchbox" tempa, cream driver, Macau casting ($7-10)
4. red body, red upper wings, "Fly Virgin Atlantic" tempa, cream driver, Macau casting ($6-8)
5. red body, red upper wings, "Fly Virgin Atlantic" tempa, cream driver, Thailand casting ($3-5)
6. white body, white upper wings, no tempa, red driver, Thailand casting ($10-15)(GF)
7. white body, white upper wings, "Circus Circus" tempa, Thailand casting ($6-8)(GS)

SB-13-A DC.10, issued 1973 (continued)

7. white fuselage, pearly silver undercarriage, "Lufthansa" tempa, Macau casting ($6-8)
8. white fuselage, pearly silver undercarriage, "Alitalia" tempa, Macau casting ($6-8)
9. white fuselage, white undercarriage, "Thai" tempa, Macau casting ($6-8)
10. white body, pearly silver undercarriage, "Swissair" tempa, Macau casting ($6-8)
11. silver & red fuselage, pearly silver undercarriage, "Aeromexico" tempa, Macau casting ($6-8)
12. silver-gray fuselage, silver-gray undercarriage, "American" tempa, Macau casting ($6-8)
13. white fuselage, white undercarriage, "UTA" tempa, Macau casting ($250-300)(SA)
14. white fuselage, white undercarriage, "Thai" tempa, Thailand casting ($3-5)
15. silver & red fuselage, pearly silver

undercarriage, "Aeromexico" tempa, Thailand casting ($3-5)
16. white fuselage, white undercarriage, "Scandanavian" tempa, Thailand casting ($3-5)
17. silver-gray fuselage, silver-gray undercarriage, "American" tempa, Thailand casting ($3-5)
18. white fuselage, silver-gray undercarriage, "Sabena" tempa, Thailand casting ($3-5)

SB-15-A PHANTOM F4E, issued 1975 (continued)

4. metallic red body, white base & wings, clear window, red/white/blue wing labels, Macau casting ($7-10)
5. gray body, base & wings, clear window, "Marines" with orange & yellow stripes tempa, Macau casting ($4-6)
6. pink body, base & wings, clear window, "Marines" with orange & yellow stripes tempa, Macau casting ($6-8)(SC)
7. gray body, base & wings, clear window, "Marines" with orange & yellow stripes tempa, Thailand casting ($3-5)

SB-16-A CORSAIR F4U-5N, issued 1975 (continued)

6. light orange body, clear window, "Navy" label on right wing, star label on left wing, Macau casting ($6-8)

SB-19-A PIPER COMMANCHE, issued 1977 (continued)

2. white body, white base & wings, silver-gray interior, "XP" tempa, Macau casting ($6-8)
3. beige body, dark blue base & wings, silver-gray interior, "Commanche" tempa,

Macau casting ($4-6)

4. beige body, dark blue base & wings, silver-gray interior, "Commanche" tempa, Thailand casting ($3-5)

SB-20-A HELICOPTER, issued 1977 (continued)

4. dark blue upper body, white lower body, "Air-Aid" tempa, Macau casting ($6-8)
5. dark blue upper body, dark blue lower body, "Gendarmarie JAB" tempa, Macau casting ($250-300)(FR)

SB-22-A TORNADO, issued 1978 (continued)

6. light gray body & wings, white base, "F132" & ornate design tempa, Macau casting ($6-8)
7. red body & wings, white base, white wings with "06" tempa, Macau casting ($6-8)(UK)
8. light purple body & wings, white base, "F132" & ornate design tempa, Macau casting ($6-8)(SC)
9. light gray body & wings, white base, "F132" & ornate design tempa, Thailand casting ($3-5)

SB-23-A SUPERSONIC TRANSPORT, issued 1979 (continued)

3. white body & base, "Air France" tempa, Macau casting ($4-6)
4. white body & base, "Supersonic Airlines" tempa, Macau casting ($4-6)
5. white body & base, "Singapore Airlines" tempa, Macau casting ($175-250)
6. white body & base, "British Airways" tempa, Thailand casting ($3-5)
7. white body & base, "Air France" tempa, Thailand casting ($3-5)

8. white body & base, "Heinz 57" tempa, Thailand casting ($35-50)(US)
9. white body & base, no tempa, Thailand casting ($10-15)(GF)

SB-24-A F.16 FIGHTER JET, issued 1979 (continued)

4. white body, red base & wings, "USAF" label on right wing, star label on left wing, no side labels, Macau casting ($6-8)
5. red body, white base & wings, "USAF" tempa on right wing, star tempa on left wing, "United State Air Force" side tempa, Macau casting ($4-6)
6. light gray body, base & wings, "USAF" tempa on right wing, star tempa on left wing, "U.S. Air Force" side tempa with blue & dk. gray camouflage, Macau casting ($4-6)(US)
7. white body, base & wings, black tempa on both wings, "USAF XXX" side tempa, Macau casting ($4-6)(UK)
8. light gray body, base & wings, "USAF" tempa on right wing, "U.S. air force" side tempa with blue & dk. gray camouflage, Thailand casting ($3-5)(US)
9. light purple body, base & wings, "USAF" tempa on right wing, star tempa on left wing, "U.S. Air Force" side tempa with blue & dk. gray camouflage, Macau casting ($6-8)(SC)
10. red body, white base & wings, "USAF" tempa on right wing, star tempa on left wing, "United States Air Force" side tempa, Thailand casting ($3-5)(UK)
11. white body, base & wings, no tempa, red windows ($10-15)(GF)

SB-25-A RESCUE HELICOPTER, issued 1979 (continued)

5. dark blue upper body, dark blue lower body, black interior, black exhausts, "Royal Air Force Rescue" tempa, Macau casting ($5-8)
6. white upper body, white lower body, orange interior, chrome exhausts, "Shell" tempa, Macau casting ($5-8)
7. white upper body, red lower body, tan interior, chrome exhausts, "007" tempa, Macau casting ($10-15)(JB)(GS)

SB-26-A CESSNA FLOAT PLANE, issued 1981 (continued)

2. red body, white base & wings, black skis, clear window, "N264H" tempa, Macau casting ($6-8)
3. black body, white base & wings, black skis, clear window, "C210F" tempa, Macau casting ($6-8)
4. red body, red base & wings, black skis, clear window, "Fire" tempa, Macau casting ($3-5)
5. white body, white base & wings, black skis, clear window, "007 James Bond" tempa, Macau casting ($10-15)(JB)(GS)

SB-27-A HARRIER JET, issued 1981 (continued)

2. white body & wings, red base, "Marines" tempa, clear window, Macau casting ($6-8)
3. light gray body & wings, white base, dark gray camouflage tempa, clear window, Macau casting ($4-6)
4. gray body & wings, white base, "Marines" & dark gray camouflage tempa, clear window, Macau casting ($4-6)(US)

5. metallic blue body & wings, white base, "Royal Navy" tempa, clear window, Macau casting ($4-6)(UK)
6. pea green body & wings, white base, "Marines" & olive camouflage tempa, clear window, Macau casting ($6-8)(SC)
7. dark blue body & wings, white base, "Royal Navy" tempa, clear window, Thailand casting ($3-5)
8. light gray body & wings, white base, bullseye tempa, clear window, Thailand casting ($3-5)

SB-28-A A300 AIRBUS, issued 1983

1. white fuselage, pearly silver undercarriage, "Lufthansa" tempa, Macau casting ($6-8)
2. white fuselage, white undercarriage, "Alitalia" tempa, Macau casting ($6-8)
3. white fuselage, white undercarriage, "Air France" tempa, Macau casting ($6-8)
4. powder blue fuselage, pearly silver undercarriage, "Korean Air" tempa, Macau casting ($5-8)
5. white fuselage, white undercarriage, "Iberia" tempa, Macau casting ($5-8)
6. white fuselage, white undercarriage, "Air Inter" tempa, Macau casting ($125-175)
7. white fuselage, pearly silver undercarriage, "Swissair" tempa, Macau casting ($5-8)
8. white fuselage, white undercarriage, "Air France" tempa, Thailand casting ($3-5)
9. powder blue fuselage, white undercarriage, "Korean Air" tempa, Thailand casting ($3-5)
10. white fuselage, white undercarriage, "Alitalia" tempa, Thailand casting ($3-5)
11. white fuselage, white undercarriage, "Air Malta" tempa, Thailand casting ($20-30)(Malta)

12. white fuselage, white undercarriage, "Swissair" tempa, Thailand casting ($3-5)

SB-29-A SR-71 BLACKBIRD, issued 1990
1. black body, "U.S. Air Force" tempa, Macau casting ($3-5)
2. black body, "U.S. Air Force" tempa, Thailand casting ($3-5)

SB-30-A GRUMMAN F-14 TOMCAT, issued 1990
1. gray fuselage, white undercarriage, "Navy" tempa, Macau casting ($3-5)
2. gray fuselage, white undercarriage, "Navy" tempa, Thailand casting ($3-5)

SB-31-A BOEING 747-400, issued 1990
1. light gray fuselage, dark blue undercarriage, "British Airways" tempa, Thailand casting ($4-6)(GS)
2. white fuselage, pearly silver undercarriage, "Cathay Pacific" tempa, Thailand casting ($3-5)
3. white fuselage, pearly silver undercarriage, "Lufthansa" tempa, Thailand casting ($3-5)
4. white fuselage, pearly silver undercarriage, "Singapore Airlines" tempa, Thailand casting ($3-5)

SB-32-A FAIRCHILD A-10 THUNDER-BOLT, issued 1990
1. dark gray body, green camouflage tempa, Macau casting ($3-5)
2. dark gray body, green camouflage tempa, Thailand casting ($3-5)

SB-33-A BELL JET RANGER, issued 1990
1. white/blue body, white base, "Sky-Ranger" tempa, Macau casting ($3-5)
2. white/blue body, white base, "Sky-Ranger" tempa, Thailand casting ($3-5)

SB-34-A C-130 HERCULES, issued 1990
1. white body, white undercarriage, "USCG" tempa, Macau casting ($3-5)
2. white body, white undercarriage, "USCG" tempa, Thailand casting ($3-5)

SB-35-A MiL M-24 HIND-D, issued 1990
1. brown body, gray base, dark brown camouflage tempa, Thailand casting ($3-5)

SB-36-A LOCKHEED F-117A (STEALTH), issued 1990
1. dark gray body & base, "USAF" tempa, Thailand casting ($3-5)
2. white body & base, no tempa, Thailand casting ($10-15)(GF)

SB-37-A HAWK, issued 1992
1. red body, red base, "Royal Air Force" tempa, white under wings, Thailand casting ($6-8)(0P)
2. red body, red base, "Royal Air Force" tempa, plain underside, Thailand casting ($3-5)

SB-38-A BaE 146, issued 1992
1. white body, gray base, "Dan-Air" tempa, Thailand casting ($3-5)
2. white body, white base, "Thai" tempa, Thailand casting ($6-8)

SB-39-A BOEING STEARMAN, issued 1992
1. orange-yellow body & wings, "Crunchie" with printing on underside of wings tempa, Thailand casting ($20-30)(OP)(UK)
2. orange-yellow body & wings, "Crunchie" without printing on underside of wings tempa, Thailand casting ($3-5)
3. white body & wings, "Circus Circus" tempa, Thailand casting ($3-5)(MC)

SB-40-A BOEING 737-300, issued 1992
1. white fuselage, dark blue undercarriage, "Britannia" tempa, Thailand casting ($3-5)
2. powder blue body, silver-gray undercarriage, "KLM" tempa, Thailand casting ($3-5)

SUPERCHARGERS

In 1984, Matchbox Toys entered the "Monster Truck" market with an item called "High Riders". Using two different clip-on chassis, a street car could become a 4X4 or a truck could become a street type vehicle. Several different cars and trucks were used. High Riders value ($7-10). In 1986, Matchbox Toys improved upon the clip-on method by introducing Superchargers. These Monster Trucks included permanent cast bases affixed to miniature style body castings. Later on, Monster cars were used and noted as Mud Racers. A series of Monster Tractors was also released.

SC01-A Big Foot
SC02-A USA 1
SC03-A Taurus

SC04-A Rollin Thunder
SC05-A Flyin Hi
SC06-A Awesome Kong II
SC07-A Mad Dog
SC08-A Hawk
SC09-A So High
SC10-A Toad
SC11-A Mud Ruler
SC12-A Bog Buster
SC13-A Hog
SC14-A Mud Monster
SC15-A Big Pete
SC16-A Doc Crush
SC17-A Mud Slinger II
SC18-A '57 Chevy
SC19-A Drag-On
SC20-A Voo Doo
SC21-A Hot Stuff
SC22-A Showtime
SC23-A Checkmate
SC24-A 12 Pac
SCXX-A Wagon Wheels
SC Pullsleds

SC01-A BIG FOOT, issued 1986
1. metallic blue body, Macau base, with or without tow slot ($10-15)
2. metallic blue body, Macau base, with tow slot & rocker steering ($10-15)

SC02-A USA 1, issued 1986
1. white body, Macau base, without rocker steering ($10-15)
2. white body, Macau base, with rocker steering ($10-15)

SC03-A TAURUS, issued 1986
1. light red body, Macau base ($12-15)
2. dark red body, Macau base ($12-15)

SC04-A ROLLIN THUNDER, issued 1986
1. orange body, Macau base, without rocker steering ($10-15)
2. orange body, Macau base, with rocker steering ($10-15)

SC05-A FLYIN HI, issued 1986
1. white body, with spare tire, no rocker steering ($12-15)
2. white body, without spare tire, no rocker steering ($10-15)
3. white body, without spare tire, with rocker steering ($10-15)

SC06-A AWESOME KONG II, issued 1986
1. red body, with or without tow slot, no rocker steering ($10-15)
2. red body, with tow slot, with rocker steering ($10-15)

SC07-A MAD DOG II, issued 1986
1. yellow body, with or without tow slot, no rocker steering ($10-15)
2. yellow body, with tow slot, with rocker steering ($10-15)

SC08-A HAWK, issued 1986
1. black body, light gray interior ($15-20)
2. black body, dark gray interior ($15-20)

SC09-A SO HIGH, issued 1987
1. yellow body, with or without tow slot ($15-20)

SC10-A TOAD, issued 1987
1. red body, chrome exhausts, no rocker steering ($15-20)
2. red body, chrome exhausts, with rocker steering ($15-20)

3. red body, gray exhausts, with rocker steering ($15-20)

SC11-A MUD RULER, issued 1989
1. 1984 Corvette casting, no rocker steering, black tires ($10-12)
2. 1984 Corvette casting, with rocker steering, black tires ($10-15)
3. 1987 Corvette casting, with rocker steering, black tires ($10-15)
4. 1987 Corvette casting, with rocker steering, neon yellow tires ($8-12)

SC12-A BOG BUSTER, issued 1989
1. black body, no rocker steering, Macau base, black tires ($10-15)
2. black body, with rocker steering, Macau base, black tires ($10-15)
3. black body, with rocker steering, Thailand base, black tires ($8-12)
4. black body, with rocker steering, Thailand base, neon orange tires ($8-12)

SC13-A HOG, issued 1989
1. white body, no rocker steering, Macau base, black tires ($8-12)
2. white body, with rocker steering, Macau base, black tires ($8-12)
3. white body, with rocker steering, Macau base, neon orange tires ($8-12)
4. white body, with rocker steering, Thailand base, neon orange tires ($8-12)
5. white body, with rocker steering, Thailand base, black tires ($8-12)

SC14-A MUD MONSTER, issued 1989
1. yellow body, without or with tow slot, black tires ($8-12)
2. yellow body, with tow slot, neon orange tires ($8-12)

SC15-A BIG PETE, issued 1988
1. lime green body, chrome dump, chrome exhausts, Macau base ($10-15)
2. lime green body, chrome dump, gray exhausts, Macau base ($10-15)
3. lime green body, chrome dum, gray exhausts, Thailand base ($10-15)

SC16-A DOC CRUSH, issued 1988
1. red body, Macau base ($15-20)

SC17-A MUD SLINGER II, issued 1988
1. blue body, Macau base, black tires ($8-12)
2. blue body, Thailand base, black tires ($8-12)
3. blue body, Thailand base, neon yellow tires ($8-12)

SC18-A '57 CHEVY, issued 1988
1. metallic red body, Macau base, black tires ($8-12)
2. metallic red body, Macau base, neon yellow tires ($8-12)
3. metallic red body, Thailand base, neon yellow tires ($8-12)
4. metallic red body, Thailand base, black tires ($8-12)

SC19-A DRAG-ON, issued 1988
1. green body, Macau base ($8-12)

SC20-A VOO DOO, issued 1988
1. black body, Macau base ($8-12)

SC21-A HOT STUFF, issued 1988
1. red body, Macau base ($8-12)

SC22-A SHOWTIME, issued 1988
1. yellow body, Macau base ($8-12)

SC23-A CHECKMATE, issued 1988
1. yellow body, Macau base ($8-12)

SC24-A İ2 PAC, issued 1988
1. orange body, Macau base ($8-12)

SC-PULLSLEDS, issued 1988
1. powder blue plastic body ($7-10)
2. yellow plastic body ($7-10)

SC-XX WAGON WHEELS, issued 1988
1. red body, "Wagon Wheels" livery ($75-100)(OP)(UK)
NOTE: not issued with an "SC" number due to being an on-pack release!

HARLEY DAVIDSON

In 1992, Matchbox Toys received a license for Harley Davidson. As Matchbox Toys already had molds it was just a matter of reintroducing older models in new liveries. Models were released using Convoy, miniature, Superking and a larger plastic version originally from the Kidco line. The Convoy, miniatures and Superkings listings are under their specific headings denoting (HD). New item releases are scheduled for 1993. Other Harley Davidson items include the following:

76210 Stunt Cycles, issued 1992, available in silver/gray, dark blue, turquoise and met. red ($8-10) ea.)
76220 Motorized Stunt set, issued 1992 with 1 stunt cycle ($15-18)
76230 Collector's Edition, issued 1992 with 1 orange stunt cycle, 2 plastic badges and patch ($10-15)
76270 Collector's Set, issued 1992 with 1

silver/gray stunt cycle, CY-8 Kenworth Truck, 2 MB50-C Cycles (orange & blue), 1 K-83 Cycle in black, patch and poster ($35-50)

THUNDERBIRDS

Possibly one of the biggest licensing winners ever made by Matchbox Toys, Thunderbirds became an instant sellout throughout England in its fall 1992 debut. Thunderbirds, based on a 1960s TV show, were originally modeled by Dinky back at the show's original inception. In the spring and fall of 1992, the BBC TV network in England resurrected the show and with that help Matchbox Toys created a licensing winner. Only four models were released in 1992 with plastic figures released in early 1993.

TB-001 THUNDERBIRD 1, issued 1992
1. metallic blue body, gray wings and red nosecone ($7-10)

TB-002 THUNDERBIRD 2 with Thunderbird 4, issued 1992
1. large green #2 with red retractable legs and removable pod. Comes with the tiny Thunderbird #4 in yellow ($15-20)

TB-003 THUNDERBIRD 3, issued 1992
1. red body, white rocket pads ($7-10)

TB-005 PENELOPE'S FAB 1, issued 1992
1. light pink body, clear windows, cream interior, chrome China base ($7-10)

RESCUE GIFT SET, issued 1992
1. standard packaging ($20-30)
2. "Radio Times" packaging ($75-100)

SUPERFAST MINIS

Introduced for Europe in 1990, Superfast Minis were 1/90 scale miniatures. Original versions made in Macau had a fifth wheel in the center of the base. Six versions were later issued with Triple Heat sets (a World's Smallest, Superfast Mini and Miniature). Two Convoy style transporters were also released. The trailers to these trucks opened up into launchers. Both are designated as MD-250. Superfast Minis were issued in either two packs with a launcher or in sets of four models or with the transporter.

MD-250 KENWORTH TRANSPORTER, issued 1990
1. red MB41-D cab with yellow & orange stripes, gray trailer with red, orange & yellow label. Comes with red Porsche Turbo and White/orange Lumina ($25-45)
2. black MB41-D cab with flames, black trailer with flames labels. Comes with silver Porsche 959 and red Ferrari F40 ($25-45)

MD-200 CARS ASSORTMENT, issued 1990 or 1992. 2 packs ($5-8), 4 packs ($10-15)

1957 CHEVY, issued 1990
1. red body, Macau base
2. black body, Macau base
3. yellow body, China base (TH)

PORSCHE 959, issued 1990
1. silver/gray body, "Porsche" logo, Macau base (MD250)
2. silver/gray body, stripes design, China base (TH)

LAMBORGHINI COUNTACH, issued 1990
1. yellow body, Macau base
2. dark orange body, Macau base
3. white body, China base (TH)

CHEVY CAMARO, issued 1990
1. silver/gray body, Macau base
2. yellow body, Macau base
3. metallic blue body, China base (TH)

FERRARI F40, issued 1990
1. red body, "Ferrari" logo, Macau base (MD250)
2. red body, "F40" design, China base (TH)

BMW M1, issued 1990
1. silver/gray body, Macau base
2. white body, Macau base

JAGUAR XJ6, issued 1990
1. silver/gray body, Macau base
2. metallic green body, Macau base

T-BIRD TURBO COUPE, issued 1990
1. black body, Macau base
2. metallic blue body, Macau base

CHEVY LUMINA, issued 1990
1. white & orange body, "Matchbox Motorsports" livery, Macau base (MD250)

PORSCHE TURBO, issued 1990
1. red body, "Porsche" logo, Macau base

FORD LTD POLICE CAR, issued 1992
1. dark blue body, "Police" livery, China base (TH)

MOTORCITY

Motorcity, designated as "MC", was devised in 1986 with the release of six mini play environments based on molds purchased from the Tomy company of Japan. The Motorcity includes playsets, play environments and other playsets. The Gift set Motorcity sets will be discussed in the "Gift Set" section of this book which includes the models in sets. This section will cover the play environments and play sets.

MC-1-A CAR WASH, issued 1986
1. white & blue with "Car Wash" labels ($7-10)
2. gray & blue with "Matchbox" labels ($7-10)

MC-2-A PETROL STATION, issued 1986
1. white & gray with "Gasoline" labels ($7-10)
2. gray & white with "Matchbox" labels ($7-10)

MC-3-A PIT STOP, issued 1986
1. white & gray ($7-10)
2. gray & red with "Matchbox" labels ($7-10)

MC-4-A GARAGE, issued 1986
1. white with "Garage" labels ($7-10)
2. gray with "Matchbox" labels ($7-10)

MC-5-A CONSTRUCTION CRANE, issued 1986
1. white & orange ($7-10)
2. gray & orange ($7-10)

MC-6-A CONVEYOR LOADER, issued 1986
1. white & orange with "Matchbox" labels ($7-10)
2. gray & orange with "Matchbox" labels ($7-10)

MC-20-A MOTORCITY PLAYTRACK, issued 1987
1. playtrack with bridge, petrol station & signs ($15-20)

MC-30-A MOTORCITY PLAYTRACK, issued 1987
1. playtrack with bridge, petrol station, car park & signs ($20-25)

MC-40-A MOTORCITY PLAYTRACK, issued 1987
1. playtrack with garage, car hoist, bridges, car park, signs ($20-25)

MC-75-A MINI FOLD 'N' GO, issued 1991
1. plastic garage ($8-10)
2. plastic police station ($8-10)
3. plastic rescue station ($8-10)
4. plastic fire station ($8-10)

MC-100-A MOTORCITY PLAYSET, issued 1988
1. playtrack with playmat, built-in street of shops & accessories ($12-15)

MC-150-A MOTORCITY AIRPORT, issued 1990
1. aircraft hanger, runway, carpark & playmat ($15-20)

MC-200-A MOTORCITY PLAYTRACK, issued 1988
1. reissue of MC-20 ($15-20)

MC-300-A MOTORCITY PLAYTRACK, issued 1988
1. reissued of MC-30 ($20-25)

MC-400-A MOTORCITY PLAYTRACK, issued 1988
1. reissue of MC-40 ($20-25)

MC-500 MOTORCITY 500 SUPERSET, issued 1991
1. playtrack with multi-level parking garage & accessories ($25-30)

MC-510 SUPER TRANSPORT SET, issued 1991
1. playtrack with roadway, airport, railway track, accessories ($30-40)

MC-520 MOTORCITY BUILDING ZONE, issued 1992
1. playtrack with multi-storey site & accessories ($30-40)

MC-550 MOTORCITY ELECTRONIC SERVICE CENTRE, issued 1991
1. garage with playmat with 7 working features with 7 sound effects ($40-50)

MC-560 INTERCOM CITY, issued 1992
1. city environment with "talk-a-tronic" system ($90-110)
NOTE: available in different languages!

MC-610 CONTAINER PORT, issued 1991
1. playmat with container ship, crane, railtrack, models ($20-25)

MC-620 CONSTRUCTION YARD, issued 1991
1. playmat with multi-storey site, crane, lift ($20-25)

MC-630 FOLD 'N' GO GARAGE, issued 1991
1. fold-up garage with accessories ($18-25)

MC-640 FOLD 'N' GO CAR PARK, issued 1991
1. fold-up car park with accessories ($18-25)

MC-660 ELECTRONIC RESCUE STATION, issued 1992
1. playmat with rescue station featuring electronic sounds ($40-50)

MC-700 MINITRONICS, issued 1992
1. electronic sound Car Wash ($10-15)
2. electronic sound Tune-Up Center ($10-15)
3. electronic sound Gravel Pit ($10-15)
4. electronic sound Crane ($10-15)

550112 MATCHBOX MOTORS, issued 1983
1. playmat with garage ($15-20)

550117 SUPER SPIN CAR WASH, issued 1983
1. playmat with real working water carwash ($15-20)

50680 STACKEMS, issued 1992
1. service station ($10-15)
2. emergency station ($10-15)
3. truck center ($10-15)

550108 CONVOY TRUCK STOP, issued 1983
1. playmat with truckstop building ($20-30)

50957 JEEP JAMBOREE, issued 1991
1. playmat with plastic mountain & special MB5-D ($25-40)

GIFT SETS

The gift sets division of this book is a little more complicated in its numbering system since 1982 when a simple "G" was the common prefix used for Gift set. Now many different prefixes are used to denote a gift set style and at times just a serial number is given and in some cases no number at all was used to designate a set! Common prefixes used are MC (Motor City), C(for Japanese sets), CY(for Convoy sets), MB(for Matchbox), SB(for Skybuster sets), CS(for Construction), EM(for Emergency), etc. Price values for gift sets are figured by adding up prices of the individual models and adding 10-20% if in a gift box.

G-1C CAR TRANSPORTER SET (continued)
2. contains MB6-B, MB21-C, MB39-B, MB55-D, MB67-C, K-10-D (1983)
2. contains MB9-E, MB15-D, MB23-D, MB33-D, MB75-D, K-120-A (1986)

G-2D CAR TRANSPORTER ACTION PACK, issued 1986
1. contains K-120A and five assorted Super GT's (various)

G-3H JCB CONSTRUCTION SET, issued 1987
1. contains MB32-D, MB60-G, MB75-D, SB-1-A in JCB liveries

G-4E CONVOY GIFT SET (continued)
2. MB59-B, MB61-C, CY3-A, CY7-A, CY9-A, plastic truck stop (1983)
3. MB45-C, MB59-B, CY3-A, CY5-A, CY17-A, plastic truck stop (1985)
4. MB10-C, MB183 DAF cab, CY17-A, CY18-A, CY19-A (1987)

G-5E FEDERAL EXPRESS ACTION PACK, issued 1987
1. contains MB20-D, MB26-F, MB33-D, MB60-F, SB-1-A in Federal Express livery

G-6E VIRGIN ATLANTIC ACTION PACK, issued 1987
1. contains MB15-D, MB65-B, MB68-E, MB75-D, SB-10-A in Virgin Atlantic livery

G-7C EMERGENCY ACTION PACK (continued)
2. contains MB8-D, MB12-D, MB22-C, MB57-D, MB75-D, plastic buildings (1983)

G-8C TURBO CHARGED ACTION PACK, issued 1984
1. contains MB6-C, MB17-D, MB39-C, MB44-D, MB52-C, plastic launcher

G-10D PAN AM ACTION PACK, issued 1986
1. contains MB10-C, MB54-F, MB65-B, BM68-E, SB-10-A in Pan Am liveries

G-11B LUFTHANSA ACTION PACK, issued 1986
1. contains MB30-E, MB54-F, Mb59-D, MB65-B, SB-28-A in Lufthansa liveries

G-25-A COLLECTASET, issued 1984
1. contains 25 assorted miniatures (varies to selection)

C-1 SPORTS CARS, issued 1984 (Japan)
1. contains MB24-D, MB35-C, MB39-C (red), MB40-C

C-2 4WD OFF-ROAD, issued 1984 (Japan)
1. contains MB7-D, MB13-D, MB44-D, MB49-D (1984)

C-3 EUROPEAN CARS, issued 1987 (Japan)
1. contains MB7-D, MB43-E, Mb66-F, MB75-D

C-4 CONSTRUCTION VEHICLES, issued 1984 (Japan)
1. contains MB19-D, MB29-C, MB32-D, MB64-D

C-6 EMERGENCY SET, issued 1987 (Japan)
1. contains MB18-C, MB44-F, MB63-E, MB75-D

C-7 POLICE SET, issued 1984 (Japan)
1. contains MB10-C, MB33-C, MB61-C, MB75-D

2. contains MB10-C, MB33-C, MB44-F, MB57-D

C-8 CAR TRANSPORTER SET, issued 1984 (Japan)
1. contains MB3-C (white), MB24-D (white), MB39-C, CY-1-A

C-9 NASA SET, issued 1984 (Japan)
1. contains MB54-F, CY-2-A, SB-3-B

C-10 CONSTRUCTION SET, issued 1987 (Japan)
1. contains MB29-C, MB30-E, MB32-D, CY-203 Lowloader

C-11 AIRPORT SET, issued 1987
1. contains MB54-F, MB67-F, SB-10-A, SB-28-A in special liveries

C-15 CAR TRANSPORTER SET, issued 1987
1. contains MB7-F, MB9-E, MB23-D, MB31-E, MB22-D, K-120-A

CY-201-A FIRE RESCUE SET, issued 1985
1. contains MB54-F, MB75-D, CY-13-A

CY-202-A POLICE SET, issued 1985
1. contains MB10-C, Mb61-C, CY-11-A

CY-203-A CONSTRUCTION SET, issued 1985
1. contains MB28-C, MB30-E, MB32-D, CY-203 Lowloader

CY-204-A NASA SET, issued 1986
1. contains MB54-F, MB75-D, CY-2-A

CY-205-A FARM SET, issued 1987
1. contains MB46-C, MB51-C, CY-20-A

CY-206-A TELECOM SET, issued 1987
1. contains MB48-F, MB60-G, CY-15-A in British Telecom liveries

MC-7-A FARM SET, issued 1988
1. contains MB35-F, 40-C, MB43-A, MB46-C, MB51-C, TP-103-A, CY-20-A

MC-8-A CONSTRUCTION SET, issued 1988
1. contains MB19-D, MB29-C, MB30-E, MB42-E, MB64-D, MB70-E, CY-203 Lowloader

MC-9-A RACING SET, issued 1989
1. contains MB6-H, MB7-F, MB30-E, MB58-E, MB65-D, Team Transporter

MC-10-A 10 MATCHBOX MINIATURES, issued 1988
1. contains 10 assorted miniatures (varies to contents)

MC-11-A CAR TRANSPORTER SET, issued 1988
1. contains MB17-F, MB39-D, CY-10-A

MC-12-A AEROBATIC TEAM SET, issued 1988
1. contains MB46-F, MB75-D, CY-21-A in Red Rebels liveries

MC-13-A POLICE SET, issued 1988
1. contains MB10-C, MB61-C, CY-11-A

MC-15-A FIRE SET, issued 1990
1. contains MB16-E, Mb30-F, MB54-F, MB63-E, SB-26-A, CY-130-A

MC-16-A MOTORCITY PLAYPACK, issued 1989 (U.K.)
1. contains assortment of eight miniatures & Super GT's with playmat

MC-17-A BRITISH AIRWAYS SET, issued 1991 (not issued in U.S.)
1. contains MB35-F, MB68-E, MB72-E, SB-23-A, SB-31-A British Airways livery

MC-18-A FERRARI SET, issued 1991
1. contains MB24-H, MB70-D, MB74-H. MB75-D, CY-24-A

MC-20-B GIFT PACK, issued 1988
1. contains assortment of 20 miniatures (contents varies)

MC-23-A PORSCHE SET, issued 1991
1. contains MB3-C (yellow), MB7-F (white), MB55-F (red), Mb71-F (black), CY-25-A

MC-24-A RED ARROWS SET, issued 1992 (U.K.)
1. contains MB35-F, MB68-E, Mb75-D, 2X SB-37-A in Royal Air Force liveries

MC-50-A MOTORCITY CARRY PACK, issued 1991
1. contains assortment of miniatures, Super GT's & plastic assessories

MC-801 GARAGE PLAYPACK, issued 1991 (U.K.)
1. contains MB21-E, 2 Super GTs, CY-105-A (Shell) & plastic garage
2. contains MB21-E, 2 Super Gts, CY-17-A (Shell) & plastic garage

MC-803 CIRCUS SET, issued 1992
1. contains MB31-B, MB35-F, MB72-I & CY-25-A

MC-804 CIRCUS SET, issued 1992
1. contains MB31-B, MB35-F, MB40-C, MB72-I, MB73-C, CY-25-A and SB39-A

MC-805-A AROUND LONDON TOUR, issued 1992
1. includes MB4-E, MB8-G, MB17-C, MB31-G, Mb63-E & playmat

MC-963-A 30 PIECE SPECIAL COLLECTION, issued 1992 (U.S.-Sam's Club)
1. includes assortment of 30 pieces (includes 15 specially colored models)
2. includes assortment of 30 pieces (all in normal colors)

MB-824 BUY 5 GET 5 FREE, issued 1991 (U.K.)
1. includes assortment of 10 models (contents varies)

MB-828 STAR CARS, issued 1990 (Holland)
1. includes assortment of 5 cars mounted on star shaped cards

MB-830 RALLY PACK, issued 1991 (U.K.)
1. includes 2 Super GT's, 2 launchers & stop watch

MB-835 CHRISTMAS WORLD RALLY, issued 1991 (German)
1. contains four miniatures in special colors with playmat (3 different)

SB-801 SKYBUSTERS VALUE PACK, issued 1991
1. contains three assorted Skybusters (contents varies)

EM-75 ACTION FIRE, issued 1992
1. contains MB16-E, Mb18-C, MB30-E, MB75-D, CY-13-A, TP-110-A raft

EM-90 30 PIECE CARRY PACK, issued 1992
1. contains assortment of miniatures & plastic assessories

CS-75 HEAVY DUTY SQUAD, issued 1992
1. contains MB19-D, MB29-C, MB30-E, MB64-D, MB70-E, CY-30-A

CS-90 30 PIECE CARRY PACK, issued 1992
1. contains assortment of miniatures & plastic assessories

0138 SUPER GT 38 PIECE GIFT SET, issued 1989
1. contains assortment of 38 Super GTs

76270 HARLEY DAVIDSON COLLECTOR'S EDITION, issued 1992
1. contains 2X MB50-D (blue & orange), CY-8-A, K-83-A (black), Stunt Cycle

030653 THREE PIECE YESTERYEAR SET, issued 1984
1. contains three Yesteryear (contents vary)

DRAGSTER SET, issued 1984 (Japan)
1. contains MB4-D, MB46-D, Mb69-D, MB74-D

CLASSIC SET, issued 1984 (Japan)
1. contains MB4-D, MB42-D, MB69-D, MB71-D

BOOTS FARM PLAYPACK, issued 1991 (U.K. for Boot's stores only)
1. contains MB35-F, MB43-A, MB46-C, MB51-C, TP-103-A with playmat

BOOTS ROUND TOWN PLAYPACK, issued 1991 (U.K. for Boot's stores only)
1. contains MB17-C, MB21-E, Mb26-F, MB48-F, MB60-G, MB67-F with playmat

JAMES BOND 007 LICENSE TO KILL, issued 1989 (U.K.)
1. contains MB58-D, SB-25-A, SB-26-A, CY-105-A in special liveries

MCDONALD'S HAPPY MEAL SET, issued 1987
1. contains 16 assorted Super GTS (this set was never released to the consumer. It was issued in the U.S. Southwest as a display unit to selected McDonald's restaurants) VERY RARE ($150-250 if display intact)

CONNOISSEUR SET, issued 1985
1. contains Y-1-B, Y-3-B, Y-4-B, Y-11-B, Y-13-B, Y-14-B in special colors with special wooden display box

CLASSIC SPORTS OF THE THIRTIES, issued 1983 (Australia)
1. contains Y-14-C, Y-16-B, Y-19-A, Y-20-A

STARTER KIT 5 FOR 4 OFFER, issued 1983

1. contains four assorted Yesteryears (contents vary)

CAP CARS

First introduced in 1983, this series became one of the first new introductions into the diecast category of Universal's new Matchbox Toys Ltd. Although, not part of the miniature series, these miniature sized vehicles featured an opening base hatch in which a small cap could be inserted. The model was then placed inside a launcher. Once launched the cap would ignite making a sound like a cap gun.

Models include:
Thunderbolt
Atom
Nitro
Trail-Traker
Rocky Mountain-Ridge Runner
Mazda RX7 (yellow)*
Fireball
Starfire
Bomb
Sheriff Patrol
Trail Tramp

*this model is the same color as the Trickshifter model but with a cap car base. Value $15-20 on this model. Other Value: $8-12

FLASH FORCE 2000

This series is an extension of the Cap Car series. Introduced in 1984 it included three

Flash Fighters (good guys) and three Rampagers (bad guys). The models were futuristic in appearance. Several large plastic vehicles were also introduced. The models include:

Flash Fighter Mazda ($10-15)
Flash Fighter Corvette ($10-15)
Flash Fighter Pickup ($10-15)
Rampager Datsun ($10-15)
Rampager Pickup ($10-15)
Rampager Firebird ($10-15)

651602 Cyclone Chopper ($50-75)
651601 Dark Seeker Battle Van ($35-50)
650105 Flash Force Base ($35-50
650106 Rampage Rock ($25-35)

PARASITES

Introduced in 1985, these set of six miniature size vehicle included a fold-up "monster" which came from the tail of Halley's Comet (an interesting concept!). Each monster inhabited a model by folding itself up and disguising itself inside the vehicle.

Terrorsite (Chevy Van) ($15-25)
Extermasite (Dodge Caravan) ($15-25)
Nemisite (Corvette) ($15-25)
Gammasite (Firebird) ($15-25)
Spectersite (Pickup) ($15-25)
Destructite (Blazer) ($15-25)

TRICKSHIFTERS

Trickshifters, introduced in 1985 included three paired miniature-size vehicles for a total of six models. The

models performed seven different stunts-wheelies, special turns and flips based on the positioning of a mounted stick shift which controlled a fifth wheel.

Yellow Mazda ($8-10)
Red Datsun ($8-10)
Silver Firebird ($8-10)
Blue Mazda ($8-10)
Black Datsun ($8-10)
White Firebird ($8-10)

TURBO 2

Issued in 1987, Turbo 2's were miniature size diecast vehicles with 2 speed friction powered motors with pullback action. Six styles in two colors each were available ($10-15)

AM-2601 Pontiac Fiero-yellow
AM-2607 Pontiac Fiero-white
AM-2602 Peugeot 205-pearly silver
AM-26-8 Peugeot 205-yellow
AM-2603 Ford Supervan II-dark pink
AM-2609 Ford Supervan II-white
AM-2604 Racing Porsche-orange
AM-2610 Racing Porsche-red
AM-2605 Toyota MR2-red
AM-2611 Toyota MR2-lime green
AM-2606 Group C Racer-blue
AM-2612 Group C Racer-white

CODE SYSTEM

In the 1970s, a code system was developed by Ray Bush of U.K. Matchbox. This code has been adopted and is in wide use today to denote models manufactured, approved or not produced by Matchbox

Toys. The code system is as follows:

CODE 1: any model wholly produced by Matchbox Toys
CODE 2: any model altered by private means but with the written consent of Matchbox Toys.
CODE 3: any model altered by private means without permission of Matchbox Toys.

Of these coded models, this book covers only code 1 and 2 produced models. Code 3 models exist in the hundreds and carry no intrinsic value other than what an individual may wish to spend for one. Code 2 models have become very valuable because of the limited quantities issued. In book 2, Yesteryears were basically the only models subject to code 2 status. Since 1982, only one more Yesteryear was ever granted code 2 status and this didn't happen until 1992! Most of the other code 2's are based on miniature, Convoy and Dinky models and for simplicity's sake, all code 2's in this book are grouped together. Two code 2's not depicted with the rest due to size restriction are the following:
A. PC-1 Model A Ford Van in Framed Cabinet Unit: less than 200 made. (Black & white photo in front of book)($100-125)
b. Complete set of 1990 Miniatures in chrome plating in a wooden display unit-only TWO complete sets were authorized. One set is displayed in the "Matchbox & Lesney Toy Museum" in Durham, CT. The other was won in a club raffle and is displayed by a private collector in Massachusetts. (No value noted as only two sets exits!)

All other code 2's are cataloged under their respective series with a (C2) denoting its code 2 status. The following though is a list of code 2's with quantity of available pieces produced.

Model	Quantity
MB 7-F Porsche 959-chrome plated body,	less than 500
MB17-D London Bus-chrome plated, "Celebrating A Decade"	250
MB24-H Ferrari F40-chrome plated body, less than 500	
MB32-G Modified Racer-chrome plated body,	less than 150
MB38-E Model A Van-silver, "Matchbox Collectors Club 1989"	576
MB38-E Model A Van-neon green/orange, "Matchbox Collectors Club 1990"	714
MB38-E Model A Van-white/pink, "Matchbox Collectors Club 1991"	625
MB38-E Model A Van-brass/black, "Matchbox Collectors Clue 1992"	525
MB38-E Model A Van-"Welcome Ye To Chester"	4000
MB38-E Model A Van-in framed cabinet unit (PC1)	less than 200
MB52-D BMW M.1-chrome plated body	less than 500
MB54-H Chevy Lumina-chrome plated body	less than 150
MB61-E Nissan 300ZX-chrome plated bodY	less than 500
MB65-D F.1 Racer-chrome plated body	less than 150
MB66-E Sauber Group C Racer-chrome plated body	less than 150
MB67-F Lamborghini Countach-chrome	

plated body	less than 500
MB74-H G.P. Racer-chrome plated body	less than 150
CY-8-A Kenworth Box Truck-"Ski Fruit Yogurt"	2000
DY-8-A Commer Van-dark blue, "Motorfair 91"	2300
DY-21-A Mini Cooper-cream, small side labels	1000
DY-28-A Triumph Stag-white, small hood & trunk labels	1900
Y-5-A Leyland Titan Bus-"MICA Collectors Club"	5000

The MICA club redeveloped the code 2 status in the late 1980's to also include models that are"....code 1 but not issued to the general public". If following their definition of code 2, the following models would also be noted as code 2, but this author still considers these code 1 based on the original Ray Bush definition as being "100% produced by Matchbox Toys".

Y-12-D Model T Van-blue body, "Hoover"	500
Y-21-E Model TT Van-blue body, "Jenny Kee"	1000
Y-21-E Model TT Van-green body, "Pro Hart"	1000
Y-21-E Model TT Van-black body, "Antiques Road Show 1992"	24
Y-47-A Morris Van-black body, "Antiques Road Show 1991"	24

THE BOX AND THE BLISTERPACK

In 1982, generic boxes replaced the standard "picture" box. By 1983, only the United States had retained the standard

small closed box being of a yellow generic type with red grid which is still retained today. Most closed-in style boxes are now specialty boxes especially those used for the collector's clubs including MICA, the Junior Collectors Club and Matchbox Collectors Club.

In 1983, England and other countries opted for the "window box". This larger style box featured a cellophane front and rear for easy model viewing. Many different window boxes have been produced with just a small sampling shown in this book. Window boxes were produced for different markets and in the case of the United States for special orders especially those produced for White Rose Collectibles. For 1993, England phased out the window box for a more "environment-friendly" blisterpack.

The blisterpack become the main means in the United States for obtaining miniatures in the last decade. Although blisterpacks were also used occasionally for other countries, America was still the blishterpack "king". Some of the dozens of blisterpacks are depicted in this book.

FAKES AND COPIES

Fakes and copies still ran rampant in the last decade. Eastern Bloc countries like Poland and Hungary are infamous for producing copies of both 1970s and later 1980s vintage miniatures. A company from Spain called Guisval produced several Yesteryear copies. Kresge, of K-Mart fame, even got into the Yesteryear copy act producing models based on 1960s vintage

Yesteryears! A copy of the Y-3-A Tram was produced in 1985 for a "Typhoo Tea" promotion. One of the worst licensing violations against the Matchbox trademark occurred in 1983. A company from Hong Kong produced cheap trailer trucks and placed them in "Matchbox" packaging. Since, Matchbox Toys were now made in the Orient, they thought no one would be the wiser! This was soon stopped and packages were relabeled "Speedburners". Another serious violation of the Matchbox company included the marketing of stolen goods in generic packaging! This included Yesteryear and Dinky either in purple/yellow or white grid pattern boxes. Some miniatures in purple/yellow blisterpacks were also noted. Other brands were also stolen-Racing Champions and Corgi-and these too were put in generic boxes. This was stopped after an investigation by Matchbox Toys.

CATALOGS

POCKET CATALOGS

The year 1983 was the last for a catalog for the United States until 1993. Most catalogs after this date were of International varieties which included from two to five languages in each. Completely different catalogs were made for Australia for the first time and Japan also issued some of their own type catalogs for a short time.

1983-depicts blue cover with red & yellow band for United States & International editions. Australian edition depicts young boy with in front of rows of Matchbox.

($3-5)U.S. & Int'l; ($5-7) Australia
1. United States, International, Australia

1984-three different covers issued. International cover depicts six cars & trucks from miniatures & Superkings range. Australian cover depicts 4 different London Bus miniatures. Japanese covers depicts rows of Matchbox with Japanese text in upper left. ($3-5)International, ($5-8)Aus. & Japan
1. International (English/German/French test), International (English/Italian/ Spanish text), Australia, Japanese

1985-two different covers issued. International & German cover depicts checkered flag. Australian depicts boy clutching toys (with or without "Special Edition" on cover). ($3-5) Int'l ($7-10) German; ($5-8) Australia; ($5-8) Japan
1. International; German (5-3/4"X8-1/4"), Australia, Japan

1986-three covers issued. All editions but Australian & Japan depicts small Live-N-Learn helicopter towing "1986" with several "Matchbox" logos at bottom. Japanese is small foldout depicting. Porsche 959 & Toyota MR2. Australian edition depicts construction scene with 6 vehicles. ($3-5) all but Australia; ($7-10) German; ($5-8) Australia; ($5-8) Japan
1. International (English), German (5-3/4"X8-1/4"), French, Dutch, Australian, Japan

1987-three covers issued. English & German depicts black upper cover with striped lower with array of models in

center. Japanese, another foldout, depicts rows of models in center running diagonally. Australian depicts boy playing with Carguantua playset. ($3-5) English; ($7-10) German; ($5-8) Australian; ($5-8) Japan
1. English, German (5-3/4"X8-1/4"), Australian, German

1988-two covers issued. English & German with blue grid design with closeout view of K-70 Porsche. Japanese is black cover with white lettering. Australian is now just a foldout "growth chart". ($2-4) English; ($7-10) German; ($5-8) Japan; ($2-3) Australian
1. English, German (5-3/4"X8-1/4"), Australian (chart), Japanese

1989-depicts white cover with small inset of playset. ($2-4)
1. English, German, French/Dutch, Italian/ Spanish

1990-same cover design as 1989 but date to 1990 ($2-4)
1. English, German

1991-depicts large playset ($2-4)
1. English/German

1992-depicts 8 "Matchbox" logos from top to bottom with five vehicles on front. Special edition for Toys R Us in England features extra cover overlayed in black with "Toys R Us". ($2-4)
1. English/German, English/German (Toys R Us)

SPECIALTY CATALOGS

In 1985 some specialty catalogs were produced especially for those depicting Yesteryear or Dinky.

Yesteryear Catalogs (measure 5-3/4"X8-1/4")-1985 German, 1986 German (2-types-with Y-20 or red racing car), 1987 German, 1988 German, 1989 German & Italian, 1990 German, 1991 German, 1992 German & English ($5-10)

Dinky Catalogs (measure 5-3/4"X8-1/4") 1991 German, 1992 German & English ($6-8)

Superkings Catalogs (measure 5-3/4"X8-1/4") 1986 German, 1987 German ($5-8)

Yesteryear/Dinky Catalogs (measure 5-3/4"X8-1/4") quarterly editions for 1991 & 1992 Australia ($5-8)

Other-from Czechoslovakia (date unknown-measures 4-3/4"X7") depicts black background with Burago brand model on cover with both "Burago" & "Matchbox" logos on front. First half of catalog for Burago band, second half for Matchbox brand. Distributed by Tuzex. ($20-25)

TRADE OR DEALERS CATALOGS

Catalogs produced for U.S.A., England, German. A few editions exist for Spain or Italy. Add 5-20% for foreign editions.

1983 ($5-10)
1984 ($5-10)
1985 ($5-10)

1986 ($5-10)
1987 ($5-10)
1988 ($5-10)
1989 ($5-10)
1990 ($5-10)
1991 ($5-10)

PINS AND BADGES

There have been an array of pins and badges issued in the last ten years especially by collectors clubs such as M.I.C.A., A.I.M. and Matchbox U.S.A. Others were issued for swapmeets in England including the Yesteryear Farnham Maltings badges. However, only those specifically produced by Matchbox Toys for their clubs, toy fairs and such are listed here. Many of the pins do not denote "Matchbox" but were issued at trade fairs.

"I'm A Matchbox Collectors Club Member", issued 1980 ($3-5)(UK)
"Get Your Hands On a Matchbox", issued 1982 ($3-5)(UK)
"Matchbox" with Live N Learn bear, issued 1982 ($3-5)(UK)
"Matchbox A New Beginning", issued 1983 ($4-6)(US)
"Matchbox On The Move in 1984", issued 1984 ($3-5)(US)
"Matchbox Building New Roads To Success", issued 1985 ($3-5)(US)
"Connectables What's Yours Called-Matchbox", issued 1989 ($2-4)(US)
"Junior Matchbox Club", issued 1987 $2-4)(CN)
"JMC Gang Member Year 3", issued 1989 ($2-4)(CN)
"Get Your Free Mighty Matchbox Truck-

Tandy", issued 1989 ($5-8)(AU)

"My Pet Monster", issued 1987 ($3-5)(UK)

"Rubik's Magic", issued 1988 ($3-5)(UK)

"Monster In My Pocket", issued 1990 ($2-4)(US)

"Robotech", issued 1986 ($2-4)(US)

"Babycise" tie tack, issued 1987 ($15-20)(US)

"Days of Thunder", issued 1989 ($5-8)(US)

"Big Top Pee Wee", issued 1988 ($2-4)(US)

VOLTRON

The Voltron series, based on a cartoon TV series, featured a series of robots. The robots came in sets which could be combined to make up larger robots. These were actually manufactured by Bandai of Japan but were licensed in the United States for Matchbox to distribute. The line includes:

700001 Voltron III Miniature Lion Space Robot ($30-50)

700002 Voltron I Miniature Space Warrior Robot ($30-50)

700100 Voltron II Miniature Red Gladiator Space Robot ($30-50)

700110 Voltron II Miniature Blue Gladiator Space Robot ($30-50)

700120 Voltron II Miniature Black Gladiator Space Robot ($30-50)

700200 Voltron III The Deluxe Lion Set ($125-175)

700201 Voltron III Giant Black Lion Robot ($30-50)

700202 Voltron III Yellow & Green Mighty Lion Robots Set ($30-50)

700203 Voltron III Blue & Red Mighty Lion Robots Set ($30-50)

700210 Voltron I The Deluxe Warrior Set ($125-175)

700211 Voltron I Air Warrior Set ($30-50)

700212 Voltron I Space Warrior Set ($30-50)

700213 Voltron I Land Warrior Set ($30-50)

700220 Voltron II The Deluxe Gladiator Set ($90-150)

700401 Voltron Blazing Sword Set ($18-25)

700402 Voltron Minature Blazing Sword Set ($10-15)

ROBOTECH

In the not too distant future, the Earth becomes a final refuge for a damaged alien space vessel, the SDF-1, containing the secrets of an amazing technology far advanced from that known to man. Unbeknownst to the people of Earth, a vast armada of warships manned by the evil Zentraedi discover the SDF-1, the Earth's Robotech Defense Force is called upon to defend their planet against the attacks of this merciless alien onslaught.

This is how Matchbox Toys described the Robotech story. Robotech was a cartoon show with 85 TV episodes. Matchbox Toys received the license to market and produce much of the Robotech line. Although most of the items were manufactured by Matchbox, a few products were actually made by Japanese firms Bandai and Godaken and marketed in Matchbox packaging. The line featured robots, action figures and dolls.

7100 MECHA LAUNCHERS WITH ROBOTECH VEHICLES, issued 1986

1. RDF Recon ship ($8-10)
2. RDF Destroyer ($8-10)
3. Zentraedi Cruiser ($8-10)
4. Zentraedi Officer's Attack Ship ($8-10)

7110 MONSTER DESTROID CANNON, issued 1986

1. dark green ($15-20)

7120 MINIATURE SOF-1, issued 1986

1. white & dark blue ($15-20)

7130 ALPHA FIGHTER, issued 1986

1. green & cream ($20-25)
2. blue & white ($20-25)
3. maroon & cream ($20-25)

7131 MINI ALPHA FIGHTER, issued 1986

1. green & cream ($15-18)
2. blue & white ($15-18)
3. maroon & cream ($15-18)

7201 3-3/4" ACTION FIGURES, issued 1986

1. Rick Hunter ($3-5)
2. Scott Bernard ($8-12)
3. Lisa Hayes ($3-5)
4. Dana Sterling ($3-5)
5. Rand ($3-5)
6. Robotech Master ($3-5)
7. Micronized Zentraedi Warrior ($3-5)
8. Bioroid Terminator ($3-5)
9. Corg ($3-5)
10. Zor Prime ($3-5)
11. Roy fokker ($3-5)
12. Lunk ($18-25)
13. Max Sterling ($3-5)

14. Lynn Minmei* ($75-125)
15. Rook Bartley ($18-25)
16. Miriya ($3-5)

*only available in Harmony Gold blistercard

7203 GIANT ZENTRAEDI ACTION FIGURES, 1986

1. Breetai ($8-10)
2. Exedore ($8-10)
3. Khyron ($8-10)
4. Dolza ($8-10)
5. Armoured Zentraedi Warrior ($8-10)
6. Miriya ($15-25)

7301 ATTACK MECHA, issued 1986

1. Excaliber Mg.IV in brown ($18-25)
2. Gladiator in green ($18-25)
3. Zentraedi Power Armour Botoru Batallion in green ($18-25)
4. Zentraedi Powered Armour Quadrono Batallion in blue ($18-25)
5. Spartan in brown ($18-25)
6. Raida X in tan ($18-25)
7. Invid Scout Ship in purple ($18-25)
8. Bioroid Invid Fighter in maroon ($18-25)

7351 ARMOURED CYCLE, issued 1986 ($12-18)

7352 BIOROID HOVERCRAFT, issued 1986 ($10-15)

7353 OFFICER'S BATTLEPOD, issued 1986 ($20-25)

7354 TACTICAL BATTLEPOD, issued 1986 ($20-25)

7355 INVID SHOCK TROOPER, issued 1986 ($20-25)

7356 VERITECH HOVER TANK, issued 1986 ($20-25)

7357 VERITECH FIGHTER, issued 1986 ($50-75)

7395 SDF-1 ACTION PLAYSET, issued 1986 ($75-100)
7700 ROBOTECH WARS VIDEO PLAYSET, issued 1986 ($95-125)

850001 SDF-1, issued 1986
1. white & blue body ($40-50)

850100 VERITECH FIGHTERS, issued 1986
1. Max Sterling's fighter in blue ($15-18)
2. Miriya's fighter in red ($15-18)
3. Rick Hunter's fighter in white ($15-18)

850201 BATTLOIDS, issued 1986
1. Civil Defense Unit Spartan Battloid in tan ($10-15)
2. Tactical Corps Spartan Battloid in white ($10-15)
3. Civil Defense Unit Gladiator in green ($10-15)
4. Tactical Corps Gladiator in red ($10-15)
5. Civil Defense Unit Raidar X in brown ($10-15)
6. Tactical Corps Raidar X in gray ($10-15)
7. Civil Defense Unit Excalibar Mk.IV in brown ($10-15)
8. Tractical Corps Excalibar Mk.IV in white ($10-15)

5101 LYNN MINMEI 11-1/2" FASHION DOLL, issued 1986 ($20-30)
5102 LISA HAYES 11-1/2" FASHION DOLL, issued 1986 ($20-30)
5103 DANA STERLING 11-1/2" FASHION DOLL, issued 1986 ($20-30)
5104 RICK HUNTER 11-1/2" FASHION DOLL, issued 1986 ($20-30)

5201 EXERCISE OUTFIT, issued 1986 ($5-8)
5202 STAR DISGUISE OUTFIT, issued 1986 ($5-8)
5203 STREET CLOTHES, issued 1986 ($5-8)
5204 PARTY DRESS, issued 1986 ($5-8)
5251 STAGE DRESS, issued 1986 ($5-8)
5252 FANCY DRESS CLOTHES, issued 1986 ($5-8)
5253 MISS MACROSS OUTFIT, issued 1986 ($5-8)
5254 EVENING GOWN, issued 1986 ($5-8)
5255 FASHION ACCESSORIES, issued 1986 ($5-8)

5410 DANA'S HOVER CYCLE, issued 1986 ($50-75)

PEE WEE HERMAN

Matchbox Toys hit a licensing winner with Pee Wee Herman, who's Pee Wee's Playhouse Saturday morning TV show was very popular in 1989 and 1990. The series was introduced in 1989 with further introductions in 1990. The movie Big Top Pee Wee was supposed to have spawned a series of action figures from the movie but only one item, Vance the Talking Pig was issued.

ACTION FIGURES
1. Pee Wee Herman, issued 1989 ($8-12)
2. King of Kartoons, issued 1989 ($10-15)
3. Chairry, issued 1989 ($10-15)
4. Cowboy Curtis, issued 1989 ($8-12)
5. Miss Yvonne, issued 1989 ($8-12)
6. Jambi & Puppet Band, issued 1989 ($15-20)

7. Globey & Randi, issued 1989 ($12-15)
8. Conkey, issued 1989 ($12-15)
9. Pterri, issued 1989 ($12-15)
10. Magic, Screen, issued 1989 ($8-12)
11. Ricardo, issued 1990 ($15-20)
12. Reba, issued 1990 ($20-25)

Large Toys
3500 Talking Pee Wee Herman, issued 1989 ($35-50)
3510 Non-Talking Pee Wee Herman, issued 1989 ($35-50)
3520 Pterri, issued 1989 ($35-50)
3530 Chairry, issued 1989 ($35-50)
3540 Billy Baloney, issued 1989 ($35-50)
3550 Playhouse Playset, issued 1989 ($75-100)
3568 Miniature Pee Wee Herman & Scooter ($25-35)
3590 Pee Wee's Scooter (full size ride-on), issued 1989 ($350-500)
3710 Vance the Talking Pig, issued 1990 ($35-50)
-Ventriloquist Pee Wee Herman, issued 1990 ($75-100)

OTHER TOYS

BIG MOVERS

Big Movers were introduced in 1989 and discontinued in 1991. This line included Matchbox's introduction into the press steel market. Made similar to Tonka and Nylint, these Korean manufactured toys found little favor on the market and were soon discontinued. The line consisted of several sized models from small to the very large.

BM-3400 10" dump truck ($20-35)
BM-3500 Stubbies Assortment ($10-15 ea.)
a. cement mixer
b. dump truck
c. snorkel truck
d. tanker
e. wreck truck
BM-3550 Stubbie Transporter
a. yellow, "IPEC" livery ($15-20)
b. white, "Matchbox" livery ($15-20)
BM-3600 4X4 Road King pickup ($25-35)
BM-3800 car carrier ($50-75)
BM-3810 container truck, "Midnight X-Press" ($50-75)
BM-3850 wreck truck ($25-35)
BM-3900 14" super dump truck ($40-60)

RAILWAYS

In 1991, Matchbox made its first introduction into the train layout world. Although the first Matchbox train was made by Lesney in the 1950's in the Yesteryear series, Matchbox never made any real train sets that operated in any manner until 1991. Two sets were produced with a third to be introduced in 1992 but this was never released.

TN-50 BIG BOXED SET, issued 1991
This set includes green locomotive, green coach and gray coal tender with two section layout ($75-100)
TN-100 GIANT BOXES SET, issued 1991
This set includes blue locomotive, red coach, red caboose, red diesel locomotive and gray coal car with three section layout ($100-150)

SCREAMIN STOCKS

Introduced in 1992, Screamin Stocks consists of two miniature sized cars, the Chevy Lumina and Ford T-Bird stock cars in Goodwrench and Texaco liveries respectively. A key is inserted into the back as a key car but the "key" utilizes sounds of a revving engine. ($8-12 each)

HOT FOOT RACERS

Another 1992 release is Hot Foot Racers. Each of the four styles available come with a foot launcher that is strapped to a child's foot. The model is then launched by clicking the launch mechanism at one's foot. ($8-12 each)

BLAZIN TURBOS/COLANI CARS

Issued in 1987, this set of four plastic models consisted of pullback friction cars based on designs by the world famous designer Luigi Colani. These were released as Colani cars in Europe but as Blazin' Turbos in the United States.

AM-4050 Colani ($7-10 each)
a. gold plated
b. silver plated
c. red plated
d. blue plated

POCKET ROCKETS

Issued in 1987 was a series of "chunky" pullback friction motored plastic vehicles. "Chunky" styles of actual cars like Mercedes, Lamborghini and Porsche are among the 12 styles available in the colors of red, blue or yellow (4 each color). Issued under assortment AM-4060 ($4-6 each)

POCKET ROCKET BUGGIES

Issued in 1987 were four dune buggies in plastic also using friction pullback action. Designated as AM-4080, four styles were available-two each in black or maroon. ($5-8 each)

POWER LIFTERS

Powerlifters were issued in 1988. These plastic models featured a twin speed pull back motor with a windback spoiler to create a wheelie. Twelve models available based on six castings. Issued under AM-4090. ($8-10 each)

Corvette
a. Vette
b. Viper

Stingray Vette
a. Stingray
b. White Heat

BMW
a. Mauler
b. Thunderbolt

Toyota
a. Green Demon
b. Burner

Camaro
a. Snaker
b. Blackstar

Pickup
a. Typhoon
b. Trailblazer

FLASHBACKS

Flashbacks were issued in 1988. The six plastic trucks were actually launchers. Inside each launcher vehicle was a key car. When the launcher impacted a wall, the key car emerged from the rear of the launcher. Many of the Flashbacks proved to be defective and this resulted in a premature discontinuation of the line. Designated AM-5010 ($8-12 each)

1. Rescue Truck and Porsche 959
2. Desert Camouflage Truck and 300ZX
3. Pickup and Ferrari
4. Off Roader and Porsche 959
5. 4X4 Pickup and Ferrari
6. Forest Camouflage and 300ZX

SPEED RIDERS

Originally released by LJN, Matchbox modified the base molds to include the Matchbox trademark. Three categories of three models each from this 1986 introduction ran on pen-light batteries.

Street Bird ($7-10)
Street Vette ($7-10)
Baja Beast (($7-10)
Super Stocker ($7-10)
Racin' Rebel ($7-10)
Corvette ($7-10)
Sand Blaster ($7-10)
Street Heat ($7-10)
Go-4-It ($7-10)

BURNIN' KEY CARS

Originally marketed by Kidco and introduced by Matchbox Toys in 1984, 1987 featured all new castings. Models feature plastic bodies and metal bases. A key is inserted into the rear of the models to launch them. Twelve different models based on six castings were produced. These include Nissan 300ZX, Porsche 959, Corvette, Firebird, Ferrari and Ford Mustang. Value ($3-5)

DUELIN' DRAGSTERS

Eight Duelin' Dragsters were introduced in 1987 based on four castings. Each featured a stubby version of a Corvette, Firebird, Camaro and Mustang. These are pull-back action models. Models issued in pairs. Value $5-8 (as pair)

LIGHTNING KEY CARS

In 1992, Matchbox reissued their Key Cars in brighter neon colors under the name Lightning Key Cars. In England these were also issued with a sound launcher; same as used for Screamin' Stocks, as these were issued as "Thunder Machines". Twelve styles available base on six casting ($3-5 each)

CONNECTABLES

Connectables were first released in 1989. Models were plastic vehicles which could be split apart and connected to each other making up hundreds of possible models and combinations. Models came in two, three and five piece multipacks. Although discontinued in the U.S. in 1991, the product continues to be available in Europe.

Two Piece Connectables, issued 1989 & 1990 ($3-5 each)

NOTE: some models have two variations.
1. Chevy Prostocker
2. Racing Porsche 935
3. V.W. Dragster
4. Plane
5. Jeep 4X4
6. Peterbilt Wrecker
7. Flareside Pickup
8. Road Roller
9. Model A Hot Rod
10. Chevy Wagon
11. Racer
12. Dodge Dragster

There were three two piece connectables which were specially offered through a Kelloggs offer in England. These were available only by sending in proof of purchase plus money through the mail. Price guide value ($15-20 each.)
1. V.W. Dragster- "Snap"
2. Plane- "Crackle"
3. Road Roller- "Pop"

Three Piece Connectables, issued 1989 & 1990 ($4-7 each)

1. Kenworth Race Truck
2. Helicopter
3. Dragster
4. Half-Track
5. Lincoln Limousine
6. Airplane

Three Piece Connectables Gold Plated Series, issued 1992 ($8-12 each)

1. Kenworth Rack Truck
2. Lincoln limousine
3. Plane
4. Dragster

Other Connectables

Micro Connectables, issued 1990 ($8-12 each)
Convoy Connectables, issued 1990 ($7-10 each)
5 Piece Themed Connectables, issued 1990 ($12-15 each)
Motorized Connectables, issued 1990 ($8-12 each)
Extender Connectables, issued 1991 ($10-12 each)
Crazy Limos Connectables, issued 1992 ($5-20 each)(depending on size)
Mini Playsets, issued 1990 (4 different) ($10-15 each)
CN-800 Truck Transporter, issued 1989 ($50-75)

Another Connectable type item released in 1992 in Europe were Connectors. These are figures which feature interlocking and interchangeable body parts. Again, only distributed by Matchbox Toys. In the U.S.A., Ertl Toys marketed these are Socket Poppers. Price guide value ($8-12 each)

1. football player & swamp monster
2. indian & robot
3. rock star & dinosaur
4. wrestler & monster fly
5. vampire & cyclops

6. soldier & mutant
7. skateboarder & pteradactyl

CAROUSEL

In 1989, Matchbox Toys introduced a line of girls toys known as Carousel. The line of Carousel included horses and animals as well as large Carousel & display stand (see playsets). Price guide value on each is $15-20 except the "rare" Lion which is valued $50-75.

Classic Horses, issued 1989 includes nine different styles Bright Beauties, issued 1990 includes six different styles.

Fancy Trotter, issued 1990 includes six different styles, Fashion Parade issued 1990 includes six different styles. Classic Animals, issued 1991 includes four different styles.

MONSTER IN MY POCKET

In 1991, Matchbox Toys was one of many companies to receive the license to produce Monster In My Pocket. Matchbox got to do the figures which included dozens of monsters in assorted colors. Contents vary to each set.

MT-250 12 Pack Scary Monsters, issued 1992 ($12-15)
MT-260 6 Pack Scary Monsters, issued 1992 ($7-10)
MT-270 Howlers-4 different, issued 1992 ($6-8 each)
MT-810 4 Pack Monsters, issued 1991 ($4-5)

MT-820 12 Pack Monsters, issued 1991 ($10-12)
MT-850 Monster Mountain display, issued 1991 ($10-12)
MT-860 Monster in My Pocket Battle Card Game, issued 1991 ($4-6)
MT-900 Monster Clash Game, issued 1992 ($15-18)

WORLD'S SMALLEST MATCHBOX

In 1990, Matchbox answered back to Galoob's Micro Machines with their own line called World's Smallest Matchbox. These small models were 1/200 scale and were possibly too small as these didn't sell well. In 1992, the series was resurrected in England as Micro Motor World.

INTRODUCTORY PLAYSETS, issued 1990 ($3-5 each)

Sets numbered 1-5.

35400 Mini-Playsets, issued 1990 ($5-8 each)
1. Fire Station
2. Bank
3. Car Dealership
4. Grocery Store
5. Auto Parts Store
6. Hotel
7. Truck Stop
8. Hospital
9. Airport
10. Car Wash
11. Marina
12. Drag Strip

35320 DELUXE PLAYSETS, issued 1990 ($8-10)
1. Car Rental
2. Restaurant
3. Service Station
4. Police Station
5. Off-Road
6. Bridge

35370 MAGNA-WHEEL CARRY CASE, issued 1990 ($8-12)
1. Exotics
2. Emergency
3. Sports
4. Leisure

MATCHBOX 2000

Matchbox 2000 was a new concept toy based on futuristic vehicles that moved by magnetic power. Each small plastic vehicle contained a magnet which interacted with each other or with two playsets. Models were packaged in three packs and most were released in two color varieties. Price guide value of each pack of three ($7-10)

NA-821 Vertical Service Center, issued 1991 ($15-20)
NA-822 robotic Positioning Module, issued 1991 ($15-20)

RING RAIDERS

In 1989, Matchbox Toys obtained a license from "Those Characters From Cleveland" and introduced Ring Raiders. These were small aircraft that attached to a ring to fit upon your finger. The series included a number of aircraft in multi-packs

with additional playsets. The line was supposed to have been supported by a TV series which never materialized.

8100 Assortment, issued 1989 includes 4 aircraft ($10-15)
1. Freedom Wing
2. Rescue Wing
3. Rebel Wing
4. Victory Wing
5. Valor Wing
6. Bandit Wing
7. Bravery Wing
8. Havoc Wing
9. Vicious Wing
10. Hero Wing
11. Ambush Wing
12. Vulture Wing

8120 Air Award Medal series, issued 1989 1 chrome model & medal ($8-12)
8130 sky Base Courage, issued 1989 ($12-15)
8140 Sky Base Freedom, issued 1989 ($12-15)
8150 Battle Blaster, issued 1989 ($8-12)
8160 Skull Squadron Mobile Base, issued 1989 ($15-20)
8170 Ring of Fire Video, issued 1989 ($12-18)
8190 Wing Command Display Stand, issued 1989 ($8-12)
RG570 Electronic Freedom Fighter, issued 1990 ($35-50)

BABYCISE

Babycise was introduced in 1986 as a Shared Development System, a specially designed infant exercise program developed

by a board of certified pediatricians and therapists. The series lasted only one year and was then discontinued. The following items were issued:

4100 Baby Bells ($10-15)
4140 Triangle Play Block ($15-20)
4110 Clutch Ball ($8-12)
4150 Mirror ($10-15)
4120 Balance Beam ($15-20)
4200 Video Gift Set ($25-40)
4130 Bolster ($18-25)
4225 Starter Set ($25-40)

LINKITS

Linkits were introduced in 1986 with line introductions in 1987. Linkits were a type of building system which linked sections together to form different toys. Price guide values as $5-15 for smaller sets up to $35 for larger sets.

LK-1 Good Robug
LK-2 Bad Robug
LK-3 Radions
LK-4 Stridants
LK-5 Espions
LK-6 Terratrek
LK-7 Artillius Unit
LK-8 Robugs on blistercard
LK-9 Expander Pack
LK-10 Cube
LK-11 Linkon LK-1-11
LK-22 Agrobot
LK-23 Micro Robot
LK-25 Terratrek
LK-26 Megadrom
LK-27 Maxar
LK-28 Vector & Agrobot

LK-29 QT-40
LK-30 Astromil
LK-31 Kraniodome
LK-34 Workshop II
LK-37 Assortment
LK192/101 Space Station
LK-50 Frog
LK-51 Agrippa
LK-52 Lizzod
LK-53 Reptopods
LK-54 Cranio Jet
LK-56 Mercurion
LK-57 Battloid
LK-58 Ballistix
LK-16/101 Shark
LK-16/102 Dragon
LK-32/101 Green Beast
LK-32/102 Robot
LK-32/103 Space Monkey
LK-32/104 Star Warriors
LK-64/102 Transmission Team
LK-64/103 Robot Racers
LK-64/104 Space Walkers
LK-64/105 Chicken Monsters
LK-64/201 Puzzle
LK-128/101 Transport Team
LK-128/102 Space Creatures
LK-128/103 Insects
LK-20/21
LK-192/102 Dinosaurs
LK-192/103 Master Robots

OH JENNY

Oh Jenny was a licensed product actually made by another company. The other company sold the product in England under its own name and the product was called Oh Penny. The series includes playsets and figurines. Oh Jenny was introduced in 1989. Spanish packaged playsets exist.

4300 Complete playset (includes 4301, 4302, 4303, & 4304), issued 1989 ($85-100)
4301 Family Home, issued 1989 ($30-40)
4302 Stable, issued 1989 ($20-30)
4303 Swimming Pool, issued 1989 ($10-15)
4304 Tree House, issued 1989 ($15-18)
4305 Figures assortment, issued 1989 ($4-6)
1. Farm Animals
2. Treehouse Kids
3. Jenny's Pets
4. Poolside Family
5. Sweet Stables Family
6. House Family

4306 Play assortment, issued 1989 ($8-10 each)
1. Pony & Cart set
2. Wagon Set
3. Car & Family
4340 Camper, issued 1990 ($25-40)
4350 Shopping Mail, issued 1990 ($40-50)

RUBIK'S

Back in the early 1980's, the Rubik's Cube became one of the greatest toy phenomena the world had ever seen. Over half a billion cubes were sold. There were competitions to see who could solve the cube the fastest. There was even a Guiness Book entry. In 1987, Erno Rubik signed a license with Matchbox Toys to bring his new puzzle on to the market. Although never reaching the sales of the cube, "Link The Rings" began a license with other Rubik's puzzles for Matchbox.

MA-012 Rubik's Magic Puzzle-Unlink The Rings, issued 1987 ($10-15)
MA-016 Rubik's Magic Strategy Game, issued 1987 ($81-25)
MA-018 Rubik's Clock, issued 1989 ($15-18)
MA-040 Rubik's Magic Picture Game, issued 1988 ($5-8)
1. Monster Sports
2. Octopuss Garden
3. Dinosaur Days
4. Crazy Orchestra
MA-080 Link The Rings, issued 1987 ($10-15)
MA-400 Rubik's Cube, issued 1989 ($8-12)
MA-650 Rubik's Illusion Game, issued 1989 ($15-20)
MA-660 Rubik's Dice, issued 1990 ($15-20)
MA-670 Rubik's Triamid, issued 1990 ($15-20)
MA-680 Rubik's Fifteen, issued 1990 ($15-20)
MA-690 Rubik's Tangle, issued 1990 ($15-20)

BUBBLEHEADS

This was another "made by someone else" but marketed by Matchbox product. Bubbleheads, introduced in 1991, included three sets of two balls with faces on them. The balls were constructed in such a way that the face on the ball stayed always right side up. Price guide value ($5-10 each)

1. Silly Sports-four different
2. Mad Monsters-four different
3. Funny Faces-four different

MY PRECIOUS PUFFS

My Precious Puffs, based on the looks of a dandelion puff, was a series of toys for girls. There were three themes-Beauty, Music and Party. Each set of Puffs included eight different puffs. These were introduced in 1987 and were discontinued the following year. Price guide value $5-8 each.

ZILLION

Although originally introduced in 1987 in answer to the Laser Tag phenomena, the United States market canceled the release of "Laser Flash". The series received limited release in England under the name "Zillion". The product was actually made by Sega for Matchbox toys. The only item in the series seen by this author is the Pistol & Electronic Target set. The original price in 1987 was $35.00. Price guide value $50-75.

TUFF WHEELS, issued 1989

In 1989, Matchbox Toys introduced the largest Matchbox models ever manufactured-Tuff Wheels. These were manufactured by Mack Plastics for Matchbox Toys. Tuff Wheels were ride-on, battery operated vehicles. Each model supported either one or two children. Only three Tuff Wheels were ever manufactured although catalogs depicted many more including different volt sizes as well as non-battery powered versions.

7000 4X4 Pickup, issued 1989 ($350-450)
7010 Turbo Racer, issued 1989 ($350-450)
7020 Ferrari, issued 1990 ($500-750)

WIZZZER TOPS, issued 1989

In 1989, Matchbox Toys received the license to produce Wizzzer Tops. At one time Wizzzers were manufactured by Mattel and are still produced today, but under another company's brand. The top was designated as WX-8470. With different decorations and combinations of different colored halves, there are at least one hundred or more Matchbox Wizzzer varieties! Price guide value $5-8

RADIO CONTROL/L.A. WHEELS

There were two series of Radio Control cars-those marketed through Matchbox Toys U.S.A. and those marketed by the Universal Associated Company. U.A.C. was a Matchbox subsidiary that marketed Matchbox Toys not offered through the normal U.S. distributor Matchbox Toys U.S.A. One of the products issued by them were L.A. Wheels radio control vehicles. Radio Control Matchbox were first introduced in 1987.

6000 R.C. Radskate, issued 1987 ($75-100)
6010 R.C. Ripskate, issued 1987 ($75-100)
6020 Revvin' 427 Cobra, issued 1987 ($50-75)
6030 Revvin' Pro Street Mustang, issued 1987 ($50-75)
6060 R.C. Snoopy Skateboard, issued 1988 ($75-100)

L.A. Wheels models range from ($50-150) depending on size.

HOT ROD RACERS

Hot Rod Racers, introduced in 1986, were 1/20th scale muscle cars in plastic that featured a rear stick shift that made the model make revving sounds. The series included only four models. Price guide value $20-25 each.

2101 Corvette, issued 1986
2103 1986 Firebird, issued 1986
2103 1957 Chevy, issued 1986
2204 Ferrari, issued 1987

HEAD GAMES

An interesting concept "game" for 1988 was the release of Head Games. This game is strapped to your head like a hat. Head movement puts a ball in a cup or basket! Price guide value $12-18 each.

86001 Basketball
86002 Golf
86003 Tennis
86004 Football

W.A.C.K.O.

Another game from 1988 is W.A.C.K.O., abbreviated from Wild And Crazy Kids Only. These are skill games based on arcade-type fun. Two assortments of four games each. Price guide value $10-15 each.

8650 Assortment-Sports
1. Baseball
2. Hoops
3. BMX Racer
4. Gobblers

8660 Assortment-Action
1. Shoot The Rapids
2. Alligator Alley
3. Jungle Safari
4. Escape From Devil's Castle

SUPERFAST MACHINES

Superfast Machines was a range of four vehicles that were actually plastic snap-together kits which included battery powered motors. The series was introduced in 1989 and featured two different styles of boxes for each model introduced. Price guide value at $8-10 each. Race track set value $15-25.

1. Tomcat
2. Speed King
3. Streaker
4. Aero-Shot

LIVE N LEARN

Still being produced, the Live-N-Learn line is now over twenty years old being first introduced in 1972. The series includes quality plastic toys for infants. New models issued since 1982 are listed below.

LL-101 Town Set, issued 1991 ($25-30)
LL-102 Farm Set, issued 1991 ($25-30)
LL-103 Railway Set, issued 1991 ($25-30)
LL-120 Mushroom Playhouse, issued 1987 ($35-50)

LL-150 Playboot Playhouse, issued 1984 ($35-50)
LL-210 Activity Bear Cot Bumper, issued 1989 ($25-35)
LL-220 Rock-A-Bear Buggy, issued 1989 ($25-35)
LL-230 Fold & Play Garage, issued 1989 ($35-45)
LL-300 Peg-A-Picture, issued 1983 ($25-35)
LL-310 Jigsaw Puzzle Playmat, issued 1989 ($20-25)
LL-330 My Soft Story Books, issud 1991 ($12-15)
LL-410 Cuddly Activity Bear, issued 1991 ($25-30)
LL-500 Activity Quilt, issued 1987 ($35-50)
LL-600 Snap-A-Tune, issued 1987 ($25-35)
LL-910 Smart Bear, issued 1987 ($20-30)
LL-1000 Natter Phone, issued 1987 ($25-35)
LL-1300 Squeek-N-Seek, issued 1987 ($20-25)
LL-1500 Tony The Tubby Tooter, issued 1983 ($15-18)
LL-3000 Shufflie Zoo, issued 1986 ($50-75)
LL-4500 Activity Bear Play Centre, issued 1988 ($100-125)
LL-4510 Big Band Bear, issued 1988 ($35-40)
LL-4520 Bathtime Bear, issued 1988 ($35-45)
LL-4600 Activity Rattle, issued 1983 ($15-18)
LL-4700 Stacking Anchor, issued 1984 ($18-25)
LL-4800 Cloud, issued 1984 ($18-25)
LL-4900 Dozey Daisy, issued 1984 ($15-25)

LL-5000 Mini-Mate, issued 1984 ($25-40)
LL-5100 Pop-Up Till, issued 1984 ($35-45)
LL-5200 Baby Bear Rattle, issued 1985 ($10-15)
LL-5300 Elephount, issued 1985 ($20-25)
LL-5400 Squeeze-N-Go Sam, issued 1985 ($15-25)
LL-5500 Squeeze-N-Go Froggy, issued 1985 ($12-15)
LL-5800 Pop-Up Peter, issued 1986 ($10-15)
LL-5900 Squeeze-N-Go Racer, issued 1986 ($15-25)
LL-6000 Learning Bear, issued 1986 ($35-40)
LL-6100 Vanity Bear, issued 1986 ($20-25)
LL-6200 Squeaky Bear, issued 1986 ($20-25)
LL-6300 Tick Tock Bear, issued 1986 ($20-25)
LL-6400 Trumpet Teether, issud 1986 ($10-15)
LL-6500 Squeaky Star, issued 1986 ($8-12)
LL-6600 Sing Song Shell, issued 1986 ($8-12)
LL-8000 Baggie Bunnie, issued 1988 ($12-15)
LL-8010 Baggie Bunnie Activity Bunnie, issued 1988 ($15-20)
LL-8020 Baggie Bunnie Activity Blanket, issued 1988 ($12-15)
LL-8030 Baggie Bunnie Activity Soft Blocks, issued 1988 ($12-15)
LL-8040 Baggie Bunnie Activity Chime Ball, issued 1988 ($10-15)
LL-8050 Baggie Bunnie Activity Soft Rattle, issued 1988 ($10-15)
LL-8060 Baggie Bunnie Activity Soft Touch Book, issued 1988 ($10-15)

TWINKLETOWN

In 1986, Matchbox Toys introduced a side series to their Live-N-Learn line called Twinkletown. The series included six plastic toy trucks and four playsets. Each vehicle included a small storybook. The series was unnumbered. The following were released:

Freddy Ready Fire Truck ($12-15)
Tony's Tow Truck ($12-15)
Sheldon's Shovelier ($12-15)
Carrier Camper ($12-15)
Melvin Mail Van ($12-15)
Stevie's School Bus ($12-15)

Twinkletown Gas Station ($20-25)
Twinkletown School House ($20-25)
Twinkletown Fire House ($20-25)
Twinkletown Construction ($20-25)

PLASTIC KITS

After being introduced in 1976, plastic kits remained in the Matchbox line through 1989. Then the division was sold to Revell in 1990. AMT was sold in the early 1980's to Ertl. AMT versions not noted below, only those with "Matchbox" only trademark.

PK-39 Northrop F-5B (1986) ($10-15)
PK-40 Starfighter TF-104 G (1986) ($10-15)
PK-41 Mikoyan MiG-21 MF (1986) ($10-15)
PK-42 T-2C Buckeye (1987)($10-15)
PK-43 Bell OH-58D (1987) ($10-15)
PK-44 Mirage IIIB (1987)($10-15)
PK-45 Bae Harrier GR Mk.3 (1987) ($10-15)

PK-46 Bae Hawk 200 (1987) ($10-15)
PK-47 Mystere IV.A (1987)($10-15)
PK-48 MBB BK.117 (1988)($10-15)
PK-126 BAC Lightning (1985)($15-20)
PK-127 Twin Otter (1985)($15-20)
PK-128 Jaguar T.MK2 (1986)($15-20)
PK-129 Armstrong Whitworth Meteor (1987)($15-20)
PK-130 Panavia Tornado F.Mk.3 (1987)($15-20)
PK-131 Saab SK-37 Viggen (1988)($15-20)
PK-132 Heinkel He 70 (1988)($15-20)
PK-133 Westland Wessex (1988)($15-20)
PK-134 Douglas Skynight (1988)($15-20)
PK-177 Churchill Bridgelayer (1983) ($15-20)
PK-178 Challenger (1988)($15-20)
PK-312 Auto Union Type D (1985)($15-20)
PK-412 Phantom FG1/FGR2 (1986) ($20-25)
PK-413 Boeing Vertol Chinook (1987) ($20-25)
PK-452 Rolls Royce Phantom (1983) ($20-25)
PK-551 Victor K.2 (1983)($25-35)
PK-5009 NATO Troops (1984)($10-15)

MAXX F-X/FREDDY KREUGER

This was to have been a series of 12" figures which featured a normal male doll who could dress himself up into some of the most popular monsters in history including Dracula, Frankenstein, the Mummy, Wolfman, the Creature, Alien, Jason and Freddy Kreuger. Only the Freddy Kreuger doll was ever released and was almost immediately discontinued due to its larger talking counterpart receiving so much bad publicity from religious protest. Both

Freddy dolls were withdrawn from the market.

Maxx F-X Freddy Kreuger Doll, issued 1989 ($25-35)
Talking Freddy Kreuger Doll, issued 1989 ($65-75)

PLUSH TOYS

Matchbox Toys experimented in plush toys several times with their first try in 1986 with Need-A-Littles. This was followed by Razzcals in 1987, Pooch Troop, Denver the Dinosaur and the Noid all in 1989. Matchbox Toys also did marketing for Amtoy in Europe of My Pet Monster, Brush-A-Loves, Little Brush-A-Loves and Kiss-A-Loves in 1988.

Need-A-Littles, issued 1986 ($30-40 each)
1. Lamb
2. Puppy
3. Bunnie
4. Teddy Bear
5. Kitten

Pooch Troop, issued 1989 ($20-30 each)
1. Colonel Ollie Collie
2. Top Dog
3. Sergeant Barker
4. Doc Bernard

Razzcals, issued 1987 ($10-15 each) four varieties (also called "Grumples")
Denver The Dinosaur, issued 1989 ($20-30)
The Noid, issued 1989 ($20-25)

DOLLS

Although Matchbox Toys had sold the Vogue Doll line in 1981. They continued to dabble in the doll category but with little success. From 1989 to 1991 they introduced Baby Secrets, Sweet Treats, Baby Baby, Baby Cheer-Up, Real Models, Popsicle Kids, Butterfly Princess and Hush-A-Bye Baby. The last three noted were released through U.A.C. (Universal Associated Company) rather than through the normal Matchbox Toys USA distributor. Matchbox Toys also marketed two LJN dolls in Europe-Michael Jackson and Boy George in 1985. In 1987, they were the European license for Play-Mates' Cricket doll.

SWEET TREATS, issued 1989 (dolls at $12-15 each, Trike at $25-35)
1. Raspberry Sherbert
2. Blueberry Cheesecake
3. Banana Split
4. Sugar Plum Pudding
5. Coconut Cupcake
6. Cotton Candy
7. Cocoa Pop
8. Treatmaker Trike

REAL MODELS, issued 1990 ($18-25 each)
1. Beverly Johnson
2. Christie Brinkley
3. Cheryl Tiegs

BABY SECRETS, issued 1989 ($25-35)
BABY CHEER-UP, issued 1990 ($20-25)
BUTTERFLY PRINCESS, issued 1990
1. single pack ($3-5)
2. four pack ($10-15)

POPSICLE KIDS, issued 1990 ($10-15 each)
1. Berry Blue
2. Merry Cherry
3. Grape Cakes
4. La De Lime
5. Lotta Lemon
6. Ooh LaOrange
BABY BABY, issued 1991 ($35-45)
HUSH-A-BYE BABY, issued 1991 ($25-35)

PLAYSETS & MISCELLANEOUS TOYS & GAMES

This section covers playsets and toys and games which were issued that don't fit into any other category or they include specialty models.

0950 Commando Playset, issued 1990 ($15-20)
1050 Laser Wheels Speed of Light Stunt Set, issued ($12-15)
1310 Competition Arena, issued 1988 ($15-20)
1960 World Class Collectors Display Stand, issued 1990 ($10-15)
4450 Carousel Musical Playset, issued 1989 ($50-75)
4480 Carousel Display Stand, issued 1989 ($15-20)
8050 Aircraft Carrier Casse, issued 1988 ($15-20)
8450 Boola Ball, issued 1987 ($18-25)
MG-6 Deluxe Garage, issued 1984 ($20-30)
MG-7 Compact Garage, issued 1984 ($20-30)
MG-9 Gearshift Garage, issued 1987 ($20-30)

S-250 Streak Racing, issued 1984 ($15-25)
S-450 Streak Racing, issued 1984 ($20-30)
CC-1500 Carguantua, issued 1986 ($50-75)
LR-730 Lightning Flipout Drag Strip, issued 1991 ($15-25)
LR-740 Electronic Drag Strip, issued 1991 ($15-25)
500101 Play & Pack Hat Construction Yard, issued 1984 ($20-25)
500102 Play & Pack Hat Rescue Center, issued 1984 ($20-25)
500103 Play & Pack hat 4X4 Test Center, issued 1984 ($20-25)
560109 Speed Shooter, issued 1984 ($35-50)

KIDCO AND LJN

Kidco and LJN were independent toy companies, separate from Matchbox Toys. In 1984, Matchbox Toys bought Kidco in a merger agreement. In the United States the Kidco products were still packaged "Kidco" but in England were packaged as "Matchbox". Castings were never modified and retained Kidco on the baseplates.

LJN used Matchbox in Europe to market some of their products. The packaging retained the "LJN" logo and the "Matchbox" logo was added to packaging. The extent of Kidco and LJN Matchbox products is too broad to be covered fully in this book. The following lines were offered by Matchbox.

LJN:
Dungeons & Dragons
Stunt Riders
Dune

Tri-Ex Rough Riders 4X4
Michael Jackson Doll
Power Blaster Cycles
Omni Force Rough Rider 4X4
A-Team Rough Riders 4X4
Rough Riders 4X4
Boy George Soft Sculpture Doll

KIDCO:
Key Cars	Burnin' Key Car Demolition Cars
Key Car Trucks	Key Cars motorcycle
Lock-up cars	Gold Key Lock-ups
Glowin' Burnin' Key Cars	

Another company, Sanrio also released their "Hello Colour" products with Matchbox logos in England in 1985. Odd-Zon, Tiger Toys and Wilson Sports have also released products with the "Matchbox" logo in Europe including popular toys like Skip-It, Koosh Balls & Moon Shoes to name but a few.

PREPRODUCTIONS & PROTOTYPES

As with the two previous books, preproductions and protypes are depicted in this book. Because of space restrictions, not all series or as many preproductions can be shown in this volume. Again, it is beyond the scope of this book to value preproduction models. They have a high collectible value to certain collectors. Series covered in this volume include miniatures, Convoy and Yesteryears.

PLACES TO VISIT-SOURCES

Many persons have requested contacts on places to visit or places to purchase their models. The places to visit include Matchbox Toys museums with perhaps the largest collections in the world. Sources include both mail order and shops in which to purchase your models.

Museums:

MATCHBOX & LESNEY TOY MUSEUM
home of Charles Mack, 62 Saw Mill Rd.
Durham, Connecticut

MATCHBOX ROAD MUSEUM
Pearl Street
Newfield, New Jersey

CHESTER TOY MUSEUM
(MATCHBOX ROOM)
13A Lower Bridge Street
Chester, England

Main Sources:

Midwest Die-Cast Miniatures
681 Paxton Pl.
Carol Stream, IL 60188
(mail order only)

Harold's Place
532 Chestnut St.
Lynn, Mass. 01904
(store & mail order)

Diecast Toy Exchange
401 Carlisle Ave.
(P.O. Box 268)
York, Penn. 17405
(store & mail order)

Kiddie Kar Kollectibles
1161 Perry St.
Reading, Penn. 19604
(store & mail order)

Neil's Wheels, Inc.
P.O. Box 354
Old Bethpage, N.Y. 11804
(store & mail order)

These are the largest dealers in the United States and all specialize in Matchbox Toys. Other smaller dealer also carry the Matchbox line. It is suggested you join a collectors club to see who in your area may be able to help you in your collecting needs.

CLUBS

For those who would like more information on models covered in these books or would like information on newer models as well, a collectors club can be an additional source of information for the Matchbox enthusiast. The following clubs can help you through their newsletters which are published on quarterly, bi-monthly or monthly offerings. Please send a self-addressed, stamped envelope when writing to the clubs for information.

U.K. MATCHBOX

Founded in 1977. Although no longer in existence, if you are able to get copies of this club's magazine you are in for a wealth of information as supplied by its the editor Ray Bush.

MATCHBOX U.S.A.

Founded in 1977. This club offers a monthly publication with the editor being the author of this book. Charles Mack. Write to: Matchbox U.S.A., 62 Saw Mill Road, Durham, CT 06422. On the web at http://www.matchbox-usa.com

MATCHBOX INTERNATIONAL COLLECTORS ASSOCIATION

Founded in 1985. This club offers a bi-monthly publication. Editors are Stewart Orr and Kevin McGimpsey. for information in North America contact: M.I.C.A. North America, P.O. Box 28072, Waterloo, Ontario, N2l 6J8, Canada.

AMERICAN INTERNATIONAL MATCHBOX

Founded in 1970 by Harold Colpitts, this monthly newsletter is now being edited by Mrs. Jean Conner. Contact the club at 522 Chestnut Street, Lynn, MA 01904.

PENNSYLVANIA MATCHBOX COLLECTORS CLUB

Founded in 1980, this monthly newsletter and club offers to those basically in the Pennsylvania area, but membership is open to all Matchbox collectors. Write to the club secretary: Mike Appnel, 1161 Perry St., Reading, PA 19604

BAY AREA MATCHBOX COLLEC-TORS CLUB

Founded in 1971. Basically geared to those collectors in California area. Contact: BAMCA, P.O. Box 1534, San Jose, CA 95109.

CHESAPEAKE MINIATURE VEHICLE COLLECTORS CLUB

Founded in 1978. A collectors club devoted to all miniature cars including Matchbox. Contact: Win Hurley, 709 Murdock Road, Baltimore, MD 21212.

MATCHBOX FORUM

Founded in 1998 by John Yanouzas as an internet club. Contact: John Yanouzas, 7 North Bigelow Rd., Hampton, CT 06247

ILLINOIS MATCHBOX COLLECTORS CLUB

Founded 1997. Contact: Tom Sarlitto, 681 Paxton Place, Carol Stream, IL 60188

BIBLIOGRAPHY

Bowdidge, Philip, *Matchbox Gift Sets 1957-1988*, Philip Bowdidge, 8 Melrose Court, Ashely, New Milton, Hants BH25 5BY England, 1988

Cramer, Rock and Smith, James, *Matchbox Catalogs*, 3104 Mesa Drive, El Paso, TX, 79901 and 431 George Cross Drive, Norman, OK 73069

Mack, Charles, *Matchbox Models of Yesteryear*, Matchbox USA Publication, 62 Saw Mill Rd., Durham, CT 06422, 1989
_____, *Matchbox Skybusters*, Matchbox USA Publication, 62 Saw Mill Rd., Durham, CT 06422, 1989
_____, *Convoy Catalog-2nd Edition*, Matchbox USA Publication, 62 Saw Mill Rd., Durham, CT 06422, 1989
_____, *Matchbox 1-75 Catalog 1969 to Date-2nd Edition*, Matchbox USA Publication, 62 Saw Mill Rd., Durham, CT 06422 1990
_____, *TP Catalog*, Matchbox USA Publication, 62 Saw Mill Rd., Durham, CT 06422, 1984
_____, *Matchbox USA* newsletter 1983-1992, Matchbox USA Publication, 62 Saw Mill Rd., Durham, CT 06422,
_____, *Matchbox Toys, The Superfast Years*, Schiffer Publishing Ltd., 4880 Lower Valley Rd., Atglen, PA 19310, 1992

Martin, Lorry, *Matchbox Super GT's* L. Martin 13 Kenwood Close, Trowbridge, Wiltshire BA14 7DN, England, 1988

Various, *Collecting Matchbox Diecast Toys, First Forty Years,* Major Productions Ltd., 42 Bridge Street Row, Chester, England, 1989